THE SPICE BOOK

THE SPICE BOOK

Andrei Vladimirescu

John Wiley & Sons, Inc.
New York Chichester Brisbane Toronto Singapore

ACQUISITIONS EDITOR Steven Elliot
MARKETING MANAGER Susan Elbe
PRODUCTION SUPERVISOR Richard Blander
DESIGNER Kevin Murphy
MANUFACTURING MANAGER Inez Pettis
ILLUSTRATION COORDINATOR Anna Melhorn

This book was set in Times Roman by Publication Services and printed
and bound by Malloy Lithographing. The cover was printed by Phoenix Color Corp.

The paper in this book was manufactured by a mill whose forest management programs
include sustained yield harvesting of its timberlands. Sustained yield harvesting principles
ensure that the number of trees cut each year does not exceed the amount of new growth.

Library of Congress Cataloging in Publication Data:
Vladimirescu, Andrei.
    The spice book / Andrei Vladimirescu.
        p.   cm.
    Includes bibliographical references.
    ISBN 0-471-60926-9
    1. SPICE (Computer file)   2. Electric circuit analysis—Data
processing.   S. Electronic circuit design—Data processing.
    I. Title.
    TK 454.V58   1994
    621.319'2'028553—dc20   93-33667
                                                                CIP

Printed in the United States of America

10 9 8 7 6 5

Printed and bound by Malloy Lithographing, Inc.

To my parents,
Gheorghe and Jeana-Maria-Victoria

PREFACE

This book is written for electrical engineering students and professionals who use one of the many versions of the SPICE program to analyze and design circuits. The topics presented in this book are universally valid for SPICE users no matter which version they use. This point is reinforced in the text by using the most popular SPICE versions to run the examples developed in the chapters.

SPICE has become the standard computer program for electrical simulation, with over 40,000 copies in use worldwide. The name SPICE stands for *Simulation Program with Integrated Circuit Emphasis* and was inspired by the application to integrated circuit (IC) design, which made computer simulation mandatory. Today, SPICE in its many versions is used not only for IC design but also for analog printed circuit boards, power electronics, and other applications.

The majority of the commercial SPICE packages are based on and support the functionality of SPICE2, version G6, from the University of California at Berkeley. The current circuit simulation development at the University of California at Berkeley is devoted to the SPICE3 program. Few commercial products are based on SPICE3, but a number of these programs support SPICE3 functionality that is not available in SPICE2. A commercial version of SPICE that has gained popularity in universities is PSpice, from the MicroSim Corporation. PSpice, which was first introduced as a Personal Computer Program, has become very popular because of the wide use of PCs.

The material in this book was developed based on the SPICE2 program, whose functionality and syntax are supported by all other SPICE simulators. The SPICE netlist standard is defined by SPICE2, and all derivatives of the program accept a SPICE2 input file; functionality specific to a certain SPICE program and not available in SPICE2 is introduced as an extension to the SPICE language and is documented in the respective user's guide. Examples throughout this book are simulated alternatively on SPICE2, SPICE3, or PSpice. Functionality available only in SPICE3 is documented, and useful features proprietary to PSpice are mentioned.

This book combines in a natural progression a tutorial approach on how to advance from hand solutions of typical electrical and electronic circuit problems to using SPICE, with some reference information on the program necessary for the more advanced user.

The text should be useful to the SPICE novice as well as to the experienced user. The reader is assumed to have a basic electrical engineering background and be able to use a computer.

The approach in this book emphasizes that SPICE is not a substitute for knowledge of circuit operation, but a complement. *The SPICE Book* is different from previously published books on this subject in the approach of solving circuit problems with a computer. The solution to most circuit examples is sketched out by hand first and followed by a SPICE verification. For more complex circuits it is not feasible to find the solution by hand, but the approach stresses the need for the SPICE user to understand the results. Although the program can detect basic circuit specification errors, it cannot flag conceptual errors. It is up to the user to question the program through the various analysis modes in order to get insight into what is wrong with the circuit. Briefly stated, the results of SPICE are only as accurate as the circuit description and the component models used.

The first six chapters provide information about SPICE relevant to the analysis of both linear passive circuits and electronic circuits. Each of these chapters starts out with a linear example accessible to any new user of SPICE and proceeds with nonlinear transistor circuits.

The latter part of the book goes into more detail on such issues as functional and hierarchical models, distortion models and analysis, basic algorithms in SPICE, analysis option parameters, and how to direct SPICE to find a solution when it fails.

This book is ideally suited as a supplement to a wide range of circuits and electronics courses and textbooks. It is of special interest in junior, senior, and graduate courses, from introductory courses on electric circuits up to analog and digital integrated circuits courses.

The subject of computer-aided circuit simulation is put in a historical perspective in the Introduction to this book. The milestones of the research in the late 1960s and early 1970s that led to the SPICE program are presented first. The proliferation of SPICE versions and the salient features of the most popular programs are described. The Introduction follows the evolution of the SPICE effort at the University of California at Berkeley from the beginning to the present day. This historical perspective concludes with the current research in the area of electrical computer simulation and the possible future SPICE developments in the 1990s.

Chapter 1 is an introduction to the computer simulation of electrical circuits and the program SPICE. The approach used in SPICE to solve electrical problems is described in simple terms of the Kirchhoff voltage law, the Kirchhoff current law, and branch constitutive equations. A linear RLC circuit is used to exemplify the workings of SPICE. The reader is also introduced to the SPICE input language, the network specification, the analysis commands, and the types of result output available. The sequence of events for simulating a circuit is completed by examples on how to run SPICE on the most common computers.

Chapter 2 presents in detail the circuit specification in terms of elements, models, and the conventions used. The SPICE syntax is detailed for two-terminal elements, such as resistors, capacitors, inductors, and voltage and current sources, and multiterminal elements, such as controlled sources, switches, and transmission lines.

Chapter 3 introduces the semiconductor device elements and models available in SPICE. The dual specification as device and model is explained for semiconductor elements. Only the first-order models are described in this chapter for devices represented by several levels of complexity. The model parameters are related to the branch-constitutive equations of the device as well as to electrical characteristics. The most important physical effects and corresponding parameters are described for the five semiconductor devices supported: diodes, bipolar junction transistors (BJTs), junction field effect transistors (JFETs), metal-oxide-semiconductor field effect transistors (MOSFETs), and metal-semiconductor field effect transistors (MESFETs). This chapter does not cover the details of each model but provides references dedicated to the subject.

Chapter 4 contains an overview of the analysis modes of SPICE and a detailed description of DC analysis. In the DC mode SPICE can perform an operating point analysis, compute DC transfer curves, estimate the value of the transfer function, and perform sensitivity calculations.

Chapter 5 describes the SPICE functionality in the small-signal frequency domain. The AC mode analysis types, such as the frequency sweep, noise, and distortion analyses, are introduced by means of both linear and nonlinear circuit examples.

Chapter 6 presents the time-domain, or transient, simulation. In the time-domain analysis mode SPICE computes the transient response of a circuit and the harmonics of a signal. At least one worked-out circuit example is included for each analysis type. The reader acquires the basic knowledge of using SPICE by the end of this chapter.

Chapter 7 introduces the concept of functional simulation. Higher-level abstractions and hierarchy can be modeled in SPICE using controlled sources and sub-circuit blocks. Logic gates and operational amplifiers can be described using the macro-modeling approach. Examples demonstrate the compactness and efficiency of macro-modeling for opamp circuits.

The last three chapters of the book, Chaps. 8 to 10, are intended for the more advanced user. The material presented in the first part should be sufficient for solving most circuit problems encountered in undergraduate and graduate courses. There are three main topics in the second half, which can be be studied independently of each other. Chapter 8 covers in some detail distortion analysis, Chap. 9 contains an explanation of the solution techniques built into SPICE and the analysis options that may be necessary for solving complex circuits, and Chap. 10 uses the information in the previous chapter to steer the user on how to ensure the convergence of SPICE.

Chapter 8 offers an in-depth look at distortion analysis. The details of small-signal and large-signal distortion analysis are described with the help of several examples.

A brief overview of the algorithms and numerical methods used in SPICE is presented in Chap. 9. The purpose of this chapter is to offer some insight into the internal workings of SPICE for the user interested in taking advantage of all the available analysis controls or options, which are also described in this chapter. The main topics are solution of sparse linear equations, iterative solution of nonlinear equations and convergence, and numerical integration.

Chapter 10, which concludes this book, is a primer on convergence and the actions a user can take to overcome DC and time-domain convergence problems. Solutions to convergence problems are offered using initialization, analysis options, and nonlinear model parameters. The importance of understanding the operation of the circuit and the limitations of the models used is emphasized for obtaining accurate results.

Five appendixes are included at the end of the book. The first contains the complete equations for the semiconductor devices and the full list of model parameters. The second appendix lists the most common error messages of SPICE2 and provides guidance on corrective action. The error messages included are common to most SPICE versions, although the exact wording may differ. Appendix C summarizes all the SPICE statements introduced in this book. Appendix D contains the Gear integration formulas of orders 2 to 6. The last Appendix contains a sample SPICE deck of a circuit that requests most analyses supported by SPICE2.

This book is a result of my association with Professor D. O. Pederson, who has guided me during my academic studies as well as during my professional activity. I acknowledge Judy Lee for the graphic design and the presentation of the schematics and the simulation results. I also acknowledge the review and comments contributed by Dr. Constantin Bulucea in addition to the valuable comments made by the following reviewers for John Wiley and Sons: Kenneth Martin, UCLA; Richard Dorf, University of California at Davis; Ron Rohrer, Carnegie Mellon University; Norb Malik, University of Iowa; Bruce Wooley, Stanford University; Darrell L. Vines, Texas Tech University; James R. Roland, University of Kansas; David Drury, University of Wisconsin, Platteville; Robert Strattan, University of Oklahoma; John O'Malley, University of Florida, Gainesville; Gordon L. Carpenter, California State University, Long Beach; and Elliot Slutsky, Cal Poly, Pomona.

Together with colleagues and customers of Daisy Systems, Analog Design Tools, Valid, and Cadence, as well as University of California-Berkeley students, they have contributed to the material covered in this book.

October 1993   Andrei Vladimirescu

CONTENTS

Chapter Two

CIRCUIT ELEMENT AND NETWORK DESCRIPTION

Chapter Three

SEMICONDUCTOR-DEVICE ELEMENTS

Chapter Four
DC ANALYSIS   **114**

Chapter Five
AC ANALYSIS   **141**

Chapter Six
TIME-DOMAIN ANALYSIS   **168**

Chapter Nine

SPICE ALGORITHMS AND OPTIONS

Chapter Ten

CONVERGENCE ADVICE

APPENDIX D

GEAR INTEGRATION FORMULAS

APPENDIX E

SPICE INPUT DECK

INDEX

Introduction

SPICE—
THE THIRD DECADE

This introduction is a review of the evolution of SPICE from the initial research project at the University of California at Berkeley in the late 1960s, through the 1970s and 1980s, and into the 1990s.

A general description of SPICE techniques, analysis modes, and intended areas of application is provided first. The relation between solution algorithms, semiconductor-device models, and circuits to be characterized is explored in order to clarify the merits and limits of this program. The current trends in electrical-circuit simulation and the role of SPICE in its third decade are presented in the last part.

I.1   THE EARLY DAYS OF SPICE

SPICE in its different versions has been the main computer-aided analysis program used in analog design for over 20 years. SPICE is the result of the work of a number of talented graduate students in the Department of Electrical Engineering and Computer Science at the University of California at Berkeley, who had a mandate to produce the best computer program for the simulation of practical integrated circuits, ICs, under the guidance of D. Pederson and R. Rohrer. That this program was written by engineering students for engineers explains the simple and computationally efficient approach chosen for the network equations and the built-in semiconductor-device models.

The program known as SPICE today was first released under the name CANCER (Nagel and Rohrer 1971) in 1970 and acquired the name SPICE1 (Nagel and Pederson 1973) in 1972. SPICE2 (Cohen 1975; Nagel 1975), which in its various versions enjoys

1

the largest use worldwide today, was released in 1975. The universal acceptance is due not only to its robustness and ease of use but also to its free distribution by UC Berkeley.

It is important to note that the early years of the SPICE development were dedicated to the investigation of the most accurate and efficient numerical methods for circuit representation, input language, nonlinear equation solution, integration algorithms, sparse-matrix solutions, and nonlinear semiconductor-device modeling. The main goal of the SPICE project has been to provide an efficient computer tool for the design of the emerging ICs in the late 1960s and early 1970s. The choice of the nodal admittance representation is based on the relative ease of setting up the circuit matrix and the quick access to the DC operating point. In the design of linear ICs such as the μA 741, checking the bias point and performing a small-signal analysis were essential. A number of related programs originated from this research. First, BIAS-3 (McCalla and Howard 1971), a program for the nonlinear DC solution of bipolar circuits, was developed; it was later included in SLIC (Idleman, Jenkins, McCalla, and Pederson 1971) to address the analysis of linear bipolar ICs. The need for accurate large-signal time-domain simulation for the characterization of highly nonlinear circuits such as oscillators fueled the research for numerical integration and the development of the programs TRAC (Johnson et al. 1968) at Autonetics, a division of Rockwell Corporation, and CIRPAC (Shichman 1969) at Bell Labs. The algorithms of TRAC evolved into the programs TIME (Jenkins and Fan 1971) at Motorola and SINC at Berkeley. These efforts were continued through the 1970s with the MSINC program at Stanford and MTIME, which is still in use at Motorola.

The algorithmic research carried out during the development of these programs converged to the use of the Newton-Raphson solution of nonlinear equations, limiting techniques, implicit integration methods using fixed time steps, and reordering schemes for sparse matrices.

The emphasis on linear IC design using bipolar technology explains the priority given the implementation of bipolar device models, diodes, and transistors. CANCER (Nagel and Rohrer 1971) implemented the Ebers-Moll model (Ebers and Moll 1954) for the bipolar transistor described by 18 parameters. The circuit size was limited to 400 components, with only 100 transistors and diodes, and up to 100 nodes. The circuit decks were submitted on punched cards, and the program was developed and initially used on a CDC 6400.

An excellent review of the algorithms, techniques, and milestones in circuit simulation evolution can be found in the paper by D. Pederson (1984).

I.2 SPICE IN THE 1970s

In the early 1970s L. Nagel, continuing to develop the CANCER type of program, called the new version SPICE, Simulation Program with Integrated-Circuit Emphasis. In May 1972 SPICE1 was distributed for the first time in the public domain (Nagel and Pederson 1973).

The most important addition to this program was in the area of semiconductor device models. A better model for bipolar transistors, the integral charge-control model of H. Gummel and C. Poon (1971), was introduced in 1970. This model, available in SPICE1, included second-order effects, such as high-level injection and low-level recombination, and represented a major advancement over the Ebers-Moll model.

Models for two other semiconductor devices were added to SPICE1: the junction field effect transistor (JFET) and the metal-oxide-semiconductor field effect transistor (MOSFET). In this first implementation the two models were very similar and were based on the first-order quadratic model of H. Shichman and D. Hodges (1968).

A new approach to IC modeling, known as macromodeling (Boyle, Cohen, Pederson, and Solomon 1974), was introduced at this time to overcome the long run times required by the use of detailed transistor-level schematics. The first macromodels were developed for operational amplifiers by replacing many transistors through functionally equivalent controlled sources, which are simulated much faster. Macromodels are to this day the main approach to representing SPICE equivalent circuits for a variety of complex ICs.

A novel circuit-theoretical concept, the adjoint network, was introduced by Director and Rohrer (1969) in the late 1960s and added to SPICE very efficient computation of sensitivity, noise, and distortion.

The next major release of the program, SPICE2 (Nagel 1975), was completed in 1975 and offered significant improvements over SPICE1. A new circuit representation, known as modified nodal analysis (MNA) (Ho, Ruehli, and Brennan 1975; Idleman et al. 1971), replaced the old nodal analysis. This new representation added support for voltage-defined elements, such as voltage sources and inductors. A memory management package was developed in SPICE2 (Cohen 1975). It allowed the program to allocate dynamically the entire available memory to the solution of the circuit. This capability addressed the need to design larger ICs, because by the mid-1970s ICs had grown in complexity and the component limit of SPICE1 had become a serious limitation. The accuracy and speed of the analysis were improved by the addition of a time-step control mechanism and the stiffly stable multiple-order integration method of Gear (1967).

Independent research on circuit simulation conducted at IBM, which had introduced ECAP in 1965, led to the ASTAP program (Weeks, Jimenez, Mahoney, Mehta, Quassemzadeh, and Scott 1973). ASTAP used a different circuit representation, known as sparse tableau, which allowed access to all desired circuit state variables at the cost of run time and memory. ASTAP and SPICE2 used implicit, variable-order integration and the Newton-Raphson nonlinear solution, leading to their classification as third-generation circuit simulators (Hachtel and Sangiovanni-Vincentelli 1981).

A noteworthy event that took place in the second half of the 1970s was the introduction by NCSS of ISPICE (Interactive SPICE), the first commercially supported version. This trend of commercial SPICE derivatives grew considerably in the 1980s.

In the late 1970s all semiconductor companies used circuit simulation and most adhered to SPICE. IC technology had advanced the complexity of circuits to large-scale integration, LSI, and these circuits used mostly n-channel MOS devices. A new push was initiated by industry to improve the SPICE2 device models to keep up with technology.

The representation of MOSFETs in SPICE2 was significantly overhauled at this time. It was important to add device geometry information, such as drain and source areas, perimeters, and number of squares. In addition to the simple square-law MOS-FET model available in 1976, two more complex models were added, that described such effects as subthreshold conduction, carrier velocity-limited saturation, and short- and narrow-channel effects (Vladimirescu and Liu 1980). The need to measure sub-picocoulomb charges in memory cells also led to the implementation of a charge-based MOS model (Ward and Dutton 1978) in addition to the existing piecewise linear Meyer capacitance model (Meyer 1971).

Bipolar transistor geometries were also shrinking and the frequency of operation rising. Improvements such as base pushout, split base-collector capacitance, substrate capacitance, and transit-time modeling were added to the Gummel-Poon model. The increase in circuit size also brought about the need for increasing the accuracy of the sparse-matrix solution by allowing for run-time pivoting to correct any singularity that may occur during a long transient simulation.

1.3 SPICE IN THE 1980s

At the beginning of the 1980s, the introduction of the minicomputer gave engineering groups easier access to computer resources, and SPICE saw a tremendous increase in use. The VAX 11/780 quickly became the platform of choice for running SPICE, and engineers were able to view the results of the simulation on their terminals as soon as the analysis had been completed.

With the proliferation of the number of users, it became obvious that support for SPICE users was lacking. Large companies had internal CAD groups dedicated to support and enhancement of software packages for their engineers. Small engineering firms, however, had little help when using public-domain programs such as SPICE. This need was the driving force of new businesses with the charter to upgrade and support public-domain SPICE2. Examples include HSPICE from Meta-Software (1991).

LSI chips required electrical-simulation speeds in excess of an order of magnitude faster than SPICE. A number of approaches for speeding up electrical simulation by relaxing the accuracy or limiting the class of circuits to which it can be applied were used in several programs. Timing simulators such as MOTIS, or MOS Timing Simulator (Chawla, Gummel, and Kozak 1975), which was first introduced in the mid-1970s, led the way to a number of programs that took advantage of MOSFET characteristics, the infrequency of events in digital circuits, and the absence of feedback. Although fast, timing simulators did not have the accuracy needed in the design of sophisticated microprocessor and memory chips. Algorithmic innovation in timing simulation (Newton and Sangiovanni-Vincentelli 1984) led to the waveform relaxation technique, exemplified by the RELAX2 program (White and Sangiovanni-Vincentelli 1983), and iterated timing analysis, used in the SPLICE simulators (Kleckner, 1984; Newton 1978; Saleh, Kleckner, and Newton 1983). This class of electrical simulators achieved speedups in excess of an order of magnitude compared to SPICE for MOS digital circuits. The attempt to use these programs to characterize analog circuits required the implementation of the more accurate and time-consuming SPICE device models and often resulted in longer run times than SPICE.

A different approach to faster simulation of complex ICs is the mixed-mode, or hybrid, simulation, where individual blocks can be evaluated depending on the performed function; e.g., only analog blocks need electrical characterization, whereas digital blocks can be evaluated using logic simulation. Early efforts in this category include SPLICE from UC Berkeley, DIANA from the University of Leuven (DeMan et al. 1980), and SAMSON from Carnegie Mellon University (Sakallah and Director 1980). In spite of these first programs, good commercial mixed-mode simulators are lacking. One possible explanation is that the need for customization of mixed-mode simulators for specific applications goes against the desire of software tool companies to develop universal tools.

Yet another approach for speeding up electrical simulation was to tailor the direct-method algorithms of SPICE2 to various parallel computer architectures. Examples include CLASSIE (Vladimirescu 1982), which was developed at UC Berkeley for CRAY vector computers, which belong to the class of single-instruction multiple-data (SIMD) machines, and PACSIM (Deutsch and Newton 1984), of SIMUCAD Corporation for multiple-instruction multiple-data (MIMD) computers, such as the Sequent or Alliant machines, based on the MSPLICE project at UC Berkeley. Program SLATE (Yang, Haji, and Trick 1980), which emerged from research at the University of Illinois at Urbana, decoupled the analysis of circuit blocks and took advantage of latency to speed up the time-domain simulation. An alternate way to more speed has been to build dedicated hardware accelerators for circuit-simulation algorithms (Vladimirescu, Weiss, Danuwidjaja, Ng, Niraj, and Lass 1987; White 1986). These accelerators did not make it beyond a prototype, due to either insufficient performance or inflexibility.

Speedups of up to an order of magnitude were achieved for circuits having a regular hierarchical structure. Flat-circuit netlists, often the result of layout extractors, could be simulated only a few times faster. The lack of impressive speed returns, the customization needed for the various parallel architectures, and the emergence of RISC workstations with ever-increasing processing speeds doomed these efforts in the late 1980s.

A major development by the mid-1980s was the proliferation of the personal computer in the engineering field. By early 1984 PSPICE (MicroSim 1991), the first PC version of SPICE, was available on the IBM PC-XT. Although eight times slower than on a VAX 11/780, the de facto reference for SPICE throughput, PC-based SPICE programs have attracted many new users and considerably expanded the popularity of this electric simulator.

SPICE received an additional boost from the three companies Daisy, Mentor, and Valid, also referred to as DMV, which in 1981–1982 introduced integrated software packages for electronic design using microprocessor-powered engineering workstations. The field they developed is called computer-aided engineering. All three addressed the most lucrative aspect of digital design first. The need for an integrated analog simulation tool that would cover design entry, simulation, and the graphic display of results on the same screen/workstation had become obvious.

DMV realized this need and linked its schematic capture to a SPICE version and developed waveform display tools. Daisy took the lead in electrical simulation development by supporting an improved SPICE2 version, DSPICE, as part of its analog

Virtual Lab software, while the other two offered a user interface with SPICE2 or deferred the choice of the simulator to the end user. A new company, Analog Design Tools (ADT), emerged in 1985 with a well-integrated analog CAE product, an Apple Macintosh–like user interface called the Analog Workbench, which extended electric simulation to board-level analog engineers and power-supply designers, who historically had been reluctant to use computers. Both ADT and Daisy developed analog component libraries needed by analog system designers. The major achievement of these CAE companies was to extend the use of SPICE to the system- and board-level analog designer and to add new functionality and models to the program to serve the needs of those applications.

The SPICE technology was also advanced by the contributions of talented CAD groups at Bell Labs, Analog Devices, Texas Instruments, Hewlett-Packard, Tektronix, Harris Semiconductor, and National Semiconductor. Most of these groups had provided output display tools on graphic terminals in the second half of the 1970s. During the 1980s the effort was directed toward robust convergence, such as ADICE from Analog Devices and TekSPICE from Tektronix; accurate semiconductor device modeling, such as TI SPICE, ADVICE at Bell Labs, and HP SPICE; and additional functionality and user-friendly features. Although these proprietary developments were not available to the user at large, ideas and results of this parallel research work eventually found their way into public-domain or commercial software.

University research made new contributions to SPICE technology during this decade. At the beginning of the 1980s, the widespread use of UNIX in the university research environment offered increased interaction between user and program. SPICE2, however, was a FORTRAN batch program and was difficult to modify and limited in its potential use of C-shell utilities. These limitations led to the SPICE3 project at UC Berkeley (Quarles 1989a), the goal of which was to rewrite and improve SPICE2 version 2G6 (Vladimirescu, Zhang, Newton, Pederson, and Sangiovanni-Vincentelli 1981) using the C language to produce an interactive, modular, easily understood, structured program with a graphic tool for the display of results. SPICE3 (Quarles 1989b) was released in the public domain in March 1985.

An important achievement of the concurrent SPICE work in this decade was the elimination to a large degree of convergence problems. The use of continuation methods and education of users contributed toward reliable DC analysis. Improved models and techniques for handling discontinuities resulted in robust time-domain simulation as well.

An interesting concept that gained support toward the end of the 1980s was to provide the user the capability of describing the functions that govern the operation of devices used in the simulation. This feature was first available in ASTAP and then expanded in the SABER simulator (Analogy Inc. 1987) from Analogy. SABER, initially developed as a piecewise linear electrical simulator addressed to the simulation of analog systems, promoted the behavioral representation of entire circuit blocks by time-domain or frequency-domain equations. Modeling entire circuit blocks at a functional level rather than transistor level speeds up the simulation and enables a designer to evaluate an entire analog board or system.

I.4   SPICE IN THE 1990s

Today SPICE is synonymous with analog computer-aided simulation. Every major supplier of analog CAD/CAE software offers a well-supported and enhanced version of SPICE2 or SPICE3. The major CAE companies—Cadence and Mentor—offer a proprietary SPICE version as part of their analog CAE products: Analog Workbench and Analog Artist from Cadence and Accusim from Mentor. The main emphasis for the near future is on increased functionality, higher-level modeling, and tighter integration with schematic capture, display tools, and component libraries, as well as with physical design tools, such as printed circuit boards and integrated circuit layout.

An example of a state-of-the-art analog CAE product is the Analog Workbench II (1990) from Cadence, introduced by Analog Design Tools in 1985. All information needed for simulation is entered in graphical form and through menus. The circuit is entered as an electric schematic. In order to simulate a differential amplifier, a schematic replaces the SPICE deck. Input signals are defined and checked in a Function Generator tool. Time-domain simulation is controlled by pop-up menus, and the resulting waveforms can be viewed and measured in an Oscilloscope tool; a similar setup with a Frequency Sweeper and a Network Analyzer tool is used to control and view the results of an AC small-signal analysis. Similar analog CAE packages are available today from Microsim, Intusoft, Analogy, Viewlogic, Intergraph, Mentor, and others.

Significant research will be dedicated to extending the functionality of electrical simulation beyond the established analysis modes of SPICE, that is, beyond nonlinear DC and time-domain analysis and small-signal frequency-domain analysis. A number of interesting developments started in this direction at the end of the 1980s.

One extension is exemplified by *Harmonica* (Kundert and Sangiovanni-Vincentelli 1986) a nonlinear frequency-domain analysis program developed at UC Berkeley. Solution in the frequency domain is especially useful for finding the steady-state response of circuits with distributed elements and high-Q resonators. This approach is not very efficient for nonlinear transistor circuits.

Another direction of research is steady-state analysis, which solves the above problem in the time domain. This mode is particularly important for circuits with long settling times, such as switching power supplies. Although research on this topic took place in the 1970s (Aprille and Trick 1972), no reliable program is available today. Current research is under way at MIT and UC Berkeley; an envelope-following method is used in NITSWIT (Kundert, White, and Sangiovanni-Vincentelli 1988), and S-SPICE (Ashar 1989) is a vehicle for studying various troubleshooting techniques for the steady-state solution.

SPICE also lacks capabilities for specific applications, such as filter design in general and switched capacitor filters in particular. Specialized programs such as SWIT-CAP (Fang and Tsividis 1980) have been developed to fill this need. This trend of developing specific functionality for given applications not well suited to traditional SPICE analysis will continue in this decade.

Modeling technology is an important aspect of circuit simulation and is instrumental in defining the capabilities and the accuracy of a program.

At the top level of circuit representation, more support will be developed for the behavioral/structural description of entire circuit blocks. The ability to represent entire circuit blocks by an equation or a set of equations will make simulation of complete analog systems possible. Such functionality creates the need for powerful model-generation software capable of automating the process. Also, a description language for analog behavior, Analog Hardware Descriptive Language (Kurker et al. 1990), is under development under the guidance of the IEEE Standards Coordinating Committee 30.

At the transistor level of representation, in order to keep up with ever-shrinking semiconductor devices, SPICE will probably evolve to an open architecture that would enable CAD groups of IC manufacturers to implement better device models or upgrade the default ones. Improved transistor models have been reported in technical journals during the last decade with little impact on the various SPICE releases. Examples include the MEXTRAM model (de Graaf and Klosterman 1986) for bipolar transistors, which is reported to describe quasi-saturation and high-frequency effects better than the current Gummel-Poon model and which could be a useful addition to SPICE.

Whereas for the last two decades circuit simulation has been used mostly for analyzing fully-specified circuits, in this decade more emphasis will be put on the design aspect. Research work in the area of analog synthesis has been reported by groups at Carnegie Mellon University, Centre Suisse d'Electronique et de Microelectronique (CSEM), and the University of California at Berkeley; the synthesis tools these groups have developed, OASYS (Harjani, Rutenbar, and Carley 1989), IDAC (Degrauwe et al. 1987), and OPASYN (Koh, Sequin, and Gray 1987), respectively, can be used to design well-defined circuit blocks, such as operational amplifiers, from a collection of analog cells available in the knowledge base. During the next decade analog synthesis tools will evolve to facilitate the design of complex analog and mixed analog-digital systems. In conjunction with other software modules, SPICE will form the analytic core of analog optimization and synthesis software tools.

I.5 CONCLUSION

The new developments in circuit simulation do not make SPICE obsolete but rather complement it. SPICE will continue to be the main electrical simulator, because it solves the fundamental equations of an electrical system. In a recent report on PC-based analog simulation published in the magazine *EDN* (Kerridge 1990), the author concludes that "for the foreseeable future nothing will supplant SPICE as the industry standard for analog simulation." SPICE will probably add a number of analysis modes, such as nonlinear frequency-domain analysis and higher-level modeling capabilities, by supporting blocks described by integro-differential or algebraic equations.

Advances in computer technology will also increase the applicability of circuit simulation. Over the next few years the power of engineering workstations will increase to 1000 MIPS, 1000 megabytes of memory, and 1000 gigabytes of disk storage, according to Bill Joy's forecast at the 1990 Design Automation Conference (Joy 1990). This translates into a 50,000-transistor circuit simulation capability. One important upgrade needed in SPICE to make such a large simulation a reality is the decoupling of the analysis of circuit blocks at the level of the differential or nonlinear equations.

ACKNOWLEDGMENT

The author would like to thank D. Pederson for the inspiring discussions and suggestions that helped identify the various trends and the chronology of circuit simulation over the past three decades.

REFERENCES

"Analog Workbench II Adds Framework Features." *High Performance Systems* 11 (March).

Analogy, Inc. 1987 (December). *Saber: A Design Tool for Analog Systems.* Beaverton, OR: Author.

Aprille, T. J., and T. N. Trick. 1972. "Steady-State Analysis of Nonlinear Circuits with Periodic Inputs." *Proceedings of the IEEE* 60: 108–116.

Ashar, P. N. 1989. "Implementation of Algorithms for the Periodic Steady-State Analysis of Nonlinear Circuits." Univ. of California, Berkeley, Research Memo (March).

Boyle, G. R., B. M. Cohen, D. O. Pederson, and J. E. Solomon. 1974. "Macromodeling of Integrated Circuit Operational Amplifiers." *IEEE Journal of Solid-State Circuits* SC-9 (December): 353–363.

Chawla, B. R., H. K. Gummel, and P. Kozak. 1975. "MOTIS–An MOS Timing Simulator." *IEEE Transactions on Circuits and Systems* CAS-22 (December): 901–909.

Cohen, E. 1975. "Program Reference for SPICE2." Univ. of California, Berkeley, ERL Memo No. UCB/ERL M75/520 (May).

dè Graaff, H. C., and W. J. Klosterman. 1986. "Compact Bipolar Transistor Model for CACD, with Accurate Description of Collector Behavior." *Proceedings Ext. Conference on Solid-State Devices and Materials,* Tokyo: 287–290.

Degrauwe, M., et al. 1987. "IDAC: An Interactive Design Tool for Analog CMOS Circuits." *IEEE Journal of Solid-State Circuits* SC-22 (December): 1106–1116.

DeMan, H., et al., 1980. "DIANA: Mixed Mode Simulator with a Hardware Description Language for Hierarchical Design of VLSI." *IEEE ICCC '80 Conference Proceedings,* Rye Brook, NY (October): 356–360.

Deutsch, J. T., and A. R. Newton. 1984. "A Multiprocessor Implementation of Accurate Electrical Simulation." *21st ACM/IEEE DAC Conference Proceedings,* Albuquerque.

Director, S. W., and R. A. Rohrer. 1969. "The Generalized Adjoint Network and Network Sensitivities." *IEEE Transactions on Circuit Theory* CT-16 (August): 318–323.

Ebers, J. J., and J. L. Moll. 1954. "Large Signal Behavior of Bipolar Transistors." *Proceedings IRE* 42 (December): 1761–1772.

Fang, S. C., and Y. P. Tsividis. 1980. "Modified Nodal Analysis with Improved Numerical Methods for Switched Capacitive Networks." *IEEE ISCAS Conference Proceedings*: 977–980.

Gear, C. W. 1967. "Numerical Integration of Stiff Ordinary Differential Equations." Report 221, Dept. of Computer Science, Univ. of Illinois, Urbana.

Gummel, H. K., and H. C. Poon. 1970. "An Integral Charge-Control Model of Bipolar Transistors." *Bell System Technical Journal* 49 (May): 827–852.

Hachtel, G. D., and A. L. Sangiovanni-Vincentelli. 1981. "A Survey of Third-Generation Simulation Techniques." *Proceedings of the IEEE* 69 (October): 1264–1280.

Harjani, R., R. A. Rutenbar, and L. R. Carley. 1989. "Analog Circuit Synthesis for Performance in OASYS." *IEEE ICCAD Conference Proceedings,* Santa Clara, CA (November): 492–495.

Ho, C., A. E. Ruehli, and P. A. Brennan. 1975. "The Modified Nodal Approach to Network Analysis." *IEEE Transactions on Circuits and Systems* CAS-22 (June): 504–509.

Idleman, T. E., F. S. Jenkins, W. J. McCalla, and D. O. Pederson. 1971. "SLIC—A Simulator for Linear Integrated Circuits." *IEEE Journal of Solid-State Circuits* SC-6 (August): 188–204.

Jenkins, F. S., and S. P. Fan. 1971. "TIME—A Nonlinear DC and Time-Domain Circuit Simulation Program," *IEEE Journal of Solid-State Circuits* SC-6 (August): 182–188.

Johnson, E. D., et al., 1968 (June). "Transient Radiation Analysis by Computer Program (TRAC)." Technical Report issued by Harry Diamond Labs., Autonetics Div., North American Rockwell Corp., Anaheim, CA.

Joy, W. 1990. "Engineering the Future." Keynote address at the *27th ACM/IEEE Design Automation Conference,* Orlando, FL (June).

Kerridge, B. 1990. "PC-Based Analog Simulation." *EDN* (June): 168–180.

Kleckner, J. E. 1984. "Advanced Mixed-Mode Simulation Techniques." Univ. of California, Berkeley, ERL Memo No. UCB/ERL M84/48 (June).

Koh, H. Y., C. H. Sequin, and P. R. Gray. 1987. "Automatic Synthesis of Operational Amplifiers Based on Analytic Circuit Models." *IEEE ICCAD Conference Proceedings,* Santa Clara, CA (November): 502–505.

Kundert, K. S., and A. S. Sangiovanni-Vincentelli. 1986. "Simulation of Nonlinear Circuits in the Frequency Domain." *IEEE Transactions on Computer-Aided Design of Integrated Circuits and Systems.* CAD-5 (October): 521–535.

Kundert, K. S., J. White, and A. S. Sangiovanni-Vincentelli. 1988. "An Envelope-Following Method for the Efficient Transient Simulation of Switching Power and Filter Circuits." *IEEE ICCAD Conference Proceedings,* Santa Clara, CA (November): 446–449.

Kurker, C. M., et al. 1990. "Development of an Analog Hardware Description Language." *Proceedings of the IEEE 1990 CICC,* Boston (May): paper 5.4.

McCalla, W. J., and W. G. Howard, Jr. 1971. "BIAS-3: A Program for the Nonlinear DC Analysis of Bipolar Transistor Circuits." *IEEE Journal of Solid-State Circuits* SC-6 (February): 14–19.

Meta-Software. 1991. *HSPICE User's Guide.* Campbell, CA: Author.

Meyer, J. E. 1971. "MOS Models and Circuit Simulation," *RCA Review* 32 (March): 42–63.

MicroSim, 1991. *PSpice, Circuit Analysis User's Guide,* Version 5.0. Irvine, CA: Author.

Nagel, L. W. 1975. "SPICE2: A Computer Program to Simulate Semiconductor Circuits." Univ. of California, Berkeley, ERL Memo No. UCB/ERL M75/520 (May).

Nagel, L. W., and D. O. Pederson. 1973. "SPICE (Simulation Program with Integrated Circuit Emphasis)." Univ. of California, Berkeley, ERL Memo No. ERL M382 (April).

Nagel, L. W., and R. A. Rohrer. 1971. "Computer Analysis of Nonlinear Circuits, Excluding Radiation (CANCER)." *IEEE Journal of Solid-State Circuits.* SC-6 (August): 166–182.

Newton, A. R. 1978. "The Simulation of Large Scale Integrated Circuits." Univ. of California, Berkeley, ERL Memo No. UCB/ERL M78/52 (July).

Newton, A. R., and A. L. Sangiovanni-Vincentelli. 1984. "Relaxation-Based Electrical Simulation." *IEEE Transactions on Computer-Aided Design of Integrated Circuits and Systems.* CAD-3 (October): 308–331.

Pederson, D. O. 1984. "A Historical Review of Circuit Simulation." *IEEE Transactions on Circuits and Systems* CAS-31 (January): 103–111.

Quarles, T. L. 1989a. "Analysis of Performance and Convergence Issues for Circuit Simulation." Univ. of California, Berkeley, ERL Memo No. UCB/ERL M89/42 (April).

Quarles, T. L. 1989b. "SPICE3 Version 3C1 User's Guide." Univ. of California, Berkeley, ERL Memo No. UCB/ERL M89/46 (April).

Sakallah, K., and S. W. Director. 1980. "An Activity-Directed Circuit Simulation Algorithm." *IEEE ICCC '80 Conference Proceedings.* Rye Brook, NY (October): 1032–1035.

Saleh, R. A., J. E. Kleckner, and A. R. Newton. 1983. "Iterated Timing Analysis and SPLICE1." In *ICCAD '83 Digest.* Santa Clara, CA.

Shichman, H. 1969. "Computation of DC Solutions for Bipolar Transistor Networks." *IEEE Transactions on Circuit Theory* CT-16 (November): 460–466.

Shichman, H., and D. A. Hodges. 1968. "Modeling and Simulation of Insulated-Gate Field-Effect Transistor Switching Circuits." *IEEE Journal of Solid-State Circuits* SC-3: 285–289.

Vladimirescu, A. 1982. "LSI Circuit Simulation on Vector Computers." Univ. of California, Berkeley, ERL Memo No. UCB/ERL M82/75 (October).

Vladimirescu, A., K. Zhang, A. R. Newton, D. O. Pederson, and A. L. Sangiovanni-Vincentelli. 1981. "SPICE Version 2G User's Guide." Dept. of Electrical Engineering and Computer Science, Univ. of California, Berkeley (August).

Vladimirescu, A., D. Weiss, K. Danuwidjaja, K. C. Ng, J. Niraj, S. Lass. 1987. "A Vector Hardware Accelerator with Circuit Simulation Emphasis." *24th ACM/IEEE DAC Conference Proceedings,* Miami (June).

Vladimirescu, A., and S. Liu. 1980. "The Simulation of MOS Integrated Circuits Using SPICE2." Univ. of California, Berkeley, ERL Memo No. UCB/ERL M80/7 (February).

Ward, D. E., and R. W. Dutton. 1978. "A Charge-Oriented Model for MOS Transistor Capacitances." *IEEE Journal of Solid-State Circuits* SC-13 (October): 703–707.

Weeks, W. T., A. J. Jimenez, G. W. Mahoney, D. Mehta, H. Quassemzadeh, and T. R. Scott. 1973. "Algorithms for ASTAP—A Network Analysis Program." *IEEE Transactions on Circuit Theory* CT-20: 628–634.

White, J. 1986. "Parallelizing Circuit Simulation—A Combined Algorithmic and Specialized Hardware Approach." *IEEE ICCD '86 Conference Proceedings,* Rye Brook, NY (October).

White, J., and A. L. Sangiovanni-Vincentelli. 1983. "RELAX2: A New Waveform Relaxation Approach for the Analysis of LSI MOS Circuits." *Proceedings—IEEE International Symposium on Circuits and Systems,* Newport Beach, CA (May).

Yang, P., I. N. Hajj, and T. N. Trick. 1980. "SLATE: A Circuit Simulation Program with Latency Exploitation and Node Tearing." *IEEE ICCC '80 Conference Proceedings,* Rye Brook, NY (October).

One

INTRODUCTION TO ELECTRICAL COMPUTER SIMULATION

1.1 PURPOSE OF COMPUTER SIMULATION OF ELECTRICAL CIRCUITS

Knowledge of the behavior of electrical circuits requires the simultaneous solution of a number of equations. The easiest problem is that of finding the DC operating point of a linear circuit, which requires one to solve a set of equations derived from Kirchhoff's voltage law, KVL, and current law, KCL, and the branch constitutive equations, BCE. For a small circuit with linear elements, described by linear branch voltage-current dependencies, the exact DC solution is readily available through hand calculations. For larger linear circuits the DC solution and especially the frequency-domain or time-domain solutions are very complex. The analysis of circuits that contain elements described by a nonlinear relation between current and voltage adds another level of complexity, requiring the solution of the nonlinear branch equations simultaneously with the equations based on Kirchhoff's laws. Only small circuits can be solved by hand calculations, which yield only approximate results. Engineers learn in electronics courses to make certain approximations in order to predict the DC operation of small circuits by hand.

Another level of complexity is added when one has to predict the behavior in time or frequency of an electrical circuit. The nonlinear equations become integro-differential equations, which can be solved by hand only under such approximations as small-signal approximation or other limiting restrictions.

For many years designers working with discrete components have used breadboards to analyze and test the behavior of electronic circuits. To this day there are designers who use this approach for building analog circuits. But the breadboarding approach became inadequate with the breakthrough of integrated-circuit fabrication and associated novel circuit techniques in the years 1964–1965. Not only did transistors integrated on the same silicon chip behave differently from discrete transistors on a breadboard, but so did circuit elements in ICs differ from their discrete equivalents; for example, a transistor became the standard load device in an IC as opposed to a resistor on a breadboard. The fabrication of ICs on silicon wafers was and still is an expensive process both in cost and time. The fabrication of an IC requires several stages: first, a process needs to be defined, after which the electrical design is carried out and a circuit layout generated; photomicrographic plates used at each step during fabrication to obtain the desired circuit equivalent in silicon are produced from the layout; and, finally, the wafers are tested for correct operation. This costly fabrication flow required a correct electrical design the first time through. Since the electrical design engineer did not have the luxury of a trial-and-error approach in silicon to verify the correctness of the design, a *virtual breadboard* was needed. This could be produced on a digital computer by means of an electrical-analysis or simulation program. Programs intended for the electrical analysis of networks without taking any shortcuts in the solution of the KCL, KVL, and BCE equations are called *circuit simulators*.

Another important factor contributing to the development of computer programs for the analysis of electrical circuits was the advance in digital computers that occurred in the same years. General-purpose computers, such as the IBM 360 series, based on solid-logic technology (SLT), a hybrid technology using silicon semiconductor diodes and transistors and a precursor of monolithic ICs, were introduced offering capabilities similar to those of present-day computers, with mean time between failures (MTBF) drastically improved over the previous generation. The CDC 6400/6600 scientific computer was also introduced at that time, with architectural innovations that preceded the principles of today's supercomputers.

These two technological factors, ICs and powerful computers, defined both the need and the tool for automating the design process of electronic circuits. A number of researchers started studying the best techniques and algorithms for automating the prediction of the behavior of electric circuits (Pederson 1984). The first-generation programs, such as ECAP I in 1965 (IBM 1965) could only solve piecewise linear networks, and their scope was fairly limited. Advances in numerical techniques led in the late 1960s to the development of nonlinear analysis programs, such as SPICE1, initially named CANCER (Nagel and Rohrer 1971), and ECAP II (Branin, Hogsett, Lunde, and Kugel 1971).

A number of so-called *third generation* circuit simulation programs available today have their roots in the above second-generation nonlinear programs. The efforts of two decades ago have crystallized in the two circuit simulators now most often used, SPICE2 (Nagel 1971) and ASTAP (Weeks et al. 1973), currently called ASX. A detailed overview of the evolution of circuit simulation in general and SPICE in particular is presented in the Introduction.

1.2   WHAT IS SPICE?

SPICE is a general-purpose circuit-simulation program for nonlinear DC, nonlinear transient, and linear AC analysis. As outlined above, it solves the network equations for the node voltages. The program is equally suited to solve linear as well as nonlinear electrical circuits. Circuits for various applications, from switching power supplies to RAM cells and sense amplifiers, can be simulated with equal accuracy by SPICE. Circuits can contain resistors, capacitors, inductors, mutual inductors, independent voltage and current sources, dependent voltage and current sources, transmission lines, and the most common semiconductor devices, diodes, bipolar junction transistors (BJTs), junction field effect transistors (JFETs), metal-oxide-semiconductor field effect transistors (MOSFETs), and metal-semiconductor FETs (MESFETs).

The *DC analysis* part of the program computes the bias point of the circuit with capacitors disconnected and inductors short-circuited. SPICE uses iterations to solve the nonlinear network equations; nonlinearities are due mainly to the nonlinear current-voltage (I-V) characteristics of semiconductor devices.

The *AC analysis* mode computes the complex values of the node voltages of a linear circuit as a function of the frequency of a sinusoidal signal applied at the input. For non-linear circuits, such as transistor circuits, this type of analysis requires the *small-signal* assumption; that is, the amplitude of the excitation sources are assumed to be small compared to the thermal voltage ($V_{th} = kT/q = 25.8$ mV at 27°C, or 80.6°F). Only under this assumption can the nonlinear circuit be replaced by its linearized equivalent around the DC bias point.

The *transient analysis* mode computes the voltage waveforms at each node of the circuit as a function of time. This is a *large-signal* analysis: no restriction is put on the amplitude of the input signal. Thus the nonlinear characteristics of semiconductor devices are taken into account.

More types of analysis, associated with the above three basic simulation modes, are available in SPICE. They are described in Chaps. 4 through 6.

The best way to understand how SPICE works is to solve a circuit by hand; SPICE sets up and solves the circuit equations using the nodal equations (Dorf 1989; Nilsson 1990; Paul 1989) in the same manner as one writes the KVL, KCL, and BCE equations for a circuit and solves them. In the first example the DC solution of a linear resistive network is calculated by hand by an approach similar to the SPICE solution.

EXAMPLE 1.1

Calculate the node voltages in the resistive bridge-T circuit shown in Figure 1.1 and find the current flowing through the bias voltage source, V_{BIAS}. Use the source and resistor values given in the figure.

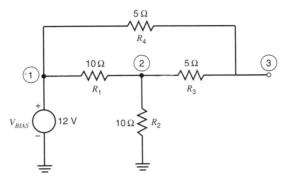

Figure 1.1 Bridge-T circuit.

Solution

The voltage at node 1 is equal to V_{BIAS}, or 12 V; the voltages at nodes 2 and 3 are found by writing the corresponding nodal equations:

$$\text{node 2}: \quad -G_1V_1 + (G_1 + G_2 + G_3)V_2 \quad\quad - G_3V_3 = 0 \quad\quad (1.1)$$

$$\text{node 3}: \quad -G_4V_1 \quad\quad\quad - G_3V_2 + (G_3 + G_4)V_3 = 0 \quad\quad (1.2)$$

where G_1, G_2, G_3, and G_4 are the conductances of resistors R_1 through $R_4 : G_1 = G_2 = 0.1$ mho and $G_3 = G_4 = 0.2$ mho. This is a system of two equations and two unknowns that can easily be solved for V_2 and V_3:

$$0.4V_2 - 0.2V_3 = 1.2$$

$$-0.2V_2 + 0.4V_3 = 2.4$$

The solution is $V_2 = 8$ V and $V_3 = 10$ V. The current through the bias source is equal to I_{R2}:

$$I_{VBIAS} = I_{R2} = G_2V_2 = 0.8 \text{ A}$$

In SPICE, Eqs. 1.1 and 1.2 are formulated from a graph of the circuit topology and are solved using Gaussian elimination (see Chap. 9). In Section 1.3.2 the results derived by hand in this example are compared with those obtained from SPICE.

The analysis of a one-transistor amplifier, a nonlinear circuit, is described in the following example. The hand derivation offers some insight on how SPICE automates the solution of the bias point for a nonlinear circuit.

EXAMPLE 1.2

Find the operating point of the one-transistor circuit of Figure 1.2; the bipolar transistor has a current gain $\beta_F = 100$ and a saturation current $I_S = 10^{-16}$ A.

Solution

The bias point of the transistor is defined by the collector and base currents, I_C and I_B, and the junction voltages, V_{BE} and V_{BC}. The two sets of equations needed for this solution are KVL and the BCEs of the bipolar transistor. The KVL equation is

$$R_B I_B + V_{BE} = V_{CC} \tag{1.3}$$

The most commonly used relation for the bipolar transistor equates the collector current I_C to the base current I_B:

$$I_C \approx \beta_F I_B \tag{1.4}$$

which is derived from the BCEs of the transistor. This approximation is valid only as long as the transistor operates in the forward active region; that is,

$$V_{BE} > 0.5 \text{ V} \quad \text{and} \quad V_{BC} < 0 \text{ V} \tag{1.5}$$

One more equation is needed in addition to Eqs. 1.3 and 1.4: the BCE that defines the current-voltage relation between I_C and V_{BE}:

$$I_C = I_S e^{qV_{BE}/kT} \tag{1.6}$$

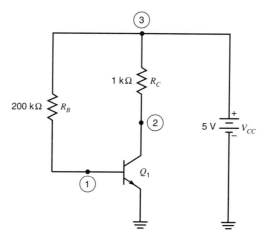

Figure 1.2   One-transistor circuit.

At this point two linear equations and one nonlinear equation must be solved to find the values defining the operating point of the transistor, I_C, I_B, and V_{BE}. The basic information on BJT equations and parameters, or *model*, is presented in Chap. 3, and the complete set of equations and parameters of the SPICE BJT model are listed in Appendix A.

To solve Eqs. 1.3–1.6, start with the assumption that $V_{BE} \approx 0.7$ V and replace it in Eq. 1.3 to yield the following value for I_B:

$$I_B = \frac{5 - 0.7}{200} \frac{V}{k\Omega} = 0.0215 \text{ mA}$$

Then I_C follows from Eq. 1.4:

$$I_C = 100 I_B = 2.15 \text{ mA}$$

The assumption for V_{BE} should be verified in Eq. 1.6 based on the value of I_C and then refined. Obviously, no one would ever go to that much trouble for hand calculations. On a computer, however, repeating, or *iterating*, the above solution until all data agree, or *converge*, is trivial. It is obvious that writing and solving by hand the KVL and BCE equations for a circuit with a few transistors is tedious.

One more detail must be checked before the above solution can be accepted: V_{BC} must satisfy the condition in Eq. 1.5. V_{BC} is calculated from the KVL equation for the BC junction mesh,

$$V_{BC} = R_C I_C - R_B I_B = (2.15 - 4.3) \text{ V} = -2.15 \text{ V}$$

and indeed it satisfies Eq. 1.5 for forward linear operation. If, however, $R_C = 2 \text{ k}\Omega$, Eq. 1.4 is no longer valid, the transistor is saturated and the two BCEs must be modified. The complete BCEs of a BJT (see Chap. 3) are coded in SPICE, guaranteeing correct analysis regardless of the operation region.

The above example has described in a nutshell the iterative procedure used in SPICE to solve a nonlinear circuit. In the next section the SPICE input file for this circuit is listed and is followed by the computer simulation and verification of the hand results.

1.3   USER INTERACTION WITH SPICE AND THE COMPUTER

This section describes the steps a user must follow for performing a SPICE analysis. First, the circuit schematic must be cast in a format that can be understood by SPICE, namely, the *SPICE input language*. For a description of this step, the type of information contained in the SPICE input file is examined in more detail by formulating the SPICE description for the bridge-T circuit in Figure 1.1 and the one-transistor amplifier in Figure 1.2. Second, several commands must be issued to the host computer to run the simulator, save the output in a file, and inspect the results graphically or in ASCII format. The two circuits from the previous section are used to acquaint the user with SPICE.

1.3.1   Electric Circuit Specification—The SPICE Input

Before running SPICE on any computer, batch or interactive, a user must create an ASCII file containing two kinds of information: the circuit description and the analysis requests. This file is referred to as the *SPICE input file*, or *SPICE deck*. In order to identify files easily, the input file is customarily named with the name of the circuit followed by a suffix, which can be `.ckt` or `.cir` for a circuit, `.spi` for SPICE, `.in` for input, and so on. Although the SPICE input file can have any name, the above naming convention is recommended:

```
circuit_name.suffix
```

The user can create the file with the editor of choice. The SPICE input language is free format and consists of a succession of statements. Most statements are a single line long, but SPICE accepts multiline statements; a *continuation line* must start with a + in the first column. A statement contains a number of fields separated by *delimiter characters*, which are a blank, a comma, an equal sign (=), or a left or right parenthesis; extra blanks are ignored.

A most important common feature of the various SPICE versions is that all accept the basic SPICE input language, that is, the SPICE2 syntax (Vladimirescu, Zhang, Newton, Pederson, Sangiovanni-Vincentelli 1981). Newer and proprietary versions of SPICE have additional functionality, and the user is advised to consult the users' guide of the specific version for the extra features. As long as one uses the functionality described in this text, with the exception of the SPICE3 (Quarles 1989) or PSpice (MicroSim 1991) extensions, the input files should be readable by a variety of SPICE versions. High-end SPICE products support a schematic-level specification. Examples include the *Analog Workbench* and *Analog Artist* from Cadence, *AccuSim* from Mentor, *SpiceNet* from Intusoft, and *Design Center* with PSpice from MicroSim.

The elements of the SPICE language can be introduced naturally by creating the SPICE decks for the bridge-T circuit in Figure 1.1 and the one-transistor circuit in Figure 1.2.

EXAMPLE 1.3

Write the SPICE input for the bridge-T circuit of Figure 1.1 and the one-transistor amplifier of Figure 1.2.

Solution

Any SPICE input file must start with a *title statement*, which identifies the circuit, and must conclude with an *end statement*, which is always the same, **.END**. These two lines must always be the first and the last, respectively. A number of lines start with an asterisk in the first column; these are *comment statements* and are used to document the circuit description and analysis requests.

The SPICE specification is component-oriented. Therefore, in order to describe the circuit in Figure 1.1 one needs to transfer the information in the figure, that is, element name, connectivity, and value, to a text file, as shown below. The circuit description consists of the element statements on lines 3 through 7, and each element can be easily identified from Figure 1.1. Other lines in the SPICE input file are the title, identifying the circuit as BRIDGE-T CIRCUIT; two comment lines, starting with an asterisk in the first column; and the **.END** line, which must always conclude a circuit and analysis description.

Another type of statement needed in a SPICE input file is the *control statement*. Control statements contain a period in the first column and define the types of analysis to be performed and the output variables to be stored. In order to verify the hand calculation of the bias point of the circuit, an **.OP** line is required, which requests the DC operating point analysis.

```
BRIDGE-T CIRCUIT
*
VBIAS 1 0 12
R1 1 2 10
R2 2 0 10
R3 2 3 5
R4 1 3 5
*
.OP
.END
```

The above circuit description is saved in the file bridget.ckt, which is used as input to SPICE. This input description is also known as a *SPICE deck* from the time that punched cards were used. Some details on how to run a SPICE simulation are presented in the next section. The circuit description is read by SPICE and compiled into an internal representation. Nodal equations identical to Eqs. 1.1 and 1.2 are set up internally in SPICE and solved as a linear system.

Next consider the one-transistor circuit of Figure 1.2. The circuit contains resistors, a bipolar transistor, and a voltage source. Following the approach used above, we transcribe the information from Figure 1.2 into the format required by SPICE: one element per line, with the name first followed by nodes and values. The resistors R_C and R_B and the voltage source, V_{CC}, can be specified with the same format used in the above deck. A new element not present in the BRIDGE-T CIRCUIT deck is the BJT. The first character for a BJT is **Q**, so the name is Q1; three nodes must be specified, for the collector, base, and emitter, respectively. Instead of a single value, such as a resistor, a

BJT is characterized by a number of parameters defined by a `.MODEL` statement; assume that the name of the model that describes Q_1 is QMOD. This name replaces the value field on a transistor line.

The model parameters for transistor Q_1, the forward gain β_F, and the saturation current, I_S, are specified in the global `.MODEL` statement for QMOD. The transistor is of type NPN and the SPICE parameter names for β_F and I_S are **BF** and **IS**, respectively.

After all the statements described so far are typed in an editor of choice, the SPICE input specification looks as listed below. We save the circuit in a file called `bjt.ckt`.

```
ONE-TRANSISTOR CIRCUIT (FIG. 1.2)
*
Q1 2 1 0 QMOD
RC 2 3 1K
RB 3 1 200K
VCC 3 0 5
*
.MODEL QMOD NPN IS=1E-16 BF=100
*
.OP
.END
```

SPICE distinguishes between *name fields* and *number fields*. A name field, such as Q1, RB, or QMOD, must start with a letter (*A–Z*) followed by additional letters or numbers. SPICE2, SPICE3, and PSpice are case insensitive, but some versions of SPICE for specific computers may require that the input file be uppercase.

A number field can contain an integer, such as 5 or −123; a floating-point number, such as 3.14159; a floating-point number in engineering notation, such as 1E-16 or 2.65E5; or either an integer or a floating-point number followed by one of the following *scale factors* recognized by SPICE: **T** = 1E12, **G** = 1E9, **MEG** = 1E6, **K** = 1E3, **MIL** = 25.4E-6, **M** = 1E-3, **U** = 1E-6, **N** = 1E-9, **P** = 1E-12, and **F** = 1E-15. The values of resistors R_C and R_B in the above example, 1K and 200K, respectively, are expressed in kilo-ohms using the scale factor **K**.

In a number field, a letter that immediately follows a scale factor or that immediately follows a number and is not a scale factor is ignored. Hence, 10, 10V, 10VOLTS, and 10HZ represent the same number, 10, and M, MA, MSEC, MMHOS represent the same scale factor, 10^{-3}. Many users like to append the units to the number fields, as in 10V, 1PF, or 10UM. These are all valid specifications as long as the physical unit is not confused with the scale factor; for example, 1F in SPICE is 10^{-15} and not 1 farad. Another common mistake is to confuse M, MEG, and MIL; 20MHZ is not 20 megahertz but 20 millihertz, and 0.1MIL is not 0.1 millimeter but 2.54 microns.

The circuit description consists of a number of *element statements*, Q1 through VCC in the above example. Each element statement contains the element *name*, the circuit *nodes* to which the element is connected, and the *values* of the parameters that determine the electrical characteristics of the element. The general format of an element statement is

Aname node1 node2 <node3 . . . ><MODELname><value1 . . . >

The first field always contains the name of the element, which must start with the letter that identifies the element type, such as **R** for resistor, **Q** for BJT, and **V** for voltage source. Except for the first letter, the rest of the element name can contain both characters and numbers. SPICE2 restricts the name to eight characters; SPICE3 and PSpice do not.

The following fields, *node1, node2, <node3 . . . >*, are lowercase to identify them as number fields; they represent the node numbers at which the element is connected. SPICE2 accepts only numbers for nodes, whereas SPICE3, PSpice, and most commercial SPICE versions allow node names as well. Throughout this text, fields whose names are uppercase or start with an uppercase character are name fields, and those whose names are lowercase are number fields. The fields enclosed in angle bracket ($<>$) are optional or need to be present depending on type. Bold characters are used in this book to identify key words and parameter names that are part of the SPICE language. A comprehensive list of all conventions used in this text can be found in Section 2.1.

Each element must be characterized by at least one *value1*. This number field is shown as optional because SPICE provides defaults such as 0 or 1, depending on context, for missing value fields. Alternatively, some elements, such as transistors, are characterized by several values, which are grouped on a separate line, called the *model* statement. Instead of referencing a value field, these elements reference the name of the model definition that contains the parameters, *MODELname*. The name of the model that defines the parameters of Q_1 in the example is QMOD. Note that either a value or a model name should end an element line.

The *model* statement allows one to specify only once a set of parameters common to a number of elements, for example, the parameters of all transistors with the same geometry integrated on one silicon chip. For each *MODELname* referenced, the circuit specification must contain a **.MODEL** statement. The **.MODEL** statement belongs to a different category of statements, the *global statements*.

The general format of a **.MODEL** statement is

.MODEL *MODELname MODELtype PARAM1=value1 PARAM2=value2 . . .*

The period in the first column differentiates global and control statements from element statements. *MODELname* uniquely identifies one set of parameters common to one or more elements, and *MODELtype* is one of the seven or eleven types of models supported by SPICE2 or SPICE3, respectively. Transistor Q_1 in Example 1.3 is an NPN transistor, one of the two *MODELtypes* supported in SPICE for BJTs. From zero up to the maximum number of model parameters supported for a specific model type can be specified. Each of *PARAM1, PARAM2, . . .* must be one of the accepted keywords for the model type, as defined in Chaps. 2 and 3 and Appendix A. Model QMOD defines only two parameters, **IS**, the saturation current, and **BF**, the forward current gain, for Q_1 in the one-transistor circuit.

MOSFETs of different geometric sizes can be characterized by the same model parameters, such as threshold voltage and thin-oxide thickness. Thus an element that references a .**MODEL** statement may require some values that are specific to it. Therefore, complex elements, such as transistors, are characterized by both *device parameters*, defined on the element line, and *model parameters*, grouped in the .**MODEL** statement. For example, the different channel widths, W, and lengths, L, of MOSFETs are defined in element statements.

The last category of statements necessary in a SPICE deck is the *control statement*. Control statements specify the analyses to be performed by SPICE as well as define initial states. All control statements start with a dot in the first column. In Example 1.3 the DC bias point is requested by the .**OP** line. Chaps. 4 through 6 describe in detail all control statements.

In summary, every SPICE input file has the following general structure:

Title statement

* Comment statements

Element statements

Global statements

Control statements

.**END** (end statement)

1.3.2   SPICE Simulation, DC Analysis

The next step is to run the simulation. SPICE2 is available on a variety of computers and operating systems worldwide, and this section does not enumerate all possibilities but is limited to the computers most often used by students and professionals.

UNIX is the operating system of choice in universities; it is presently available on a variety of computers ranging from the Personal Computer to the Cray. The simulation of bridget.ckt or bjt.ckt created in Example 1.3 can be accomplished in UNIX by typing

```
% spice2 < bjt.ckt > bjt.out
```

assuming that the executable program is called spice2 and that it is located in a directory that is in the search path of the user. The redirection signs, $<$ and $>$, define the file where the input data reside and the file where the results are stored, respectively. Upon completion of the simulation, bridget.out and bjt.out can be inspected by having it typed on the screen, viewing it through an editor, or printing it.

EXAMPLE 1.4

Print the result files bridget.out and bjt.out and compare the results with the hand calculations of Examples 1.1 and 1.2.

Solution

The output `bridget.out` produced by SPICE2 is listed in Figure 1.3, and the solution can be verified to be identical to the hand calculation. It is important to note that SPICE2 and PSpice always echo back the circuit description received. Many errors can be identified by carefully comparing the circuit description output by SPICE with the original schematic.

The contents of `bjt.out` are shown in Figure 1.4. The SPICE2 output contains several sections. First, the CIRCUIT DESCRIPTION is echoed so that the user can check for potential errors. Next, the BJT MODEL PARAMETERS defined in a **.MODEL**

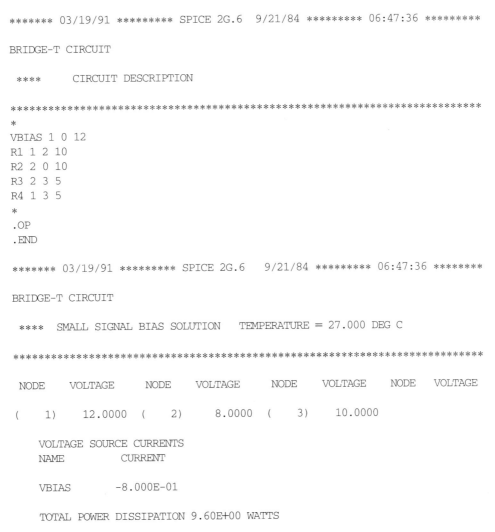

```
******* 03/19/91 ********* SPICE 2G.6  9/21/84 ********* 06:47:36 *********

BRIDGE-T CIRCUIT

 ****     CIRCUIT DESCRIPTION

***********************************************************************
*
VBIAS 1 0 12
R1 1 2 10
R2 2 0 10
R3 2 3 5
R4 1 3 5
*
.OP
.END

******* 03/19/91 ********* SPICE 2G.6   9/21/84 ********* 06:47:36 ********

BRIDGE-T CIRCUIT

 ****   SMALL SIGNAL BIAS SOLUTION    TEMPERATURE = 27.000 DEG C

***********************************************************************

  NODE    VOLTAGE    NODE    VOLTAGE    NODE    VOLTAGE    NODE   VOLTAGE

(   1)    12.0000  (    2)     8.0000  (    3)    10.0000

     VOLTAGE SOURCE CURRENTS
     NAME          CURRENT

     VBIAS        -8.000E-01

   TOTAL POWER DISSIPATION 9.60E+00 WATTS
```

Figure 1.3   SPICE2 results for DC operating point.

statement are printed. Finally, the voltages, power consumption, and small-signal characteristics computed by SPICE2 are listed.

The node voltages listed as part of the SMALL SIGNAL BIAS SOLUTION agree with the ones obtained through hand calculations in Example 1.2. The difference, of the order of tens of millivolts, is due to the assumption that $V_{BE} \approx 0.7$ V in the hand calculations. Although SPICE2 computes a first solution with a similar assumption, it continues to iterate until Eqs. 1.3, 1.4, and 1.6 are satisfied. The VOLTAGE SOURCE CURRENTS, in this case the source V_{CC}, and the TOTAL POWER DISSIPATION are also listed in this section.

The OPERATING POINT INFORMATION of transistor Q_1, consisting of I_B, I_C, V_{BE}, V_{BC}, and V_{CE}, defines the region of operation. According to these values, Q_1 is

```
******* 03/25/91 ******* SPICE 2G.6   9/21/84 ******* 23:07:40 ***********

ONE-TRANSISTOR CIRCUIT (FIG. 1.2)

 ****     CIRCUIT DESCRIPTION

******************************************************************************
*
Q1 2 1 0 QMOD
RC 2 3 1K
RB 1 3 200K
VCC 3 0 5
*
.MODEL QMOD NPN
*
.OP
*
.END
******* 03/25/91 *******    SPICE 2G.6   9/21/84 ******* 23:07:40 *********

ONE-TRANSISTOR CIRCUIT (FIG. 1.2)

 ****     BJT MODEL PARAMETERS

******************************************************************************

          QMOD
          NPN
      IS 1.00E-16
      BF 100
      NF 1
      BR 1
      NR 1
```

Figure 1.4   File `bjt.out`: DC analysis results.

```
******* 03/25/91 ******* SPICE 2G.6   9/21/84 ******* 23:07:40 ************

ONE-TRANSISTOR CIRCUIT (FIG. 1.2)

****    SMALL SIGNAL BIAS SOLUTION    TEMPERATURE = 27.000 DEG C

*************************************************************************

  NODE   VOLTAGE    NODE   VOLTAGE    NODE   VOLTAGE   NODE   VOLTAGE

(   1)    .7934  (   2)    2.8967  (   3)    5.0000

    VOLTAGE SOURCE CURRENTS
    NAME          CURRENT

    VCC          -2.124E-03

    TOTAL POWER DISSIPATION    1.06E-02 WATTS

******* 03/25/91 *******    SPICE 2G.6   9/21/84 ******* 23:07:40 *********

 ONE-TRANSISTOR CIRCUIT (FIG. 1.2)

  ****   OPERATING POINT INFORMATION    TEMPERATURE = 27.000 DEG C

*************************************************************************

**** BIPOLAR JUNCTION TRANSISTORS

NAME        Q1
MODEL       QMOD
IB          2.10E-05
IC          2.10E-03
VBE            0.793
VBC           -2.100
VCE            2.900
BETADC       100.000
GM          8.13E-02
RPI         1.23E+03
RX          0.00E+00
RO          1.00E+12
CBE         0.00E+00
CBC         0.00E+00
CBX         0.00E+00
CJS         0.00E+00
BETAAC       100.000
FT          1.29E+18
```

Figure 1.4   (*continued*)

biased in the forward active region, as predicted by hand calculations. A value of β_F, BETADC, is computed that takes into account the I_C and I_B for each transistor. The rest of the data represent the values of the elements in the linear equivalent model of the transistor and are described in more detail in Chap. 3.

The above examples are intended to show a new user of computer simulation how natural the SPICE input language is and how straightforward is the output information provided by the program for a simple DC analysis. In this and the following chapters all supported elements and analysis modes are described, as well as the best approaches for using SPICE to solve electrical circuits.

SPICE2 is a batch program. If a circuit element must be changed or an output request has been omitted, the previous steps must be repeated, namely, editing the input file, running the simulation, and viewing the output file.

SPICE3, however, is interactive, and the user is transferred inside a *spice shell* upon invocation of the program:

```
Spice 1 ->
```

SPICE3 commands must be entered at the prompt. At this point the user needs a few additional commands not available in SPICE2 in order to communicate with the program. Figure 1.5 is the transcript of an interactive SPICE3 session of the same simulation as performed above for `bjt.ckt`. As can be seen, each command is followed by the results displayed on the screen by the program. The first command, **source**, defines the input file; the second, **listing**, lists the input file for verification; the third, **op**, runs the DC operating point; and the fourth, **print all**, lists all node voltages and currents through voltage sources, that is, the same information found under the SMALL SIGNAL BIAS SOLUTION header in SPICE2.

Note that the print command does not provide the OPERATING POINT INFORMATION of SPICE2, and in release 3d2 this information can be accessed only on a device-by-device basis by using the **show q1** command. The resulting information listed on the screen is much more verbose.

For larger circuits it is useful to **display** all available output variables, but **print** or **plot** only selected ones.

While one circuit is active, the results of various analyses can be either viewed graphically on the screen or printed out. SPICE3 saves all node voltages of all simulations performed during the same SPICE3 session and provides for the interactive display of selected waveforms while the simulation is running. Input files can be edited and repeated simulations can be performed from within the SPICE3 shell. When you have finished, type **quit** to exit SPICE3. The results of the different runs have been saved in temporary files, which are displayed before the program exits, and the user has a last chance to save any desired results.

SPICE3 can also be run in batch mode through the command

```
% spice3 -b bjt.ckt > bjt.out
```

```
unix 1> spice3d2

This is a sample news file and will be printed
whenever spice or nutmeg is started.

Spice 1 -> source bjt.ckt

Circuit: ONE-TRANSISTOR CIRCUIT (FIG. 1.2)

Spice 2 -> listing
      ONE-TRANSISTOR CIRCUIT (FIG. 1.2)

    1 : one-transistor circuit (fig. 1.2)
    3 : q1 2 1 0 qmod
    4 : rc 2 3 1k
    5 : rb 1 3 200k
    6 : vcc 3 0 5
    9 : .model qmod npn is=1e-16 bf=100
   15 : .end
Spice 3 -> op
Spice 4 -> print all
v(1) = 7.934384e-01
v(2) = 2.896719e+00
v(3) = 5.000000e+00
vcc#branch = -2.12431e-03
Spice 7 -> quit
Warning: the following plot hasn't been saved:
op2   ONE-TRANSISTOR CIRCUIT (FIG. 1.2) , operating point

Are you sure you want to quit (yes)? yes
Spice-3d2 done
```

Figure 1.5 Transcript of interactive SPICE3 analysis of bjt.ckt.

At completion bjt.out contains information similar to but in a different format from what is produced by SPICE2. SPICE3 also produces a *rawfile* containing data to be displayed graphically when the -r flag is used. If no filename is specified, the display data are stored in a file called rawspice.raw. The companion postprocessor for the SPICE3 rawfiles is called *Nutmeg*.

The most common platform for running SPICE has become the IBM PC and PC clones. Although SPICE2 is not available from UC Berkeley for the PC, a number of commercial offerings, such as PSpice from MicroSim, IsSpice from Intusoft, and HSPICE from Meta Software, run under DOS or Windows. SPICE3 is distributed by UC Berkeley for the PC; the models supported range from the AT to the 486.

Although the sequence of operations remains the same, the command to simulate may differ from package to package. Some packages offer DOS shells, which facilitate the simulation sequence of creating or modifying the input circuit, running SPICE, and viewing the results. The schematic capture available in the Design Center from MicroSim and SpiceNet from Intusoft allow the circuit specification to be entered graphically rather than through the SPICE netlist language.

Although in Windows-based packages most operations can be performed from menus, it is still useful to follow file-naming conventions similar to those introduced above for circuit identification. When one runs PSpice from DOS, the input and output file names follow the program name as command arguments:

```
C > PSPICE BJT.CKT BJT.OUT
```

The output file from PSpice is of the same format and contains the same information as that from SPICE2. The name of the output file must not be specified unless a different suffix than `.OUT` is desired. Note that `BJT.OUT` as saved by PSPICE is similar in format to that saved by SPICE2. SPICE3 and Nutmeg can be run similarly on a PC. First, SPICE3 is run in batch mode, then Nutmeg is run to view the results. (The two commands can be saved in a `.BAT` file, `SPICE3.BAT`, which runs `BSPICE`, the name of batch SPICE3 on a PC, and then runs Nutmeg on the result file `RAWSPICE.RAW` created by `BSPICE`.) More on displaying results is presented in the following two sections.

Although not as common as on the PC, SPICE is also available on the Apple Macintosh. PSpice and IsSpice are offered on the Macintosh, as is MacSpice, from Deutsch Engineering, which is specifically tailored for the Macintosh interface.

Both SPICE2 and SPICE3 are distributed by the University of California, Berkeley, for several computers and operating systems. For the VMS operating system on DEC computers, the same steps as in UNIX must be followed for simulating circuits. When the input file is ready, type

```
$ run spice
```

SPICE2 will prompt for an *input file* and then for an *output file*. SPICE3 transfers to its shell, where the same commands as in UNIX are valid.

SPICE2 and SPICE3 run in batch mode on a variety of mainframe computers, such as IBM, CDC, and Cray. Typically, the jobs submitted to such computers contain job control statements in addition to the SPICE circuit description. Job control statements define the user identification, time limit, computer resources, input and output units, and the program to be run. The Electronics Research Laboratory at UC Berkeley tells where users can get a copy of SPICE for a specific mainframe computer.

1.3.3 SPICE Results for AC and TRAN Analyses

So far we have learned how to write a SPICE circuit file, what types of statements are available, how to run the analysis on the most common computers, and how to interpret the results of a DC analysis in the output file. Only one control statement, **.OP**, has been introduced so far; this was sufficient for obtaining all computed node voltages in the output files, `bridget.out` and `bjt.out`.

In addition to a DC analysis, SPICE also performs a steady-state sinusoidal analysis, invoked by the `.AC` control line, and a time-domain analysis, invoked by the `.TRAN` line. These analyses result in large amounts of data, due to the fact that each node voltage is computed for all frequencies or times. For small circuits such as the ones in Examples 1.1 and 1.2, this may not be a problem. But as the circuits grow, so will the size of the output files, making the results of interest more difficult to find. Thus the `.PRINT` and `.PLOT` control statements define the node voltages of interest or, more generally, the desired output variables. `.PRINT` and `.PLOT` define only the results saved in the output file; various versions of SPICE have additional control lines or interactive commands that let the user access results that have been saved in binary files and display them in graphical mode with the help of postprocessing programs.

The interpretation of the results of SPICE simulations is best understood if exemplified for typical applications. The following example introduces the SPICE frequency, or AC, analysis and how to display the results as a Bode plot.

EXAMPLE 1.5

Compute the node voltages and the current for the series RLC circuit shown in Figure 1.6 assuming the following periodic input signal:

$$v_{in} = 5\cos 2\pi 10t$$

Write the SPICE deck and run the program to verify the results derived from hand calculation. Then, vary the frequency of the input signal v_{in} from 1 Hz to 10 kHz and obtain the Bode plot of the magnitude and the phase of the voltage across the capacitor C_1, v_3.

Solution
First, the solution for the network is obtained using phasor calculations. In calculating the current, the resistance R_1 and the inductor reactance ωL_1 can be neglected compared to the reactance of the capacitor, $1/\omega C_1$. The current is

$$\mathbf{I} = \frac{V_{in}}{\mathbf{Z}} = \frac{V_{in}}{R_1 + 1/j\omega C_1} = \frac{5 \cdot 10^{-3}}{0.2 - j16} \approx j0.31 \cdot 10^{-3} \qquad (1.7)$$

$$|\mathbf{I}| = 0.31 \text{ mA}, \qquad \angle\mathbf{I} = 90°$$

The voltages at nodes 2 and 3, v_2 and v_3, closely follow the input signal:

$$v_3 \approx v_2 \approx v_{in}$$

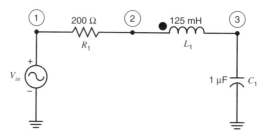

Figure 1.6   Series RLC circuit.

The SPICE deck for this circuit, `rlc.ckt`, is based on the previous examples.

```
SERIES RLC CIRCUIT
*
VIN 1 0 AC 5 0
L1 2 3 0.125
C1 3 0 1U
R1 1 2 200
.AC LIN 1 10 10
.PRINT AC IM(VIN) IP(VIN) VM(2) VP(2) VM(3) VP(3)
.END
```

Two new control statements are used, **.AC** and **.PRINT.** The first statement represents a request for a steady-state frequency-domain analysis and defines the type of frequency variation, **LIN**, the number of frequencies, 1, and the starting and ending frequencies, which in this case are the same, 10 Hz.

The second statement, **.PRINT**, is necessary if the user wants the results of the analysis to be written into the output file. Unlike the case of the DC operating point analysis, **.OP**, where SPICE automatically prints the results in the output file, for the other analyses the user must specify which data are to be saved in the output file. The .PRINT keyword must be followed by the type of analysis, AC in this case, and circuit variables to be saved in the output file. For AC variables the character V (for voltage) or I (for current) must be followed by one or more additional characters specifying whether polar or rectangular values are desired. AC voltages and currents are phasor quantities and can be expressed in terms of either real and imaginary parts, as in $VR(2)$ and $VI(2)$, or a magnitude and a phase, as in $VM(2)$ and $VP(2)$. These two statements are described in detail in Chaps. 5 and 4, respectively. In most SPICE versions, including SPICE2 and SPICE3, only the currents flowing through voltage sources can be measured; therefore, a current requested on a **.PRINT** line must always have as argument a voltage source name.

The following results are computed by SPICE for the desired phasors at 10 Hz:

```
 FREQ      IM(VIN)    IP(VIN)    VM(2)      VP(2)      VM(3)      VP(3)
1.000E+01 3.143E-04 -9.072E+01 5.000E+00 -7.203E-01 5.002E+00 -7.203E-01
```

If the amplitude of the input signal were 1 V, the voltage at each node computed by SPICE would represent the transfer function at that node referred to the input. The phase of the current, IP (VIN), is negative because the direction of flow is assumed in SPICE to be from the positive to the negative terminal through the voltage source.

The next part of this example is to obtain the Bode plot of the magnitude and phase of the voltage across the capacitor for the frequency range from 1 Hz to 10 kHz. The only two statements that need to be changed are the **.AC** statement, to reflect the desired frequency range, and the output request line. We also change the amplitude of v_{in} to 1 V in order to obtain the transfer function v_3/v_{in} and modify the value of R_1 to 50 Ω.

For the frequency variation a logarithmic scale is desirable, because it provides an even number of analysis points in all ranges. SPICE offers two choices of logarithmic intervals, the decade, **DEC**, and the octave, **OCT**. The most commonly used frequency interval is the decade, and we will use it for obtaining the Bode plot. The new **.AC** line is

```
.AC DEC 10 1 10K
```

This statement requests an AC solution of the RLC circuit at 10 frequencies per decade in the interval from 1 Hz to 10 kHz.

In addition to the **.PRINT** capability, SPICE supports a **.PLOT** command, which produces an ASCII character, or line-printer, plot. The information entered on the **.PLOT** line is identical to that entered on a **.PRINT** line. For the desired Bode plot the requesting statement is

```
.PLOT AC VDB(3) VP(3)
```

Note that the magnitude of the voltage is requested in decibels as specified by the suffix DB. The new **.AC** line and the above **.PLOT** line should replace the existing **.AC** and **.PRINT** lines in the SPICE input file rlc.ckt. The resulting Bode plot, shown in Figure 1.7, is taken from the SPICE output file, rlc.out. The amplitude peaks at 450 Hz, the resonant frequency of the circuit, and the phase changes by 180° because of the two complex poles (see also Chap. 6).

```
******* 09/08/92 ******** SPICE 2G.6 3/15/83 ******** 19:45:56 ********

SERIES RLC CIRCUIT

****  AC ANALYSIS    TEMPERATURE = 27.000 DEG C

*******************************************************************
LEGEND:

*: VDB(3)
+: VP(3)
```

Figure 1.7 SPICE2 ASCII plot of VDB(3) and VP(3).

```
        FREQ VDB(3)

*)------------- -6.000D+01  -4.000D+01  -2.000D+01     0.000D+00  2.000D+01
                - - - - - - - - - - - - - - - - - - - - - - - - - - - -
+)------------- -2.000D+02  -1.500D+02  -1.000D+02    -5.000D+01  0.000D+00
                - - - - - - - - - - - - - - - - - - - - - - - - - - - -
1.000D+00   4.288D-05 .           .           .             *            +
1.259D+00   6.695D-05 .           .           .             *            +
1.585D+00   1.067D-04 .           .           .             *            +
1.995D+00   1.694D-04 .           .           .             *            +
2.512D+00   2.678D-04 .           .           .             *            +
3.162D+00   4.247D-04 .           .           .             *            +
3.981D+00   6.726D-04 .           .           .             *            +
5.012D+00   1.066D-03 .           .           .             *            +
6.310D+00   1.689D-03 .           .           .             *            +
7.943D+00   2.677D-03 .           .           .             *            +
1.000D+01   4.245D-03 .           .           .             *            +
1.259D+01   6.727D-03 .           .           .             *            +
1.585D+01   1.067D-02 .           .           .             *            +
1.995D+01   1.691D-02 .           .           .             *            +
2.512D+01   2.682D-02 .           .           .             *            +
3.162D+01   4.254D-02 .           .           .             *            +
3.981D+01   6.751D-02 .           .           .             *            +
5.012D+01   1.072D-01 .           .           .             *            +
6.310D+01   1.706D-01 .           .           .             *            +
7.943D+01   2.719D-01 .           .           .             *            +
1.000D+02   4.348D-01 .           .           .             *          +.
1.259D+02   6.994D-01 .           .           .             *          +.
1.585D+02   1.135D+00 .           .           .            .*          +.
1.995D+02   1.873D+00 .           .           .            .*          +.
2.512D+02   3.184D+00 .           .           .            . *        + b.
3.162D+02   5.744D+00 .           .           .            .      *  +  .
3.981D+02   1.200D+01 .           .           .            .        + * .
5.012D+02   1.085D+01 .           .           .            .          * .
6.310D+02   1.336D-01 .           .      +    .            .      *      .
7.943D+02  -6.561D+00 .       +        .           .    *    .          .
1.000D+03  -1.193D+01 .       +        .           .       *    .       .
1.259D+03  -1.669D+01 .       +        .           .  *         .       .
1.585D+03  -2.114D+01 .      +         .           *.          .        .
1.995D+03  -2.542D+01 .      +         .       *    .          .        .
2.512D+03  -2.958D+01 .      +         .   *        .          .        .
3.162D+03  -3.369D+01 .      +        .  *          .          .        .
3.981D+03  -3.775D+01 .      +      . *             .          .        .
5.012D+03  -4.180D+01 .      +    *.                .          .        .
6.310D+03  -4.582D+01 .     +   *   .               .          .        .
7.943D+03  -4.984D+01 .     +*       .               .          .        .
1.000D+04  -5.385D+01 . *  +        .                .          .        .
                - - - - - - - - - - - - - - - - - - - - - - - - - - - -
```

Figure 1.7   (*continued*)

There are ways to obtain SPICE plots with high-quality graphics on a computer screen or printer. For the UC Berkeley releases, SPICE 2G6 and SPICE3, the graphic postprocessor is Nutmeg. SPICE2 plots can be viewed or printed in Nutmeg on a UNIX system by setting the rawfile option, `-r filename`, on the command line:

```
spice2 -r rlc_sp2.raw < rlc.ckt > rlc.out
```

Before a rawfile created by SPICE2 can be opened in Nutmeg, the *sconvert* utility must be run, which translates the file to the SPICE3/Nutmeg rawfile format:

```
sconvert o rlc_sp2.raw a rlc_sp3.raw
```

The Bode plot of the capacitor voltage magnitude and phase produced by Nutmeg from a SPICE2 rawfile is shown in Figure 1.8. If SPICE3 is used for simulation, the plot of any desired circuit variable can be obtained while running the program interactively from the spice3 shell, as shown in Section 1.3.2. If SPICE3 is run in batch mode, it creates by default a binary result file called `rawfile.raw`, which can then be directly loaded and viewed in Nutmeg. The rawfile created by SPICE3 can be assigned any name with the `-r filename` option on the command line, as in the following:

```
spice3 -b -r rlc.raw rlc.ckt > rlc.out
```

Note that SPICE3 does not support the **.PRINT** and **.PLOT** commands of SPICE2 and most other commercial SPICE programs. Also note that the **units** variable must be set in SPICE3 to **degrees** in order to obtain a plot compatible with SPICE2,

```
Spice -> set units = degrees
```

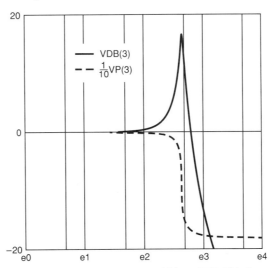

Figure 1.8   Bode plot of VDB (3) and VP (3) from Nutmeg.

Another graphics package available on UNIX machines that run X-Windows is *xgraph*. The tabular output created by the SPICE2 .PRINT command can be used as input to *xgraph*. The data produced by the .PRINT command are extracted from the SPICE output file, a few headers are added, and the results can then be viewed or printed in the xgraph tool; consult the manual page of xgraph for the details.

Commercial PC SPICE packages offer postprocessing for all popular PC graphics, including SVGA, VGA, EGA, CGA, and Hercules. Besides using the regular output file, the user can view the results of the simulation in graphic mode. MicroSim offers *Probe*, which can display and perform arithmetic operations on waveforms. In order to generate the graphics file PROBE.DAT, a control statement, .PROBE, must be added to the PSpice input file. IntuScope, the postprocessor for IsSpice, best takes advantage of PC graphics.

On Macintosh computers tabular output created by the .PRINT command on a Macintosh or another computer can be viewed and printed using plotting programs such as Cricket Graph or Kaleidograph.

The rest of this section is dedicated to introducing the third major analysis mode of SPICE, time-domain, or transient, analysis. The following example describes how to obtain the SPICE time-domain solution for the above RLC circuit.

EXAMPLE 1.6

Use SPICE to compute the time-domain response of the series RLC circuit of Figure 1.6 to a pulse with an amplitude of 5 V applied at the input at time 0 for 25 ms. Find the waveform of $v_3(t)$ between 0 and 50 ms. Use $R_1 = 50\ \Omega$.

Solution
Two new items must be introduced in the SPICE deck of the RLC circuit in order to perform a transient analysis. First, a pulse with the defined characteristics must be assigned to the voltage source VIN. Second, the control line .TRAN must be included, which defines the time interval for the analysis. The complete SPICE deck for the transient analysis is listed below.

```
SERIES RLC CIRCUIT
*
VIN 1 0 PWL 0 0 10N 5 25M 5 25.01M 0
L1 2 3 0.125
C1 3 0 1U
R1 1 2 50
.TRAN .2M 50M
.PRINT TRAN V(3) V(1)
.END
```

The first difference from the input file used in the AC analysis is found on the VIN line: the two nodes are followed by the keyword **PWL**, which identifies a piecewise linear function generated by the input source. The values following the keyword represent pairs of time-voltage values. Other waveform functions can be generated by a source; complete information on independent sources is in Section 2.2.6.

The **.TRAN** control line contains three values: the time step, which is the interval at which values are printed or plotted by a **.PRINT** or **.PLOT** command; the final time, and the starting time. SPICE uses an internally adjusted, variable time step for solving the circuit equations (see also Chaps. 6 and 9); the value on the **.TRAN** line is used only for result output purposes. All the details on specifying a time-domain analysis can be found in Chap. 6.

A **.PRINT** statement followed by an analysis type, **TRAN**, is also present in the deck. Two waveforms are requested, V(1), the input, and V(3), the voltage across the capacitor. The tabular output saved in rlc.out is used in this case as input for *xgraph*, which produces the plot of Figure 1.9. The voltage across capacitor C_1 displays an overshoot, which takes almost the entire pulse width to settle. This behavior corresponds to the peaked frequency characteristic displayed by the circuit; see Figures 1.7 and 1.8.

If PSpice is used for this analysis and the line **.PROBE** is included in the deck, upon completion of the SPICE run the user is transferred into the Probe program, which enables the results to be viewed on the screen or a hardcopy to be produced as shown in Figure 1.10.

Note that both the AC and transient analyses can be made part of a single SPICE run by the addition of the **PWL** specification to the AC characteristics on the VIN line and the inclusion of the **.AC** and **.TRAN** lines together with the relevant output request lines in the same input file.

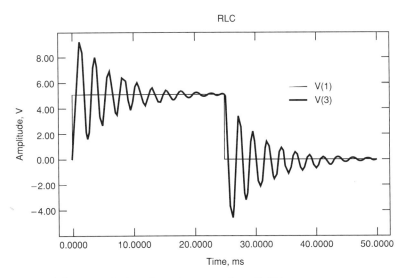

Figure 1.9 *xgraph* plot of transient waveform V(3).

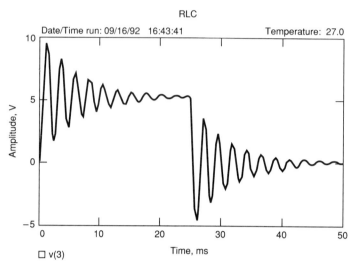

Figure 1.10   Probe plot of transient waveform $V(3)$.

The main analysis modes of SPICE as well as how to display the results of the analyses have been covered in this section. Alternatives for obtaining graphs of the results have been presented; the list of graphics packages available on computers, engineering workstations, and personal computers is intended not to be complete but only to exemplify an approach for obtaining high-quality plots from SPICE .**PRINT** data. For uniformity, the plots in the rest of the book are produced using the Oscilloscope and Network Analyzer tools of the Analog Workbench because of their superior graphic quality.

1.4   SUMMARY

This chapter introduced the basic capabilities of the electrical simulation program SPICE. The solution process of SPICE was linked to the knowledge of electric circuits necessary for using the simulation. You should be able to write a SPICE input deck for a simple linear circuit, run the basic analyses, and save the results in a output file or create a plot.

All the elements used so far can be specified in SPICE according to the following format:

Aname node1 node2 <node3> ... <MODELname><value>

Associated with complex elements, such as transistors, is the global statement .**MODEL**, which defines the parameters for a number of elements:

.**MODEL** *MODELname MODELtype <PARAM1=value1 PARAM2=value2 ...>*

The following analysis control statements are defined:

.OP
.AC *INTERVAL numpts fstart fstop*
.TRAN *TSTEP TSTOP <TSTART>*

for performing a DC, frequency-domain, or time-domain analysis, respectively. SPICE results are requested with the following control statements:

.PRINT *ANALYSIS OUT_var1 OUT_var2 ...*
.PLOT *ANALYSIS OUT_var1 OUT_var2 ...*

The general structure of SPICE deck is shown on page 22; all SPICE circuit descriptions must start with a *title* line and end with an **.END** line.

Examples of several plotting tools were presented, such as *Nutmeg*, *xgraph*, and *Probe*. The plots in the following chapters of this book are all created using the display tools of the Analog Workbench.

REFERENCES

Branin, F. H., G. R. Hogsett, R. L. Lunde, and L. E. Kugel. 1971. ECAP II—A new electronic circuit analysis program. *IEEE Journal of Solid-State Circuits* SC-6 (August): 146–165.

Dorf, R. C. 1989. *Introduction to Electric Circuits.* New York: John Wiley & Sons.

Nagel, L. 1975. SPICE2: A computer program to simulate semiconductor circuits. Univ. of California, Berkeley, ERL Memo UCB/ERL M75/520 (May).

Nagel, L., and R. A. Rohrer. 1971. Computer analysis of nonlinear circuits, excluding radiation (CANCER), *IEEE Journal of Solid-State Circuits* SC-6 (August): 166–182.

Nilsson, J. W. 1990. *Electric Circuits.* 3d ed. Reading, MA: Addison-Wesley.

Paul, C. R. 1989. *Analysis of Linear Circuits.* New York: McGraw-Hill.

IBM. 1965. 1620 electronic circuit analysis program [ECAP], [1620-EE-02X] User's Manual, IBM Application Program File H20-0170-1.

Pederson, D. O. 1984. A historical review of circuit simulation. *IEEE Transactions on Circuits and Systems* CAS-31 (January): 103–111.

Quarles, T. L. 1989. SPICE3 version 3C1 user's guide. Univ. of California, Berkeley, UCB/ERL Memo M89/46 (April).

Microsim. 1991. *PSpice: Circuit Analysis User's Guide Version 5.0.* Irvine, CA: Author.

Vladimirescu, A., K. Zhang, A. R. Newton, D. O. Pederson, and A. L. Sangiovanni-Vincentelli. 1981 (August). SPICE version 2G user's guide, Dept. of Electrical Engineering and Computer Science, Univ. of California, Berkeley.

Weeks, W. T., A. J. Jimeniz, G. W. Mahoney, D. Mehta, H. Quassenizadeh, and T. R. Scott. 1973. Algorithms for ASTAP—A network analysis program. *IEEE Transactions on Circuit Theory* CT-20 (November): 620–634.

Two

CIRCUIT ELEMENT AND NETWORK DESCRIPTION

2.1 ELEMENTS, MODELS, NODES, AND CONVENTIONS

Element statements and model statements represent the core of the circuit description, as shown in Sec. 1.3.1. Every element type accepted by SPICE2 and SPICE3 is presented in this and the following chapter. All commercial SPICE versions support the elements available in SPICE2, but not all implement the newer element types of SPICE3. SPICE elements are classified in three categories: two-terminal elements, described in Sec. 2.2; multiterminal elements, presented in Sec. 2.3; and semiconductor devices, presented in Chap. 3.

An element statement contains connectivity information and, either explicitly or by reference to a model name, the values of the defined element. Model statements are necessary for defining the parameters of complex elements. SPICE2 supports models only for semiconductor devices; these model types, common to all SPICE versions, are described in Chap. 3. SPICE3 and PSpice, however, have extended model support to most elements.

The following conventions must be observed in the SPICE2 (Vladimirescu, Zhang, Newton, Pederson, and Sangiovanni-Vincentelli 1981), SPICE3 (Johnson, Quarles, Newton, Pederson, and Sangiovanni-Vincentelli 1991), and PSpice (MicroSim 1991) circuit definitions.

The circuit nodes must always be positive integers in SPICE2 or positive integers and names in SPICE3. The circuit nodes need not be numbered sequentially. A circuit must always contain a ground node, which must always be number 0.

Every node in the circuit must have at least two elements connected to it; the only exceptions are the substrate node in MOSFETs, which has internal connections to the drain and to the source, and the nodes of unterminated transmission lines.

Every node in the circuit must have a DC path to ground. In DC, capacitors represent open circuits and inductors represent shorts. This requirement prevents the occurrence of floating nodes, for which the program cannot find a bias point.

Because SPICE2 uses modified nodal analysis (Ho, Ruehli, and Brennan 1975; McCalla 1988; see Chap. 9) to solve for both node voltages and currents of voltage-defined elements, such as voltage sources and inductors, two restrictions must be observed: the circuit cannot contain a loop of voltage sources or inductors, and it cannot contain a cutset of current sources or capacitors. The former is disallowed due to Kirchhoff's voltage law, KVL, and the latter due to Kirchhoff's current law, KCL.

Any violation of the above restrictions results in an error message and termination of the SPICE program. The possible error messages and corrective actions are described in Appendix B.

Several conventions are observed in the following sections in the presentation of element statements. In the statement format definition the characters or keywords that must be present in an actual statement are boldface, and optional keywords or values appear between angle brackets, $<\,>$.

The following description summarizes the conventions for different typesets.

1. Monotype is used for:
   - Computer (program) input and output
   - References made in the text to names, titles, or variables appearing in a computer input or output

2. Boldface monotype is used for:
   - Command names, parameter names, model and analysis types, and characters that are keywords for the program; the same type is used whether these keywords appear in a statement definition or are referred to in the text

3. Italic type is used for:
   - Variable names (subscripted characters as well)
   - Names of fields in SPICE statement definitions
   - Reference to the value of a program parameter with the same name; for example the parameter name is **L** and the variable is L
   - Highlighting new concepts

4. Uppercase versus lowercase in SPICE statement definitions:
   - Variables in uppercase or starting with uppercase in statement definitions denote a character field, such as *MODEL_name*
   - Variables in lowercase denote a numeric field in a statement as in **TC**= *tc1, tc2*
   - Exception: when a parameter name is followed by its value, the value may be denoted by the same characters as the name, in uppercase, but italic type

2.2   TWO-TERMINAL ELEMENTS

This section describes both the syntax and the branch-constitutive equations (BCEs) of all two-terminal elements except the semiconductor diode. The semiconductor diode is presented together with multiterminal semiconductor devices in Chap 3. The I-V relations of semiconductor elements are expressed by complex analytic equations, which require many parameters.

SPICE supports the following two-terminal elements:

Resistors (linear)

Capacitors (linear and nonlinear)

Inductors (linear and nonlinear)

Independent voltage sources (linear)

Independent current sources (linear)

Diodes (nonlinear)

The type of I-V branch-constitutive equation implemented in the program for each element listed above is specified in parentheses. All two-terminal elements supported by SPICE2, except the diode, are described by simple BCEs, require only one or a few parameters, and have no associated model statements.

SPICE3 supports semiconductor resistors and capacitors, which can be specified by geometric and process parameters. These special elements have an associated **.MODEL** statement for the process parameters. SPICE3 supports the following model types introduced in this chapter:

R	Diffused resistor model
C	Diffused capacitor model
URC	Uniformly distributed RC model
SW	Voltage-controlled switch model
CSW	Current-controlled switch model

PSpice also supports model statements for resistors, capacitors, inductors, nonlinear magnetic cores, and switches—for all elements except for sources.

On any two-terminal element statement the first node, *node1*, is positive, and the second node, *node2*, is negative. The branch voltage across any element is computed as

$$V_{elem} = V_{node1} - V_{node2}$$

and the current is assumed to flow from node *node1* to node *node2*.

2.2.1   Resistors

The general form of the resistor statement is

R*name node1 node2 rvalue* <**TC** = *tc1*<,*tc2*>>

R in the first column identifies a resistor labeled *Rname*, as shown in Figure 2.1, connected between nodes *node1* and *node2* of the circuit. Only the first 7 characters in *name* are used by SPICE2 to identify this resistor. SPICE3 does not restrict the length of *name*.

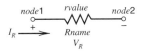

Figure 2.1   Resistor model.

The BCE of a resistor is

$$V_R = rvalue \cdot I_R \tag{2.1}$$

where the proportionality constant *rvalue* is the resistance measured in ohms, V_R the voltage across the resistor in volts, and I_R the current in amperes. The resistance may be positive or negative but cannot be zero.

SPICE models the temperature variation of the resistance by a second-order polynomial:

$$rvalue(TEMP) = rvalue(TNOM)[1 + tc1(TEMP - TNOM)$$
$$+ tc2(TEMP - TNOM)^2] \tag{2.2}$$

The keyword **TC** must be present if one or both temperature coefficients are specified; *tc1* and *tc2* are the first- and second-order temperature coefficients of the resistor specified in parts per °C or (°C)², respectively. *TNOM* is the nominal temperature, 27°C, assumed in SPICE2, and *TEMP* is a different simulation temperature specified in a `.TEMP` statement. Note that SPICE3 and PSpice require the temperature coefficients to be specified on the resistor `.MODEL` line. PSpice also supports a second temperature dependence, described by an exponential function.

Examples

```
R1 2 45 100
RC1 12 17 1K TC=0.001,0.015  (SPICE2)
RC1 12 17 RMOD 1K  (SPICE3,PSpice)
.MODEL RMOD R TC1=0.001 TC2=0.015
```

2.2.2 Semiconductor Resistors (SPICE3)

SPICE3 supports an extension of the general resistor element that allows a convenient description of a diffused resistor from geometric and process information.

The general form of a semiconductor resistor statement is

R*name node1 node2* <*rvalue*><*Mname*><**L**=*L*><**W**=*W*>

If *rvalue* is specified, this statement is equivalent to the general resistor statement and any information following the value is discarded. Note that SPICE3 does not support temperature coefficients on the resistor statement. A model statement with the general format described in Sec. 1.3.1 must be used in order to define the parameters listed in Table 2.1 for a model of type **R**.

Table 2.1 Semiconductor Resistor Model Parameters

Name	Parameter	Units	Default	Example
TC1	First-order temperature coefficient	1/°C	0.0	5E-3
TC2	Second-order temperature coefficient	1/(°C^2)	0.0	20E-6
RSH	Sheet resistance	Ω/sq.	0.0	50
DEFW	Default width	m	10^{-6}	2E-6
NARROW	Narrowing due to side etching	m	0.0	1E-7

The resistance is computed from the length L and width W of the diffusion specified in the resistor statement and the values *RSH* and *NARROW* of the model statement:

$$rvalue = RSH \cdot \frac{L - NARROW}{W - NARROW} \tag{2.3}$$

The temperature behavior is modeled the same way as for regular resistors (see Eq. 2.2). Note that the program provides default values only for W and *DEFW*, and not for L, because the width of most diffused resistors on a chip is equal to the minimum feature size; a default value for *DEFW* also prevents division by zero in Eq. 2.3 when W is omitted.

EXAMPLE 2.1

```
RDIFF1 1 2 RMOD1 L=50U W=5U
.MODEL RMOD1 R RSH=100 NARROW=.25U
```

The above statements define a resistor of resistance

$$rvalue = 100 \cdot \frac{50 - 0.25}{5 - 0.25} \, \Omega = 1047 \, \Omega$$

2.2.3 Capacitors

The general form of a capacitor statement is

Cname *node1 node2 cvalue* <**IC** = v_{C0} >

The **C** in the first column identifies a capacitor labeled *Cname* and connected between nodes *node1* and *node2* of the circuit, as shown in Figure 2.2.

The BCE of a capacitor is

$$i_C = cvalue \cdot \frac{dv_C}{dt} \tag{2.4}$$

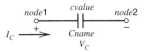

Figure 2.2 Capacitor model.

where *cvalue* is the capacitance in farads and represents the proportionality constant between the current through the capacitor, i_C, and the rate of change of the voltage across the capacitor, v_C. The integral variant of Eq. 2.4 is used in SPICE to model the capacitor:

$$v_C = \frac{1}{cvalue} \int_0^t i_C dt + v_{C0} \tag{2.5}$$

IC is optional and is used to input v_{C0}, the initial value (time $t = 0$) of the capacitor voltage. This value is used at $t = 0$ only when **UIC** (*use initial conditions*) is specified in the **.TRAN** statement (see Chap. 6).

Examples
```
C2 2 0 10P
CGS1 12 14 50F
CLOAD 31 0 20P IC=5
```

The above statement describes a linear capacitor with constant capacitance. SPICE2 and PSpice also support nonlinear capacitors whose capacitance is a nonlinear polynomial function of the terminal voltage v_C.

The general form of a nonlinear capacitor statement is

    Cname node1 node2 **POLY** *c0 c1* <*c2*...><**IC** = v_{C0} >

The keyword **POLY** identifies the capacitor *Cname* as nonlinear, and the values *c0, c1,* ... are the coefficients of the corresponding powers of v_C. The value of this capacitor is computed at each time point as

$$cvalue = c0 + c1 \cdot v_C + c2 \cdot v_C^2 + \cdots \tag{2.6}$$

The BCE for the nonlinear capacitor becomes

$$i_C = \frac{dq}{dt} = \frac{d}{dt}(cvalue \cdot v_C) = \frac{d}{dt}\left(c0 \cdot v_C + c1 \cdot v_C^2 + c2 \cdot v_C^3 + \cdots\right) \tag{2.7}$$

Thus the coefficients *c0, c1, c2,* ... should not be mistaken for a polynomial representation of the charge q.

EXAMPLE 2.2

```
C1 3 5 POLY 1N 75P 200F
```

The value of `C1` is evaluated for every new value of v_C across the capacitor according to Eq. 2.6:

$$cvalue = 10^{-9} + 75 \cdot 10^{-12}v_C + 200 \cdot 10^{-15}v_C^2 + \cdots \text{ farad}$$

2.2.4 Semiconductor Capacitor (SPICE3)

SPICE3 supports an extension of the general capacitor element to allow for a convenient description of the capacitance of a planar diffused region from geometric and process information.

The general form of a semiconductor capacitor statement is

$$C\textit{name node1 node2} <cvalue><Mname><\mathbf{L}=L><\mathbf{W}=W><\mathbf{IC}=v_{C0}>$$

If *cvalue* is specified, this statement is equivalent to the general capacitor statement and any information following the value is discarded except the initial value. If *Mname* is specified, the capacitance is calculated from the process information in model *Mname* and the given length, *L*, and width, *W*. Note that if *cvalue* is not specified, *Mname* and *L* must be provided; *W* assumes the default value in the model if not specified. Also note that either *cvalue* or *Mname*, *L*, and *W* may be specified, but not both.

A model type **C** statement must be used in order to define the parameters listed in Table 2.2. The capacitance is computed as follows:

$$cvalue = CJ(L - NARROW)(W - NARROW)$$
$$+ 2 \cdot CJSW(L + W - 2 \cdot NARROW) \tag{2.8}$$

EXAMPLE 2.3

```
CDIFF PDIFF 0 PCAP L=5U W=5U
.MODEL PCAP C CJ=100U CJSW=1N
```

Table 2.2 Semiconductor Capacitor Model Parameters

Name	Parameter	Units	Default	Example
CJ	Junction bottom capacitance	Fm^{-2}	—	5E-5
CJSW	Junction sidewall capacitance	Fm^{-1}	—	2E-11
DEFW	Default device width	m	10^{-6}	2E-6
NARROW	Narrowing due to side etching	m	0.0	1E-7

The above statements define a diffused capacitor between node `PDIFF` and node 0, with a value computed according to Eq. 2.8:

$$cvalue = 10^{-4} \cdot 5 \cdot 10^{-6} \cdot 5 \cdot 10^{-6} \text{ F} + 2 \cdot 10^{-9} \cdot (5 + 5) \cdot 10^{-6} \text{ F} = 22.5 \text{ fF}$$

Note that node names are accepted in SPICE3.

2.2.5   Inductors

The general form of an inductor statement is

L*name node1 node2 lvalue* $<$**IC** $= i_{L0} >$

The **L** in the first column identifies an inductor labeled *Lname* and connected between nodes *node1* and *node2* of the circuit, as shown in Figure 2.3.

The BCE of an inductor is

$$v_L = lvalue \cdot \frac{di_L}{dt} \tag{2.9}$$

lvalue is the inductance in henries and represents the proportionality constant between the voltage across the inductor and the rate of change of the current through the inductor. The integral variant of Eq. 2.9 is used in SPICE to model the inductor:

$$i_L = \frac{1}{lvalue} \int_0^t v_L dt + i_{L0} \tag{2.10}$$

IC is optional and is used to input the initial (time $t = 0$) inductor current, i_{L0}. This value is used at $t = 0$ only when **UIC** is specified in the **.TRAN** statement, as described in Chap. 6.

Examples
```
LXTAL 5 6 0.8
LSHUNT 23 51 10U IC=15.7M
```

The statements presented so far describe linear inductors characterized by the constant inductance *lvalue*. SPICE2 also supports nonlinear inductors the inductance of

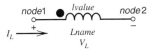

Figure 2.3   Inductor model.

which is a nonlinear polynomial function of the current i_L. The general form of a nonlinear inductor statement is

Lname node1 node2 **POLY** l0 l1 < l2 ...> <**IC** $= i_{L0}$>

The keyword **POLY** identifies the inductor *Lname* as nonlinear and the values *l0*, *l1*, ... as the coefficients of the corresponding powers of i_L. At least one other coefficient must be specified besides *l0*. The value of this inductor is computed at each time point as

$$lvalue = l0 + l1 \cdot i_L + l2 \cdot i_L^2 + \cdots \tag{2.11}$$

The BCE for the nonlinear inductor becomes

$$v_L = \frac{d\phi}{dt} = \frac{d}{dt}(lvalue \cdot i_L) = \frac{d}{dt}(l0 \cdot i_L + l1 \cdot i_L^2 + l2 \cdot i_L^3 + \cdots) \tag{2.12}$$

The coefficients *l0, l1, l2*, ... are the coefficients for a polynomial representation not of the magnetic flux ϕ but of the inductance.

EXAMPLE 2.4

```
LPAR 21 0 POLY 0 5M 1
```

The BCE for inductor LPAR is obtained from Eq. 2.12:

$$v_L = \frac{d}{dt}(0.005 \cdot i_L^2 + i_L^3)$$

which is solved at each time point.

2.2.6 Independent Bias and Signal Sources

Voltage and current sources that are independent of any circuit variables are defined by the following general statements:

Vname node1 node2 <<**DC**> dc_value> <**AC** <ac_mag <ac_phase>>>
+   <TRAN_function <value1 <value2 ...>>>
Iname node1 node2 <<**DC**> dc_value> <**AC** <ac_mag <ac_phase>>>
+   <TRAN_function <value1 <value2 ...>>>

The **V** and the **I** in the first column identify a voltage source and a current source, respectively, connected between nodes *node1* and *node2*. The polarity conventions are shown in Figure 2.4. The *dc_value* is the voltage difference between nodes *node1* and

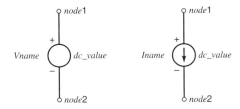

Figure 2.4 Independent voltage and current sources.

node2 for a voltage source and the current flowing from node *node1* to node *node2* through the source for a current source.

The voltage across the terminals of a voltage source is independent of the current flowing through it. Likewise, the current flowing through a current source is independent of the voltage across its terminals.

Independent sources are used to describe biases and signals for the three analytic modes of SPICE: DC, transient (time-domain), and small-signal AC. If a source definition contains no other information except the name and the nodes, the program assumes a DC source of value 0.

EXAMPLE 2.5

```
VCC 10 0 DC 5
IB 0 1 10U
IB1 1 0 -10U
VMEAS 4 5
```

All four statements define DC sources. The keyword **DC** is not necessary for defining the DC value of VCC and is used mostly for clarity when a lot of information is present on the source line. The current I_B flows from ground into node 1 of source IB. IB1 is identical to IB since both the order of the nodes and the sign of the current have been changed. VMEAS is a zero-value voltage source used in SPICE to measure currents (see also Chapter 4). The DC value of a source remains constant during a transient analysis if no other information is provided.

---

ac_mag and *ac_phase* are the magnitude and phase in degrees of an AC small-signal voltage or current. These values must be preceded by the keyword **AC** and are used only in conjunction with an AC analysis request, described in Chapter 5. If the keyword **AC** is alone, a magnitude of 1 and a phase of 0 are assumed by the program. The value of the *transfer function* at any point in the circuit referred to the input can be obtained by monitoring an AC voltage or current. The input $V_{in}(j\omega)$ is defined by the AC source. The output variables computed by the program, such as $V_{out}(j\omega)$, are identical to the transfer function, $T(j\omega)$, if the input signal has a magnitude of 1 and phase of 0:

$$V_{out}(j\omega) = T(j\omega)V_{in}(j\omega) = T(j\omega) \qquad (2.13)$$

since

$$V_{in}(j\omega) = ac\_mag \cdot \exp(j \cdot ac\_phase) = 1 \qquad (2.14)$$

For the large-signal time-domain analysis, SPICE supports five types of time-dependent signals: pulse, exponential, sinusoidal, piecewise linear, and single-frequency frequency modulated. The *TRAN_function* specification in a source statement contains a keyword that identifies one of the five functions and a set of parameters. In the following description of the five functions, the parameters and their defaults are specified.

2.2.6.1 Pulse Function
The general format of the *TRAN_function* specification of the source statement is

> **PULSE** (*V1 V2 <TD <TR <TF <PW <PER>>>>>*)

where the seven parameters have the meanings shown in Figure 2.5 and described in Table 2.3. The order of the values following the *TRAN_function* is essential for the correct specification of the signal characteristics. The parameters must be input in the given order. The initial and pulsed values, *V1* and *V2*, must be specified. If no values follow the function name, SPICE2 and SPICE3 perform the simulation and do not flag an error; PSpice, however, announces an error and aborts the analysis. The rest of the values, *TD* through *PER*, need not be input, but all values preceding the last nonzero parameter must be specified. The default values listed in Table 2.3 are used for unspecified parameters and are related to *TSTEP* and *TSTOP* of the transient analysis introduced in Chap. 1 and described in detail in Chap. 6; *TSTEP* is the output resolution of the waveforms, or time step, and *TSTOP* is the end of the time interval.

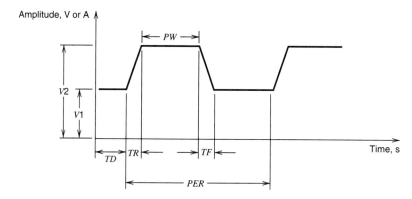

Figure 2.5   SPICE **PULSE** source function.

Table 2.3   Pulse Source Parameters

Name	Parameter	Units	Default
V1	Initial value	V or A	0.0
V2	Pulsed value	V or A	0.0
TD	Delay time	s	0.0
TR	Rise time	s	*TSTEP*
TF	Fall time	s	*TSTEP*
PW	Pulse width	s	*TSTOP*
PER	Period	s	*TSTOP*

EXAMPLE 2.6

```
VD 3 0 PULSE (1 -1 1U)
IKICK 0 2 PULSE (0 1M 1U 0 0 2U)
VIN 1 0 PULSE (0 5 0 1N 1N 99N 200N)
VSAW 3 4 PULSE (0 1 0 10U 10U 0.1U 20.1U)
```

The waveforms generated by the four statements are shown in Figure 2.6. The **PULSE** source can describe a step function, such as VD, a single pulse, such as IKICK, or a periodic signal, such as VIN and VSAW. VIN is a rectangular signal, and VSAW a sawtooth. For a step function neither *PW* nor *PER* need to be specified, and for a single pulse *PER* must not be specified.

The DC value of each of the above sources is equal to the initial value of the pulse. Neither SPICE2 nor PSpice accepts a DC value different from the first value of the PULSE function. SPICE3, however, allows the user to define a DC value different from

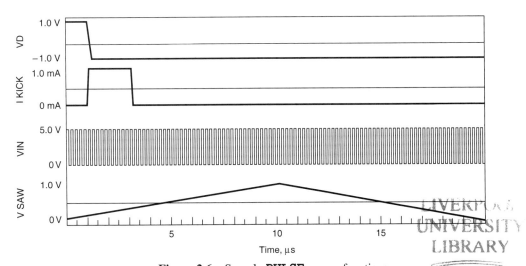

Figure 2.6   Sample **PULSE** source functions.

the starting value *V1* of the **PULSE** function. In the absence of *dc_value*, a warning message is output stating that the time zero value is used for DC.

The step function described by VD does not need any additional information besides the initial and pulsed values and the delay time of the step. SPICE uses the defaults of *TSTEP* for the fall time and *TSTOP*, the end of the transient analysis, for the pulse width.

The single pulse IKICK defines a pulse of 2 μs width and zero *TR* and *TF*. Due to the numerical integration algorithm used by SPICE, explained in more detail in Chap. 9, a voltage or current change in zero time can cause the solution not to converge. Therefore, the program substitutes the default value, *TSTEP* for *TR* and *TF*. Values smaller than *TSTEP* can be specified for *TR* and *TF*, but the faster rise and fall times cannot be seen on the line-printer plot.

For the sawtooth voltage source, VSAW, the pulse width, *PW*, should be zero. SPICE, however, does not accept a zero value for *PW* and replaces it by the default value, *TSTOP*. Thus a value one or two orders of magnitude smaller than *TR* and *TF* needs to be specified.

2.2.6.2 Sinusoidal Function

The general format of the sinusoidal function in the source statement is

SIN (*VO VA* <*FREQ* <*TD* <*THETA*>>>)

where the five parameters are illustrated in Figure 2.7 and described in Table 2.4. The offset and the amplitude must always be specified; the other values are optional.

The source assumes the offset value *VO* for times from $t = 0$ to $t = TD$ and behaves according to the following function for $t > TD$:

$$f(t) = VO + VA \sin(2\pi FREQ(t - TD))e^{-THETA(t-TD)} \qquad (2.15)$$

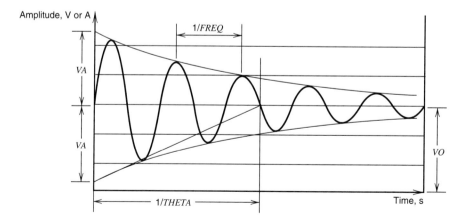

Figure 2.7 SPICE **SIN** source function.

Table 2.4 Sinusoidal Source Parameters

Name	Parameter	Units	Default
VO	Offset	V or A	0.0
VA	Amplitude	V or A	0.0
FREQ	Frequency	Hz	1/*TSTOP*
TD	Delay	s	0.0
THETA	Damping factor	s^{-1}	0.0

EXAMPLE 2.7

```
VSIN 33 34 SIN(0 1 1MEG)
I2 2 0 SIN(1M .2M 10MEG 1U 1MEG)
VCOS 5 6 SIN (0 5 100K -2.5U)
```

The first example describes a sinusoidal signal of 1 V amplitude, zero DC offset, and 1 MHz frequency. The number of signal periods depends on *TSTOP*, the length of the time interval for the transient analysis. A single-period sinusoid independent of the transient analysis interval can be specified by omitting the frequency. The default for *FREQ* according to the parameter table is 1/*TSTOP*; in other words, the period is equal to the time interval of the analysis.

The second example is a current source that supplies a sinusoidal signal of 1 mA DC current, 0.2 mA amplitude, and 10 MHz frequency that is delayed by 1 μs and decays by a factor of e, equal to 2.73, over 10 periods.

The last source, VCOS, implements a cosine signal by specifying a negative delay equal to 2.5 μs, or a quarter of a period. SPICE3 and PSpice allow a negative delay, but not SPICE2. Therefore, in SPICE2 the signal VCOS would be a sine wave.

The waveforms produced by the three sources are plotted in Figure 2.8 over a time interval of 10 μs. The number of periods that can be viewed for this interval are 10 for VSIN, 90 for I2, and 1 for VCOS.

2.2.6.3 Frequency-Modulated Sinusoidal Function

The general format for a single-frequency frequency-modulated (SFFM) transient function on a source statement is

SFFM (*VO VA* <*FC* <*MDI* <*FS*>>>)

The five parameters are defined in Table 2.5. The SFFM source is a special case of a sinusoidal source; indeed, if only the first three parameters are defined and if they are identical, the **SIN** and **SFFM** functions produce identical waveforms. The last two parameters differ between the two functions, in one case defining damping characteristics and in the other, modulation. Note that an SFFM function does not contain any delay.

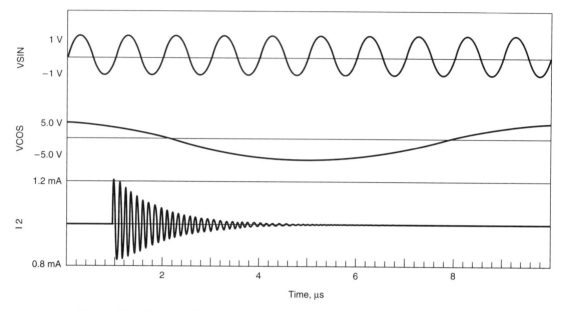

Figure 2.8   Example **SIN** source functions.

A signal described by an SFFM function has the following time behavior:

$$f(t) = VO + VA \sin(2\pi FC \cdot t + MDI \sin(2\pi FS \cdot t)) \qquad (2.16)$$

EXAMPLE 2.8

```
VIN 3 0 SFFM (0 1 1MEG 2 250K)
```

The above source produces a 1 MHz sinusoid of 1 V amplitude modulated at 250 kHz. It is shown in Figure 2.9.

Table 2.5   Frequency-Modulated Sinusoidal Source Parameters

Name	Parameter	Units	Default
VO	Offset	V or A	0.0
VA	Amplitude	V or A	0.0
FC	Carrier frequency	Hz	1/*TSTOP*
MDI	Modulation index	—	0.0
FS	Signal frequency	Hz	1/*TSTOP*

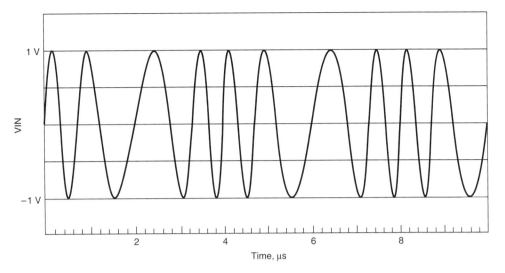

Figure 2.9 SPICE **SFFM** source function.

2.2.6.4 Exponential Function

The general format of the **EXP** *TRAN_function* in the source statement is

    EXP (*V1 V2 <TD1 TAU1 TD2 <TAU2>>*)

where the six parameters are as illustrated in Figure 2.10 and are described in Table 2.6.

The initial and pulsed values, *V1* and *V2*, must be specified, and the other parameters need not be specified; the defaults listed in the table are used for any missing parameter values.

The time behavior of an exponential source is described by the following functions of time *t*:

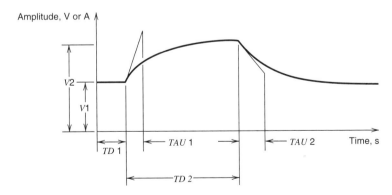

Figure 2.10 SPICE **EXP** source function.

Table 2.6   Exponential Source Parameters

Name	Parameter	Units	Default
V1	Initial value	V or A	0.0
V2	Pulsed value	V or A	0.0
TD1	Rise delay time	s	0.0
TAU1	Rise time constant	s	*TSTEP*
TD2	Fall delay time	s	*TD1 + TSTEP*
TAU2	Fall time constant	s	*TSTEP*

$$v(t) = \begin{cases} V1 & \text{for } 0 \le t \le TD1 \\ v_1(t) = V1 + (V2 - V1)\left(1 - e^{-(t-TD1)/TAU1}\right) & \text{for } TD1 < t \le TD2 \\ v_2(t) = v_1(t) + (V1 - V2)\left(1 - e^{-(t-TD2)/TAU2}\right) & \text{for } TD2 < t \le TSTOP \end{cases}$$

$$(2.17)$$

EXAMPLE 2.9

```
VEXP 1 0 EXP (0 5 0 1U 1)
IDEC1 3 1 EXP (1M 0 0 0 0 1U)
```

The first source, VEXP, represents an exponential signal that rises from 0 to 5 V with a time constant of 1 μs. The last value, 1, on the VEXP line is necessary for defining the delay *TD2* before the source starts decaying; by default the value of *VEXP* would start to fall after *TSTEP* seconds.

The second source, IDEC1, is intended to be a current that exponentially decays from 1 mA to 0 with a time constant of 1 μs. Note that since the initial value, *V1*, of IDEC1 is larger than the pulsed value, *V2*, the meanings of rise and fall are reversed; that is, *TD1* and *TAU1* become the fall delay and time constant, respectively, and *TD2* and *TAU2* become the rise delay and time constant, respectively. This is why after a short decay with a time constant equal to *TSTEP*, until *TD2* = *TSTEP*, the current resumes increasing back towards 1 mA with a time constant equal to 1 μs. Thus, the correct definition of a decaying current is the following:

```
IDEC2 3 1 EXP 1M 0 0 1U 1
```

Note that a large value, 1 second, must be specified for *TD2*; otherwise the current may start increasing before the end of the analysis interval *TSTOP*. The three waveforms are plotted in Figure 2.11, where *TSTEP* = 0.1 μs.

2.2.6.5   Piecewise Linear Function

The general format of a piecewise linear function on a source statement is

PWL $(t_1 \ V_1 < t_2 \ V_2 < t_3 \ V_3 \ldots >>)$

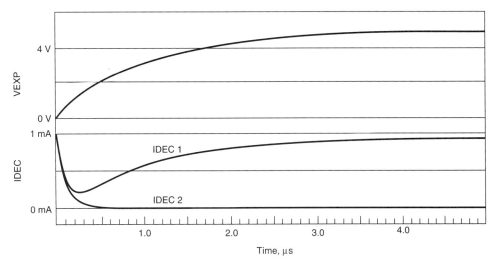

Figure 2.11 Example **EXP** source functions.

The parameters for this function are time-value coordinates. The signal described by a **PWL** statement is formed of straight lines that connect the pairs of coordinates (t_i, V_i). There is no limit to the number of coordinate pairs specified. This type of function is useful for describing sequences of pulses.

Neither t_1 nor the last defined time needs to coincide with the transient analysis limits, 0 and *TSTOP*. If $t_1 > 0$, the first coordinate pair assumed by SPICE is $(0, V_1)$; if the time of the last coordinate pair, t_n, is greater than *TSTOP*, the source assumes the last value, V_n, at *TSTOP* in SPICE2. SPICE3 and PSpice interpolate the source value at *TSTOP*:

$$V(TSTOP) = V_{n-1} + \frac{TSTOP - t_{n-1}}{t_n - t_{n-1}}(V_n - V_{n-1})$$

EXAMPLE 2.10

```
IBIT1 1 0 PWL (0 0 1U 0 1.1U 1M 2U 1M 2.1U 0 4U 0 4.1U 1M 5U 1M
+    5.1U 0 7U 0 7.1U 1M 8U 1M 8.1U 0)
VDATA 21 0 PWL (0 0 0.2U 5 3U 5 3.2U 0 6U 0 6.2U 5)
```

The two waveforms are shown in Figure 2.12; IBIT1 preserves its last specified value, 0, from 8.1 μs through *TSTOP*, equal to 10 μs. Both IBIT1 and VDATA are specified with finite rise and fall times; note that SPICE does not accept more than one **PWL** source value for a given time point.

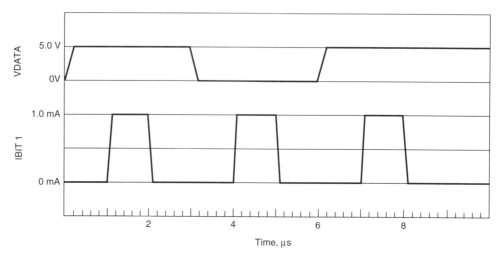

Figure 2.12   **PWL** functions.

2.3   MULTITERMINAL ELEMENTS

SPICE2 supports the following types of multiterminal elements:

Mutual inductors (linear and nonlinear)
Controlled sources (linear and nonlinear)
Transmission lines (linear)
Bipolar and field effect transistors (nonlinear)

In addition to the above element types, SPICE3 contains models for:

Switches
Uniformly distributed RC lines
Lossy transmission lines

In PSpice the switch and the lossy transmission line are also supported.

This section covers multiterminal elements with the exception of semiconductor elements, which are treated separately in Chap. 3. The elements presented in this section, generally, require only simple specifications.

2.3.1   Coupled (Mutual) Inductors

The general format of a coupled inductors statement is

Kname **L**name1 **L**name2 k

K in the first column identifies a mutual inductance specification between the two inductors *Lname1* and *Lname2*, defined somewhere else in the input file. k is the coefficient of coupling, which must be greater than 0 and less than or equal to 1. See Figure 2.13.

EXAMPLE 2.11

The SPICE specification of two coupled inductors L1 and L2 is

```
L1 1 2 1M
L2 4 3 1M
KL1L2 L1 L2 .99
```

The polarity of coupling between the two inductors is defined by the position of the dots as shown in the schematic representation in Figure 2.13. Note that the positive inductor node, containing the dot, must be defined first on the inductor statement. The value of the mutual inductance M is computed as

$$M = k\sqrt{L_1 L_2} \tag{2.18}$$

where k is the coupling coefficient and L_1 and L_2 are the inductances. The mutual inductance of the coupled inductors is $M = .99$ mH.

The BCEs of coupled inductors in SPICE are

$$v_1 = L_1 \frac{di_1}{dt} + M \frac{di_2}{dt}$$
$$v_2 = M \frac{di_1}{dt} + L_2 \frac{di_2}{dt} \tag{2.19}$$

Systems of coupled inductors are not limited to two and can be extended to a multitude of inductor pairs. The above equations must be modified accordingly to include all mutual inductances and terminal pairs.

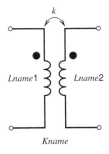

Figure 2.13
Coupled inductors.

Coupled inductors can be used to model an *ideal transformer* in SPICE. An ideal transformer has two pairs of terminals, the primary, with N_1 turns, and the secondary, with N_2 turns, and the voltages and currents obey the following relations:

$$v_2 = \frac{N_2}{N_1} v_1 \tag{2.20}$$

$$i_2 = \frac{N_1}{N_2} i_1 \tag{2.21}$$

EXAMPLE 2.12

Write the SPICE2 input for a transformer that has *turns ratio* $N_2/N_1 = 5$ and whose primary has self-inductance $L_1 = 1$ mH.

Solution
First calculate the inductance of the secondary knowing that the inductance is proportional to the square of the turns:

$$L_2 = \left(\frac{N_2}{N_1}\right)^2 L_1 = 25 \text{ mH} \tag{2.22}$$

The coupling coefficient for an ideal transformer is 1, and therefore the SPICE2 input specification for the transformer is the following:

```
LPRIM 1 2 1M
LSEC 3 4 25M
KXFRMR LPRIM LSEC 1
```

Note that PSpice restricts the value of k to less than 1. Additionally, PSpice and other commercial SPICE versions, such as SpicePLUS, implement a nonlinear magnetic core model.

2.3.2 Dependent (Controlled) Sources

Dependent sources, also known as controlled sources, supply voltages or currents that are functions of voltages or currents in other parts of the circuit. SPICE supports four types of dependent sources, a *voltage-controlled current source, VCCS; a voltage-controlled voltage source, VCVS; a current-controlled current source, CCCS*; and a *current-controlled voltage source, CCVS*. Dependent sources are useful for implementing a variety of large-signal input/output transfer functions (Epler 1987).

SPICE supports both linear and nonlinear dependent sources. A linear controlled source has two ports, and the output is equal to the input, or controlling variable, times a proportionality constant. A nonlinear controlled source is limited to a polynomial function of an arbitrary number of circuit variables in SPICE2. SPICE3 and PSpice accept an arbitrary nonlinear function of both voltages and currents in the circuit; see Sec. 7.4.1 and Sec. 7.4.4.

The general format of dependent sources is

$$CSname\ node+\ node-\ <\textbf{POLY}(ndim)>\ CTRLname/nodes\ p_0 < p_1 < p_2 \ldots >>$$

For a linear source only the information outside the brackets is specified. *CTRLname/ nodes* is a pair of nodes for voltage-controlled sources or a voltage source name that measures the controlling current for current-controlled sources. $p_0, p_1, p_2, \ldots$ are coefficients for the polynomial description. Only SPICE2 and PSpice recognize the **POLY** specification. The four types of dependent sources are as follows:

VCCS	**G**	$I_G = GV_C$	linear
		$I_G = g(V_C)$	nonlinear
VCVS	**E**	$V_E = EV_C$	linear
		$V_E = e(V_C)$	nonlinear
CCCS	**F**	$I_F = FI_C$	linear
		$I_F = f(I_C)$	nonlinear
CCVS	**H**	$V_H = HI_C$	linear
		$V_H = h(I_C)$	nonlinear

The different types listed next to the four kinds of controlled sources represent the identification character on an element statement; the symbols and controlling elements are shown in Figure 2.14. A nonlinear source can depend on more than one current or voltage. In SPICE3 only linear controlled sources are identified by **G**, **E**, **F**, or **H**; all types of nonlinear controlled sources have **B** as the first character of the element name.

The sequence of values in the dependent source statement is a function of the number of dimensions of the polynomial. Let $p_0, p_1, p_2, \ldots$ on the dependent source statement denote polynomial coefficients, x_1, x_2, x_3 be three controlling variables, and $f(x)$ be the dependent polynomial function.

The one-dimensional polynomial ($ndim = 1$) is the default in SPICE2, and neither the keyword **POLY** nor *ndim* needs to be specified. The input values represent the following coefficients:

$$f(x_1) = p_0 + p_1x_1 + p_2x_1^2 + p_3x_1^3 + p_4x_1^4 + \cdots \qquad (2.23)$$

An exception to the above assignment of coefficients is made when only one value appears on the source statement. In this case that value becomes the coefficient of the linear term, x_1. Thus, a linear controlled source statement is a special case of the general **POLY** statement, with all optional fields deleted and only one coefficient given.

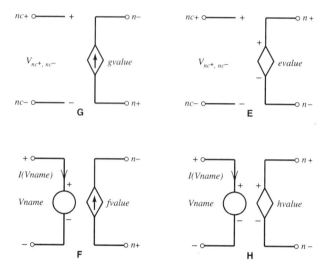

Figure 2.14　Dependent sources.

EXAMPLE 2.13

```
G1 3 2 1 2 1E-3
G1 3 2 POLY(1) 1 2 1E-3
G2 5 6 3 4 10M 1M     (SPICE2)
G2 5 6 VALUE = {0.01+1E-3*V(3,4)}      (PSpice)
```

The first two statements represent the same VCCS, or transconductance, of a current of 1 mA flowing from mode 3 through the source to node 2 and controlled by the voltage, $V_{1,2}$, between nodes 1 and 2:

$$f(x_1) = G1 = p_0x_1 = 10^{-3}V_{1,2} \text{ mho}$$

The above discrepancy in the meaning of p_0 has been deliberately chosen for one-dimensional polynomial sources, because in this way the general nonlinear dependent source statement takes the particular form of a linear controlled source for $ndim = 1$. The second statement is rejected by SPICE3.

The third example, G2, is interpreted like a regular polynomial source in SPICE2 but causes an error in PSpice. The fourth statement represents the preferred syntax for PSpice; the third statement is accepted by PSpice if the keyword **POLY**(1) is included. G2 must be described by a **B** statement in SPICE3 (see Sec. 7.4.1). The value of G2 is evaluated according to the following equation:

$$f(x_1) = G2 = p_0 + p_1x_1 = (10^{-2} + 10^{-3}V_{3,4}) \text{ mho}$$

A two-dimensional polynomial function is expressed as

$$f(x_1, x_2) = p_0 + p_1 x_1 + p_2 x_2 + p_3 x_1^2 + p_4 x_1 x_2$$
$$+ p_5 x_2^2 + p_6 x_1^3 + p_7 x_1^2 x_2 + p_8 x_1 x_2^2 + p_9 x_2^3 \qquad (2.24)$$

A three-dimensional polynomial function assigns the values on the source statement to coefficients in the following order:

$$f(x_1, x_2, x_3) = p_0 + p_1 x_1 + p_2 x_2 + p_3 x_3 + p_4 x_1^2 + p_5 x_1 x_2 + p_6 x_1 x_3$$
$$+ p_7 x_2^2 + p_8 x_2 x_3 + p_9 x_3^2 + p_{10} x_1^3 + p_{11} x_1^2 x_2$$
$$+ p_{12} x_1^2 x_3 + p_{13} x_1 x_2^2 + p_{14} x_1 x_2 x_3 + p_{15} x_1 x_3^2$$
$$+ p_{16} x_2^3 + p_{17} x_2^2 x_3 + p_{18} x_2 x_3^2 + p_{19} x_3^3 + \cdots \qquad (2.25)$$

Controlled sources are useful for emulating analog and digital circuit blocks, such as gain stages, operational amplifiers, converters, and many others.

2.3.2.1 Voltage-Controlled Current Source (VCCS)
The general format of a linear VCCS is

G*name n+ n− nc+ nc− gvalue*

A **G** in the first column followed by up to seven characters and digits define the unique name of a VCCS. $n+$ and $n-$ are the nodes between which the current source is connected, with current flowing from the positive node, $n+$, through the source to the negative node, $n-$ (see Figure 2.14). The positive and negative controlling nodes are $nc+$ and $nc-$, respectively; the voltage difference $V_{nc+,nc-}$ is the controlling variable. The BCE of a VCCS is

$$I_G = gvalue \cdot V_{nc+,nc-} \qquad (2.26)$$

where *gvalue* is the transconductance in mhos. The linear VCCS is only a special case of the more general nonlinear VCCS.

The general format of a nonlinear VCCS is

G*name n+ n−* <**POLY(** *ndim*)> *nc1+ nc1−* < *nc2+ nc2−* ... >
+ p_0 <p_1<p_2 ... >> <**IC**= $v_{nc1+,nc1-}$, $v_{nc2+,nc2-}$, ... >

The first difference between this statement and that of a linear VCCS is the keyword **POLY**, which can be present only when a nonlinear VCCS is specified; *ndim* is the number of controlling voltages, and $nc1+$ and $nc1-$ are the terminals of the first controlling voltage. The second difference is that for each additional controlling voltage a pair of controlling nodes, $nc2+$, $nc2-$, must be specified. The coefficients $p_0, p_1, p_2, \ldots$ take the meaning described above depending on the number of controlling variables. The third difference from the statement for a linear VCCS is that initial conditions can be defined for the controlling voltages. These values are used only in conjunction with the **UIC** option in the **.TRAN** statement to initialize the controlled source at the first time point. If no initial values are present, the controlling voltages are assumed to be zero, if the **UIC** option is chosen, or equal to the resulting value from the DC bias point otherwise.

EXAMPLE 2.14

```
GRPOLY 17 3 POLY(1) 17 3 0 1M 1.5M IC=2
G2DIM 23 17 POLY(2) 3 5 1 2 0 1M 17M 3.5U IC=2.5,1.3
```

The first example represents a nonlinear conductance, because the terminal nodes of the controlled current source are identical with the nodes of the controlling voltage. The BCE of the nonlinear conductance GRPOLY is

$$I_G = 0.001 \cdot V_G + 0.0015 \cdot V_G^2$$

In a transient analysis, at $t = 0$ the terminal voltage, V_G is 2 V, because the **IC** keyword and initial value are present on the GRPOLY line. Note that the keyword **POLY** is needed in PSpice for a one-dimensional source, but not in SPICE2.

The second example is a two-dimensional nonlinear VCCS of value

$$I_G = 0.001 \cdot V_{3,5} + 0.017 \cdot V_{1,2} + 3.5 \cdot 10^{-6} V_{3,5}^2$$

In a transient analysis that starts from initial conditions (that is, when the **UIC** flag is on), the two controlling voltages are initialized to 2.5 V and 1.3 V; see Chap. 6 for details.

2.3.2.2 Voltage-Controlled Voltage Source (VCVS)

The general format of a linear VCVS is

E*name n+ n− nc+ nc− evalue*

An **E** in the first column followed by up to seven characters and digits defines the unique name of a VCVS. $n+$ and $n-$ are the positive and negative nodes of the voltage source, respectively, as shown in Figure 2.14. $nc+$ and $nc-$ are the positive and negative controlling nodes, respectively; the voltage difference $V_{nc+,nc-}$ is the controlling variable. The BCE of a VCVS is

$$V_E = evalue \cdot V_{nc+,nc-} \tag{2.27}$$

where *evalue* is the voltage gain. The linear VCVS is only a special case of the more general nonlinear VCVS.

The general format of a nonlinear VCVS is

E*name* $n+$ $n-$ <**POLY**(*ndim*)> $nc1+$ $nc1-$ <$nc2+$ $nc2-$ ... >
+ p_0 <p_1 <p_2 ... >> <**IC**= $v_{nc1+,nc1-}$, $v_{nc2+,nc2-}$, ... >

The keyword **POLY** must be present when a nonlinear VCVS is specified; *ndim* is the number of controlling voltages. $nc1+$ and $nc1-$ are the terminals of the first controlling voltage; for each additional controlling voltage, a pair of controlling nodes, $nc2+$, $nc2-$, must be specified. The coefficients p_0, p_1, p_2, ... take the meaning described above depending on the number of controlling variables. **IC** is optional and defines the initial values of the controlling voltages; these values are used only in conjunction with the **UIC** option in the .**TRAN** statement to initialize the controlled source at the first time point. If no initial values are present, the controlling voltages are assumed to be zero, if the **UIC** option is chosen, or equal to the resulting value from the DC bias point otherwise.

EXAMPLE 2.15

```
E1 3 4 POLY(1) 21 17 10.5 2.1 1.75
ESUM 17 0 POLY(3) 1 0 2 0 3 0 0 1 1 1 IC=1,2,3
```

The first example, E1, defines a nonlinear voltage gain function

$$V_E = 10.5 + 2.1 \cdot V_{21,17} + 1.75 \cdot V_{21,17}^2$$

The second example represents a voltage summer, since the value of the controlled source ESUM is the sum of $V_{1,0}$, $V_{2,0}$, and $V_{3,0}$.

Exercise
Run an AC analysis of the voltage summer ESUM using SPICE and verify if this VCCS still performs the add function. Use different values for $V_{1,0}$, $V_{2,0}$, and $V_{3,0}$. For an explanation of the result, see Chaps. 5 and 7.

2.3.2.3   Current-Controlled Current Source (CCCS)
The general format of a linear CCCS is

F*name* $n+$ $n-$ V*name* *fvalue*

An **F** in the first column followed by up to seven characters and digits defines the unique name of a CCCS. $n+$ and $n-$ are the positive and negative nodes of the current source, and current flows from the positive node through the source to the negative node (see Figure 2.14). *Vname* is the voltage source through which the controlling current flows.

I(Vname) is the controlling variable. The BCE of a CCCS is

$$I_F = fvalue \cdot I(Vname) \tag{2.28}$$

where *fvalue* is the current gain. The linear CCCS is only a special case of the more general nonlinear CCCS.

The general format of a nonlinear CCCS is

F*name n+ n−* $<$**POLY**$(ndim)>$ *Vname1* $<Vname2...>$ $p_0 <p_1 <p_2...>>$
$+$ $<$**IC**$= i(Vname1), i(Vname2)...>$

The keyword **POLY** must be present when a nonlinear CCCS is specified; *ndim* is the number of controlling currents. *Vname1* is the voltage source measuring the first controlling current; for each additional controlling current a voltage source must be specified. The coefficients $p_0, p_1, p_2,...$ take the meaning described above depending on the number of controlling variables. **IC** is optional and defines the initial values of the controlling currents; these values are used only in conjunction with the **UIC** option in the **.TRAN** statement to initialize the controlled source at the first time point. If no initial values are present, the controlling currents are assumed to be zero if the **UIC** option is chosen, or equal to the resulting value from the DC bias point otherwise.

EXAMPLE 2.16

```
FCON 13 4 POLY(1) VC1 0.001 1E-4 1E-5
F2 2 3 POLY(2) VCON1 VCON2 0 0 0 0 1
```

FCON defines a current that is a quadratic function of the current flowing through the voltage source VC1:

$$I_F = 10^{-3} + 10^{-4}I_{VC1} + 10^{-5}I_{VC1}^2$$

The current of F2 is equal to the product of the currents flowing through voltage sources VCON1 and VCON2.

Exercise

Replace transistor Q_1 in Example 1.2 with a CCCS connected from collector to emitter having value *fvalue* $= \beta_F = 100$ and controlled by the current i_{RB}, and show that the bias point obtained by SPICE is close to the one obtained in Chap. 1. Explain the difference and modify the circuit to obtain the same answer.

2.3.2.4   Current-Controlled Voltage Source (CCVS)

The general format of a linear CCVS is

H*name n+ n− Vname hvalue*

An **H** in the first column followed by up to seven characters and digits defines the unique name of a CCVS. $n+$ and $n-$ are the positive and negative nodes of the voltage source, as shown in Figure 2.14. *Vname* is the voltage source through which the controlling current flows. *I(Vname)* is the controlling variable. The BCE of a CCVS is

$$V_H = hvalue \cdot I(Vname) \tag{2.29}$$

where *hvalue* is the transresistance in ohms. The linear CCVS is only a special case of the more general nonlinear CCVS.

The general format of a nonlinear CCVS is

H*name n+ n−* $<$**POLY**(*ndim*)$>$ *Vname1* $<$*Vname2* ... $>$ p_0 $<$$p_1$ $<$$p_2$ ... $>$$>$
$+$ $<$**IC** $= i(Vname1), i(Vname2)$ ... $>$

The keyword **POLY** can be present only when a nonlinear CCVS is specified; *ndim* is the number of controlling currents. *Vname1* is the voltage source measuring the first controlling current; for each additional controlling current a voltage source must be specified. The coefficients $p_0, p_1, p_2, \ldots$ take the meaning described above depending on the number of controlling variables. **IC** is optional and defines the initial values of the controlling currents; these values are used only in conjunction with the **UIC** option in the **.TRAN** statement to initialize the controlled source at the first time point. If no initial values are present, the controlling currents are assumed to be zero if the **UIC** option is chosen, or equal to the resulting value from the DC bias point otherwise.

EXAMPLE 2.17

```
HRNL 1 2 POLY(1) V12 0 0 1
HXY 13 20 POLY(2) VIN1 VIN2 0 0 0 1 -2 1 IC=0.5,1.3
```

The voltage between nodes 1 and 2 of HRNL is equal to the square of the current flowing through V12, and the value of HXY is equal to the square of the difference of the currents through voltage sources VIN1 and VIN2.

2.3.3 Switches

SPICE3 and PSpice support a nearly ideal switch model. An ideal switch has zero ON resistance and infinite OFF resistance. This behavior can be approximated satisfactorily with ON and OFF resistances that are significantly smaller or larger, respectively, than the other resistances in the circuit. The switch is therefore a resistor that toggles between a very small value, the ON resistance, and a very large value, the OFF resistance, depending on a controlling voltage or current.

Voltage- and current-controlled switches have the following general formats:

S*name n+ n− nc+ nc− Model* <ON/OFF>

W*name n+ n− V*name *Model* <ON/OFF>

where an **S** in the first column identifies a voltage-controlled switch, a **W** identifies a current-controlled switch, and the flag **ON/OFF** specifies the initial state of the switch. The switch is connected between nodes $n+$ and $n−$ and is controlled either by the voltage between nodes $nc+$ and $nc−$ or by the current flowing from the positive node to the negative node of the voltage source V*name*. *Model* is the name of the model statement that contains the parameters of the switch.

There are two types of switch models, **SW**, the voltage-controlled switch, and **CSW**, the current-controlled switch. Each model type defines four parameters, shown in Table 2.7.

The resistance of the switch as a function of the controlling variable is shown in Figure 2.15. The switch in SPICE3 displays hysteresis, switching at $V_{\text{control}} = VT$ when V_{control} is rising and at $VT − VH$ when V_{control} is falling. *GMIN* in the table is a minimum conductance, used in SPICE to protect against an ill-conditioned set of nodal equations; for other uses of **GMIN** see Chap. 3, and for modifying it see Chap. 9.

Because switches are highly nonlinear, their use in a circuit can lead to convergence problems. In order to prevent the failure of the time-domain analysis, that is, in order to avoid the situation in which the time step is too small (see Chap. 9), *RON* should be negligible compared to the smallest resistance and *ROFF* should be large enough not to affect the total value when connected in parallel to the largest resistor. It is strongly recommended that

$$\frac{ROFF}{RON} \leq 10^{12} \tag{2.30}$$

Another remedy for convergence failure of circuits with switches is the addition of capacitors across the controlling voltage and inductors in series with the controlling current, which prevent sudden changes in the controlling variables.

Note that the PSpice switch does not have hysteresis. The parameters **VT** and **VH** are replaced by **VON** and **VOFF** for an **S** switch, and **IT** and **IH** are replaced by **ION** and **IOFF** for a **W** switch. The interval between the two variables is used to provide for a smooth transition between *RON* and *ROFF*.

Table 2.7 Switch Model Parameters

Name	Parameter	Units	Default	Model
VT	Threshold voltage	V	0.0	SW
IT	Threshold current	A	0.0	CSW
VH	Hysteresis voltage	V	0.0	SW
IH	Hysteresis current	A	0.0	CSW
RON	ON resistance	Ω	1.0	SW, CSW
ROFF	OFF resistance	Ω	1/*GMIN*	SW, CSW

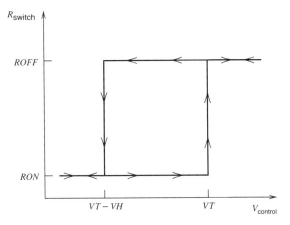

Figure 2.15 Switch resistance as a function of controlling voltage.

EXAMPLE 2.18

Use switches to model a NOR gate in SPICE3.

Solution

A straightforward implementation is shown in Figure 2.16; the circuit is equivalent to the MOS implementation of a NOR gate with the transistors replaced by switches and a load resistor connected to 5 V. The value of the load resistor is 1 kΩ. For best results in SPICE, *RON* and *ROFF* are selected such that they are much smaller and much larger, respectively, than 1 kΩ. The values are $RON = 1\ \Omega, ROFF = 1\ M\Omega$.

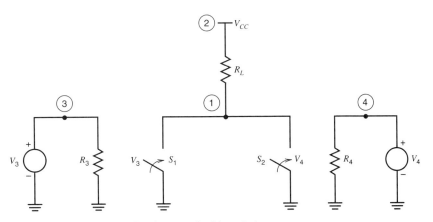

Figure 2.16 NOR gate implemented with switches.

The SPICE3 input description is listed below. The controlling terminals, nodes 3 and 4, represent the two inputs of the NOR gate. The two input signals are a sequence of logic 0 and 1 described as **PWL** voltage sources, introduced in Sec. 2.2.6.5.

```
NOR GATE WITH SWITCHES
*
* NOR GATE
*
RL 2 1 1K
S1 1 0 3 0 SW
S2 1 0 4 0 SW
VCC 2 0 5
*
* INPUT SIGNALS
*
V3 3 0 PWL 0 0 1U 0 1.01U 5 2U 5 2.01U 0 3U 0 3.01U 5
R3 3 0 1
V4 4 0 PWL 0 0 2U 0 2.01U 5
R4 4 0 1
*
.MODEL SW SW RON=1 ROFF=1MEG VT=1 VH=0
*
.TRAN .02U 4U
.PLOT TRAN V(1) V(3) V(4)
.END
```

Exercise

Run this circuit in SPICE3 and study the impact of various values for the hysteresis width, *VH*, as well as for different values of *RON* and *ROFF*. PSpice users must replace **VT** and **VH** with **VON** and **VOFF** and the model type by **VSWITCH**; for current-controlled switches, model type **ISWITCH**, and **ION** and **IOFF** must be specified. See also Chap. 7 for more details on this subject.

2.3.4 Transmission Lines

The general form of a transmission line statement is

T*name n1 n2 n3 n4* **Z0**=*Z0* <**TD**=*TD*><**F**=*freq* <**NL**=*NL*>>
+   <**IC**=$v_{n1,n2}, i_1, v_{n3,n4}, i_2$>

A **T** in the first column identifies a transmission line element, *n1* and *n2* are the nodes at port 1, and *n3* and *n4* are the nodes at port 2. A transmission line together with the SPICE equivalent model is shown in Figure 2.17. Only lossless lines are supported in SPICE2; SPICE3 and PSpice support also lossy transmission lines. *Z0* is the characteristic impedance of the line; one additional characteristic

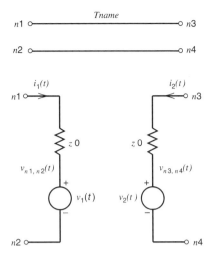

Figure 2.17 Transmission line and equivalent model.

must be specified about the line, which can be either the line delay, TD, or the normalized electrical line length, NL, expressed in wavelength units, at a given frequency $freq$. If only a frequency is specified and NL is omitted, NL defaults to 0.25; that is, $freq$ is the quarter-wave frequency. The three parameters are described by the following equation:

$$TD = \frac{NL}{freq} \qquad (2.31)$$

where

$$NL = \frac{l}{\lambda} \qquad (2.32)$$

with l the physical length of the line and λ the wavelength.

Initial Conditions, **IC**, are optional and consist of the currents and voltages at the terminals of the two ports. These values are used only in conjunction with the **UIC** option on the **.TRAN** statement.

The BCEs of the transmission line are expressed as the following functional forms of the voltage sources $v_1(t)$ and $v_2(t)$ in the equivalent model (Branin 1967) of Figure 2.17:

$$v_1(t) = v_{n3,n4}(t - TD) + Z0 \cdot i_2(t - TD) \qquad (2.33)$$
$$v_2(t) = v_{n1,n2}(t - TD) + Z0 \cdot i_1(t - TD) \qquad (2.34)$$

The two equations represent the incident and reflected waves along a lossless transmission line.

EXAMPLE 2.19

The following three definitions of line T1 are equivalent; see Eq. 2.31 and 2.32.

```
T1 1 0 2 0 Z0=50 TD=10N
T1 1 0 2 0 Z0=50 F=100MEG NL=1
T1 1 0 2 0 Z0=50 F=25MEG
```

A coaxial cable is described by two transmission lines, the first representing the inner conductor with respect to the shield, and the second the shield with respect to the outside world:

```
TINT 1 2 3 4 Z0=50 TD=1.5N
TEXT 2 0 4 0 Z0=100 TD=1N
```

Study the difference in the response of line TLINE to a pulse when terminated with a load resistor R_L of 50 Ω and 100 Ω. The SPICE input is listed below:

```
TRANSMISSION LINE EXAMPLE
*
VIN 1 0 PULSE 0 5 0 0.1N 0.1N 5N 50N
RIN 1 2 50
TLINE 2 0 3 0 Z0=50 TD=10N
RL 3 0 50
.TRAN .25N 50N
.PLOT TRAN V(2) V(3)
```

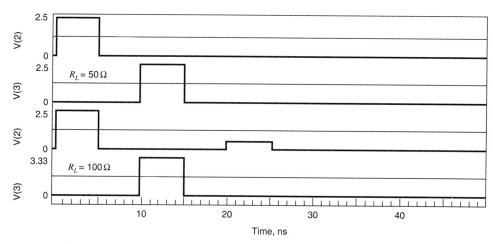

Figure 2.18   Transmission line response for $R_L = 50\ \Omega$ and $R_L = 100\ \Omega$.

```
.PRINT TRAN V(2) V(3)
.END
```

The waveforms at the input and output of `TLINE` for the two values of the load resistor R_L are shown in Figure 2.18. The difference is that a matched line with $R_{in} = R_L = Z0$ delays the input pulse by TD while the unmatched line reflects the pulse between input and output with an attenuation corresponding to the values of R_{in}, R_L, and $Z0$.

2.4 SUMMARY

This chapter describes all the elements supported in SPICE with the exception of the semiconductor devices. Both the syntax and the relation between the branch-constitutive equations and SPICE statement parameters were explained and exemplified.

First, the two-terminal passive elements, resistor, **R**, capacitor, **C**, and inductor, **L**, were defined. These elements use the following SPICE syntax:

Rname node1 node2 rvalue $<$**TC**$= tc1, <tc2>>$

Cname node1 node2 cvalue $<$**IC**$= v_{C0}>$

Lname node1 node2 lvalue $<$**IC**$= i_{L0}>$

Capacitors and inductors can be also defined as nonlinear, polynomial expressions of voltage and current, respectively.

Independent voltage and current bias and signal sources can be specified using the following generic SPICE statement:

V/I*name node1 node2* $<<$**DC**$> dc\_value><$**AC** $<ac\_mag <ac\_phase>>>$

$+$ $<TRAN\_function <value1 <value2...>>>$

where *TRAN_function* can be one of the following:

PULSE $(V1\ V2\ <TD <TR <TF <PW <PER>>>>>)$

SIN $(VO\ VA\ <FREQ <TD <THETA>>>)$

SFFM $(VO\ VA\ <FC <MDI <FS>>>)$

EXP $(V1\ V2\ <TD1 <TAU1 <TD2 <TAU2>>>>)$

PWL $(t_1\ V_1\ <t_2\ V_2\ <t_3\ V_3...>>)$

For each of the optional parameters SPICE provides default values.

Multiterminal elements supported in SPICE include coupled inductors, controlled (dependent) sources, switches, and transmission lines. Coupled inductors have the following syntax:

Kname Lname1 Lname2 k

Four types of controlled sources are supported in SPICE, a voltage-controlled current source, type **G**, a voltage-controlled voltage source, type **E**, a current-controlled current source, type **F**, and a current-controlled voltage source, type **H**. A linear dependent source has the following syntax:

$$\textbf{G/E}name\ node+\ node-\ nc+\ nc-\ value$$

$$\textbf{F/H}name\ node+\ node-\ Vname\ value$$

Vname is the name of the voltage source used to measure the controlling current. Nonlinear polynomial controlled sources use the following general syntax:

$$\textbf{G/E/F/H}name\ node+\ node-\ \textbf{POLY}(ndim)\ Vname/ncnodes\ p_0\ \ p_1$$

$$+\quad <p_2\ <p_3\dots>>$$

More complex nonlinear dependent sources supported in SPICE3 and PSpice are defined in Chap. 7.

A nearly ideal switch is implemented in SPICE3 and PSpice; the operation of the switch can be controlled by a voltage (type **S**) or a current (type **W**). The syntax for the two types of switches is

$$\textbf{S}name\ n+\ n-\ nc+\ nc-\ Model\ <\textbf{ON/OFF}>$$

$$\textbf{W}name\ n+\ n-\ Vname\ Model\ <\textbf{ON/OFF}>$$

One or more switch elements reference a **.MODEL** statement that defines the necessary parameters.

The last element introduced in this chapter is the ideal transmission line defined by the following statement:

$$\textbf{T}name\ n1\ n2\ n3\ n4\ \textbf{Z0}=Z0\ <\textbf{TD}=TD>\ <\textbf{F}=freq\ <\textbf{NL}=NL>>$$

$$+\quad <\textbf{IC}=v_{n1,n2},\ i_1,\ v_{n3,n4},\ i_2>$$

This chapter has detailed all SPICE circuit elements except semiconductor devices, which are the subject of the following chapter.

REFERENCES

Branin, F. H. 1967. Transient analysis of lossless transmission lines. *Proc. IEEE* 55: 2012–2013.

Epler, B. 1987. SPICE2 application notes for dependent sources. *IEEE Circuits and Devices Magazine* 3 (September): 36–44.

Ho, C. W., A. E. Ruehli, and P. A. Brennan. 1975. The modified nodal approach to network analysis. *IEEE Transactions on Circuits and Systems.* CAS-22 (June): 504–509.

Johnson, B., T. Quarles, A. R. Newton, D. O. Pederson, and A. L. Sangiovanni-Vincentelli. 1991. (April). *SPICE3 Version 3E User's Manual.* Berkeley: Univ. of California.

McCalla, W. J. 1988. *Fundamentals of Computer-Aided Circuit Simulation.* Boston: Kluwer Academic.

MicroSim. 1991. *PSpice, Circuit Analysis User's Guide, Version 5.0.* Irvine, CA: Author.

Vladimirescu, A., K. Zhang, A. R. Newton, D. O. Pederson, and A. L. Sangiovanni-Vincentelli. 1981 (August). *SPICE Version 2G User's Guide.* Department of Electrical Engineering and Computer Science, Univ. of California, Berkely.

Three

SEMICONDUCTOR-DEVICE ELEMENTS

3.1 INTRODUCTION

Semiconductor elements are presented together as a group because they have a common specification methodology. Unlike the elements described in Chap. 2, semiconductor elements are defined by complex nonlinear BCEs characterized by a large number of parameters. These parameters are specified in a `.MODEL` statement. As mentioned in the beginning of Chap. 2, SPICE2 supports four different semiconductor-device models, and SPICE3 and PSpice accept five different semiconductor devices.

The element statement for any semiconductor device has the following general format:

DEVname node1 node2 <*node3 . . .*> *MODname* <*value1* <*,value2*>*, . . .* >

+   <**OFF**> <**IC=***v1*<*,v2, . . .* >>

The device name *DEVname* starts with one of four characters **D**, **Q**, **J**, and **M** corresponding to the four kinds of semiconductor elements accepted by SPICE2, diode, bipolar transistor (BJT), junction field effect transistor (JFET), and metal-oxide-semiconductor field effect transistor (MOSFET), respectively. The additional semiconductor device supported in SPICE3 and PSpice, the metal-semiconductor field effect transistor (MESFET), is identified by a **Z** in SPICE3 and a **B** in PSpice. Up to seven characters can follow to identify the element. Two to four node numbers specify the connection of the semiconductor element terminals. *MODname* is the name of a model statement that contains the parameter values; one or more semiconductor devices can reference the same model. SPICE2 was initially developed for integrated circuits (ICs), and the commonality of model parameters can be traced back to the common process parameters. Besides the process parameters, which are general model parameters, each semiconductor device has certain geometric characteristics, *value1, value2,* and so on, that are unique and differentiate it from other devices having the same model parameters. Diodes, BJTs, JFETs, and MESFETs use only one parameter, *area,* to scale their geometry with respect to a *unit device.* MOSFETs require more detailed geometric information, such as channel dimensions and areas. More complete geometric information is available for MOSFETs because of the importance of this device type in LSI and VLSI circuits.

The rest of the information in a semiconductor element statement specifies the initial state of the device. These initial conditions are optional and are seldom used. The **OFF** flag is a toggle that initializes the device in the nonconducting state for the DC solution. The iterative DC solution process assumes that at the start all semiconductor elements are either conducting or on the verge of conduction. **OFF** devices are held in the cutoff state until convergence is reached, at which point the constraint is removed and iterations continue until the first converged solution is confirmed or a new solution is found. In other words, a device initially assumed to be **OFF** may turn out to be conducting at the completion of the SPICE DC solution.

The optional **IC** specification defines the values of the terminal voltages of a semiconductor device at time zero. This specification is used only in conjunction with a transient analysis, which precludes the computation of a DC bias solution. The values defined as part of the **IC** option are used only when the **UIC** option is specified in the **.TRAN** statement; see Chap. 6. In all other circumstances these values are ignored.

The companion model statement for semiconductor-element statements has the following general format:

.MODEL *MODname MODtype* <*PARAM1=value1* <*PARAM2=value2* ...>>

MODname is a unique eight-character name in SPICE2 and a name of arbitrary length in SPICE3 associated with the parameter values defined in this statement. This name is referenced by the device statements. *MODtype* is one of seven recognized device types in SPICE2 and one of twelve in SPICE3. SPICE2, SPICE3, and PSpice support the following model types:

D	Diode model
NPN	npn BJT model
PNP	pnp BJT model
NJF	n-channel JFET model
PJF	p-channel JFET model
NMOS	n-channel MOSFET model
PMOS	p-channel MOSFET model

SPICE3 supports the following additional semiconductor model types:

R	Diffused resistor model
C	Diffused capacitor model

URC	Uniformly distributed RC model
NMF	n-channel MESFET model
PMF	p-channel MESFET model

PSpice also has a built-in model for an n-channel gallium-arsenide FET, called a GAS-FET, and an additional BJT model type for a lateral pnp, called an LPNP.

Four sets of semiconductor parameter names occur in SPICE2, eight in SPICE3, and five in PSpice. More than one model type can share the same parameters, for example, npn and pnp bipolar transistors. Only those parameter names must appear on the model statement that are assigned different values from the defaults built into the program. A model statement with no values assigns the defaults of the specific type to that model name. In this chapter each element is described together with its model parameters.

The following sections describe the supported semiconductor elements. The descriptions of both the element and the companion model are presented. The parameters and supporting equations for the first-order models are introduced. For details on the semiconductor-device physics underlying the models described in this chapter, several reference texts (Grove 1967; Muller and Kamins 1977; Sze 1981) are suggested. The complete equations and list of parameters for semiconductor devices are contained in Appendix A and in more detail in *Semiconductor Device Modeling with SPICE* (Antognetti and Massobrio 1988).

3.2 DIODES

The general format of a diode element statement is

D*name n+ n− MODname* <*area*> <**OFF**><**IC**=v_{D0} >

The letter **D** must be the first character in *Dname;* **D** identifies a diode and can be followed by up to seven characters in SPICE2. $n+$ is the positive node, or anode, and $n−$ is the negative node, or cathode. The schematic symbol of a diode is presented together with its pn-junction SPICE implementation in Figure 3.1.

MODname is the name of the model defining the parameters for this diode element. The variable *area* is a scale factor equal to the number of identical diodes connected in parallel; *area* defaults to 1. For the initial iterations of the DC bias solution, the keyword **OFF** initializes the diode in the cutoff region; otherwise diodes are initialized at the limit of turn-on, with $V_D \approx 0.6$ V, for the DC solution. The keyword **IC** defines the voltage v_{D0} at time $t = 0$ for a time domain analysis. v_{D0} is used as initial value only when the **UIC** option is present in the **.TRAN** statement.

The BCE of the diode in DC is described by an exponential function:

$$I_D = IS(e^{qV_D/NkT} - 1) \qquad (3.1)$$

where *IS* is the saturation current, *N* is the emission, or nonideality coefficient, and *T* is the temperature in degrees Kelvin, K. *q* and *k* are the electron charge and Boltzmann's

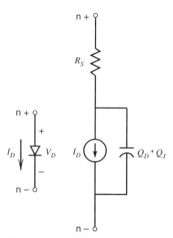

Figure 3.1   Diode element.

constant, equal to 1.6×10^{-19} C and 1.38×10^{-22} JK^{-1}, respectively; these two constants and the diode temperature define the *thermal voltage*, $V_{th} = kT/q$.

The I-V characteristic of the diode described by Eq. 3.1 is shown in Figure 3.2. The two model parameters, **IS** and **N**, can be easily derived as the slope and the intercept at the origin of the log I_D versus V_D plot, that is, of a semilogarithmic plot of Eq. 3.1. The breakdown current occurring when the diode is reverse-biased is also modeled if the value of the breakdown voltage parameter, **BV**, is defined. The current at the breakdown voltage is set to the value of parameter **IBV**; see Appendix A for complete equations.

The variation in time of the diode current of a short-base diode (Muller and Kamins 1977) is controlled by the two types of charge storage in a semiconductor pn junction, the diffusion charge and the depletion charge.

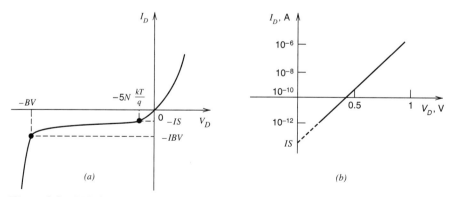

Figure 3.2   I-V characteristics of a diode: (*a*) I_D versus V_D; (*b*) log I_D versus V_D.

The diffusion charge Q_D is defined by

$$Q_D = TT \cdot I_D \tag{3.2}$$

where TT is the average transit time of minority carriers through the narrow region of a short-base diode.

The depletion charge Q_J at the pn junction is stored in a voltage-dependent junction capacitance, C_J:

$$C_J = \frac{CJO}{(1 - V_D/VJ)^M} \tag{3.3}$$

$$Q_J = \int_0^{V_D} C_J \, dV \tag{3.4}$$

CJO, VJ, and M are the zero-bias junction capacitance, built-in voltage, and grading coefficient, respectively.

In the small-signal AC analysis, the diode is modeled as the conductance in the operating point g_d and two capacitances C_D and C_J corresponding to the two charges, respectively,

$$g_d = \frac{1}{r_d} = \frac{dI_D}{dV_D} = \frac{qIS}{NkT}e^{qV_D/NkT} \approx \frac{I_D}{NV_{th}} \tag{3.5}$$

$$C_D = TT \cdot g_d \tag{3.6}$$

Two additional parameters needed for a first-order definition of the diode model are the parasitic series resistance, **RS**, and the energy gap, **EG**. The latter is used to differentiate between different types of diodes, such as silicon pn-junction diodes, other semiconductor junctions, and Schottky diodes.

The general format of the diode model statement is

.MODEL *MODname* **D** $<$**IS** $= IS$ $<$**N** $= N \ldots >>$

Table 3.1 summarizes the diode model parameters introduced so far along with the default values assigned by SPICE2. The scale factor column indicates whether and how the parameter is scaled by the factor *area* appearing in the device statement.

EXAMPLE 3.1

Following are the SPICE descriptions for two diodes.

```
DIN 0 1 DMOD OFF
.MODEL DMOD D
```

Table 3.1 Diode Model Parameters

Name	Parameter	Units	Default	Example	Scale Factor
IS	Saturation current	A	10^{-14}	1E-16	*area*
N	Emission coefficient	—	1	1.5	—
RS	Ohmic resistance	Ω	0	100	*1/area*
TT	Transit time	s	0	0.1N	—
CJO	Zero-bias junction capacitance	F	0	2P	*area*
VJ	Junction potential	V	1	0.6	—
M	Grading coefficient	—	0.5	0.33	—
EG	Activation energy	eV	1.11	1.11 Si	—
				0.69 Sbd	
				0.67 Ge	
BV	Breakdown voltage	V	∞	40	—
IBV	Current at breakdown voltage	A	10^{-3}	10U	*area*

DIN describes a protective diode at the input of another device that is normally off. Only default parameters are used to model DIN.

```
DSBD 11 17 DS1 IC=0.4
.MODEL DS1 D IS=1P CJO=0.2P VJ=0.7 M=0.5 EG=0.69
```

The above two statements define a Schottky-barrier diode. The junction is initialized at 0.4 V, the value of the voltage drop when conducting; also note that $TT = 0$, because of the absence of minority carriers. The value of *EG* corresponds to an aluminum-silicon contact.

3.3 BIPOLAR JUNCTION TRANSISTORS

The general form of a bipolar junction transistor (BJT) statement is

Q*name nc nb ne <ns> MODname <area><***OFF***><***IC***= v_{BE0}, v_{CE0} >*

The letter **Q** must be the first character in *Qname;* **Q** identifies a BJT and can be followed by up to seven characters in SPICE2. *nc, nb,* and *ne* specify the collector, base, and emitter nodes, respectively. The fourth number, *ns,* is the substrate node, and its specification is optional. If *ns* is not present, the substrate is assumed to be connected to ground.

MODname is the name of the model defining the parameter values for this transistor. Two BJT device types are supported, **NPN** and **PNP**. The schematic symbols for the two types of BJTs are shown in Figure 3.3.

The scale factor *area* is equal to the number of identical transistors connected in parallel; *area* defaults to 1. The keyword **OFF** initializes the transistor in the cutoff region for the initial iterations of the DC bias solution. By default, BJTs are initialized in the forward active region, with $V_{BE} = 0.6$ V and $V_{BC} = -1.0$ V for the DC solution. The keyword **IC** defines the values of the junction voltages, v_{BE0} and v_{CE0} at time $t = 0$ for the transient analysis. v_{BE0} and v_{CE0} are used as initial values only when the **UIC** option is present in the **.TRAN** statement.

3.3.1 DC Model

The basic DC model used in SPICE to describe the BCEs of a bipolar transistor is the Ebers-Moll model (Muller and Kamins 1977). The model shown in Figure 3.4 is the *injection version* of the Ebers-Moll model, which uses diode currents I_F and I_R as reference:

$$I_F = I_{ES}(e^{qV_{BE}/kT} - 1)$$

$$I_R = I_{CS}(e^{qV_{BC}/kT} - 1)$$

(3.7)

where the emission coefficients (see Eq. 3.1) have been assumed to equal 1. The three terminal currents of the transistor, I_C, I_E, and I_B can be expressed as functions of the two reference currents and the forward and reverse current gains, α_F and α_R, of the common-base (CB) connected BJT:

$$I_C = \alpha_F I_F - I_R$$

$$I_E = -I_F + \alpha_R I_R$$

$$I_B = (1 - \alpha_F)I_F + (1 - \alpha_R)I_R$$

(3.8)

I_{ES} and I_{CS} are the saturation currents of the BE and BC junctions, respectively. These two currents satisfy the reciprocity equation

$$\alpha_F I_{ES} = \alpha_R I_{CS} = IS$$

(3.9)

where IS, a SPICE BJT model parameter, is the saturation current of the transistor.

The SPICE implementation of the Ebers-Moll model is a variant known as the *transport version* and is shown in Figure 3.5. The injection version is commonly documented in textbooks and has been repeated above for comparison with the transport version. The currents flowing through the two sources, which represent the transistor effect of the two back-to-back pn junctions, are chosen as reference:

$$I_{CC} = IS(e^{qV_{BE}/NF \cdot kT} - 1)$$

$$I_{CE} = IS(e^{qV_{BC}/NR \cdot kT} - 1)$$

(3.10)

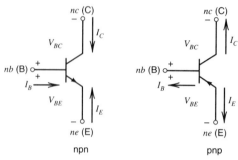

Figure 3.3   npn and pnp bipolar transistor elements.

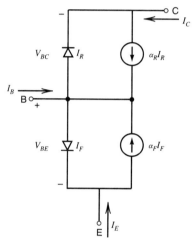

Figure 3.4   Ebers-Moll injection model of an npn transistor.

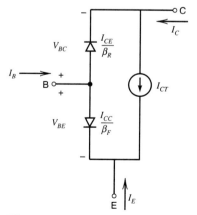

Figure 3.5   Ebers-Moll transport model of an npn transistor.

The three terminal currents assume the following expressions:

$$I_C = I_{CC} - \frac{\beta_R + 1}{\beta_R} I_{CE} = I_{CT} - \frac{1}{\beta_R} I_{CE}$$

$$I_E = I_{CE} - \frac{\beta_F + 1}{\beta_F} I_{CC} = -I_{CT} - \frac{1}{\beta_F} I_{CC} \tag{3.11}$$

$$I_B = \frac{1}{\beta_F} I_{CC} + \frac{1}{\beta_R} I_{CE} = I_{BC} + I_{BE}$$

where

$$I_{CT} = I_{CC} - I_{CE}$$

β_F and β_R in the above equations are the forward and reverse current gains, SPICE parameters **BF** and **BR**, of a bipolar transistor in the common-emitter (CE) configuration.

Depending on the values of the two controlling voltages, V_{BE} and V_{BC}, the transistor can operate in the following four modes:

Forward active	$V_{BE} > 0$ and $V_{BC} < 0$
Reverse active	$V_{BE} < 0$ and $V_{BC} > 0$
Saturation	$V_{BE} > 0$ and $V_{BC} > 0$
Cutoff	$V_{BE} < 0$ and $V_{BC} < 0$

In most applications the transistor is operated in the forward active, or linear, region, and in some situations in the saturation region. The suffixes F and R in many SPICE parameter names indicate the region of operation.

The I-V characteristics described by Eqs. 3.11 are shown in Figure 3.6 for positive values of V_{BE} and V_{CE}. These characteristics are ideal, ignoring the effects of finite output conductance in the forward and reverse regions and the parasitic series resistances associated with the collector, base, and emitter regions; these resistances are modeled by parameters **RC**, **RB**, and **RE**, respectively.

The finite output conductance of a BJT is modeled in SPICE by the Early effect (Muller and Kamins 1977) implemented by two parameters, **VAF** and **VAR**. The *Early voltage* is the point on the V_{BC} axis in the (I_C, V_{BC}) plane where the extrapolations of the linear portions of all I_C characteristics meet. This geometric interpretation of the Early effect and its SPICE implementation is shown in Figure 3.7 for the $I_C = f(V_{CE})$ characteristics: $V_{CE} = V_{BE} - V_{BC}$. The reverse Early voltage, *VAR,* has a similar interpretation for the reverse region. For most practical applications *VAF* is important and *VAR* can be neglected.

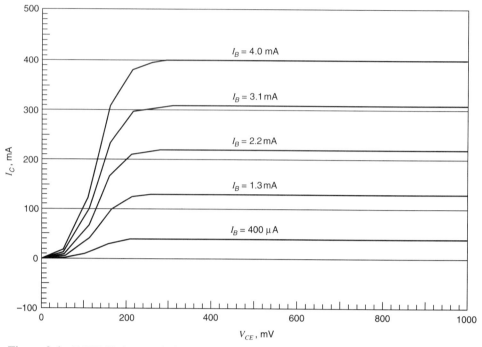

Figure 3.6   BJT I-V characteristics described by the Ebers-Moll model.

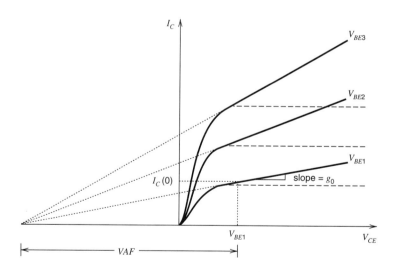

Figure 3.7   Early voltage parameter, **VAF**.

With the addition of the Early voltage, I_C and I_{CT} in Eqs. 3.11 are modified as follows:

$$I_C = (I_{CC} - I_{CE})\left(1 - \frac{V_{BC}}{VAF} - \frac{V_{BE}}{VAR}\right) - \frac{1}{\beta_R}I_{CE} = I_{CT} - I_{BC} \qquad (3.12)$$

3.3.2 Dynamic and Small-Signal Models

The dynamic behavior of a BJT is modeled by five different charges. Two charges, Q_{DE} and Q_{DC}, are associated with the mobile carriers. These are the diffusion charges represented by the current sources I_{CC} and I_{CE} in the Ebers-Moll model. The other three charges model the fixed charges in the depletion regions of the three junctions: base-emitter, Q_{JE}; base-collector, Q_{JC}; and collector-substrate, Q_{JS}.

The diffusion charges are modeled by the following equations in the large-signal transient analysis:

$$\begin{aligned} Q_{DE} &= TF \cdot I_{CC} \\ Q_{DC} &= TR \cdot I_{CE} \end{aligned} \qquad (3.13)$$

TF and TR are the forward and reverse transit times, respectively, of the injected minority carriers through the neutral base.

The depletion charges can be derived using the nonlinear equation that defines the depletion capacitance, C_J, of a pn junction, Eq. 3.3. The SPICE large-signal implementation of the three depletion charges is according to Eq. 3.4, which defines the charge Q_J. The three voltage-dependent junction capacitances are described by the following functions:

$$\begin{aligned} C_{JE} &= \frac{CJE}{(1 - V_{BE}/VJE)^{MJE}} \\[2mm] C_{JC} &= \frac{CJC}{(1 - V_{BC}/VJC)^{MJC}} \\[2mm] C_{JS} &= \frac{CJS}{(1 - V_{CS}/VJS)^{MJS}} \end{aligned} \qquad (3.14)$$

Each junction can be characterized in SPICE by up to three parameters: CJX, the zero-bias junction capacitance; VJX, the built-in potential; and MJX, the grading coefficient. X stands for E, C, or S, denoting the emitter, collector, or substrate junction, respectively.

The nonlinear BJT model in SPICE including charge storage and parasitic terminal resistances is depicted in Figure 3.8. The five charges are consolidated into three: Q_{BE}, which includes Q_{DE} and Q_{JE}; Q_{BC}, which includes Q_{DC} and Q_{JC}; and Q_{CS}, modeled by

C_{CS}, the collector-substrate capacitance. Figure 3.8 is a first-order representation of the complete Gummel-Poon BJT model available in SPICE and is sufficiently accurate for many applications. The complete model includes second-order effects, such as β_F and τ_F dependency on I_C, base push-out, and temperature effects. The complete equations and model parameters are summarized in Appendix A.

The linearized small-signal model of a BJT, also known as the hybrid-π model, is shown in Figure 3.9. The nonlinear diodes and the current generator I_{CT} in Figure 3.5 are replaced by the following linear resistances (conductances) and transconductances:

$$g_\pi = \frac{1}{r_\pi} = \frac{\partial I_B}{\partial V_{BE}} = \frac{1}{\beta_F}\frac{dI_{CC}}{dV_{BE}}$$

$$g_\mu = \frac{1}{r_\mu} = \frac{\partial I_B}{\partial V_{BC}} = \frac{1}{\beta_R}\frac{dI_{CE}}{dV_{BC}}$$

$$g_{mF} = \frac{\partial I_{CT}}{\partial V_{BE}} \tag{3.15}$$

$$g_{mR} = g_o = \frac{1}{r_o} = -\frac{\partial I_{CT}}{\partial V_{BC}} = -\frac{\partial I_C}{\partial V_{BC}} - \frac{\partial I_B}{\partial V_{BC}} = -\frac{\partial I_C}{\partial V_{BC}} - g_\mu$$

$$g_m = g_{mF} - g_{mR} = \frac{\partial I_C}{\partial V_{BE}} - g_o$$

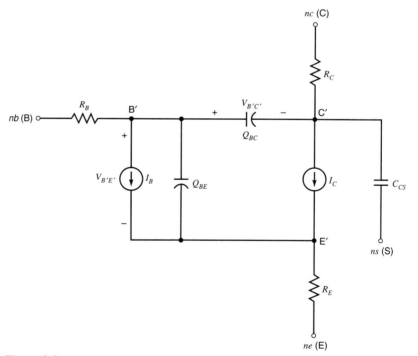

Figure 3.8 Large-signal SPICE BJT model.

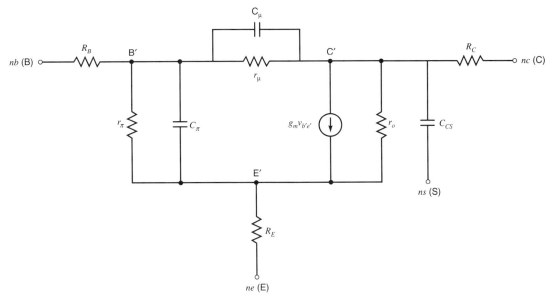

Figure 3.9 Small-signal SPICE BJT model.

The above small-signal parameters have been derived assuming no parasitic terminal resistances RC, RB, and RE; if these resistances are present, terminal voltages $V_{B'E'}$ and $V_{B'C'}$ replace V_{BE} and V_{BC}.

The small-signal AC collector current i_c, can be expressed from Eqs. 3.12 and 3.15 and the hybrid-π model (see Figure 3.9) as

$$i_c = \frac{\partial I_{CT}}{\partial V_{BE}} v_{be} + \frac{\partial I_{CT}}{\partial V_{BC}} v_{bc} - \frac{\partial I_{BC}}{\partial V_{BC}} v_{bc} = g_m v_{be} + g_o v_{ce} - g_\mu v_{bc} \quad (3.16)$$

where v_{bc} has been replaced by $v_{be} - v_{ce}$ in the second term. In the forward active region the small-signal equations assume the more commonly known expressions (Gray and Meyer 1993):

$$i_c = g_m v_{be} = g_{mF} v_{be}$$

$$g_m = g_{mF} = \beta_F g_\pi = \frac{q I_C}{NF \cdot kT}$$

$$r_\pi = \frac{1}{g_\pi} = \frac{\beta_F}{g_m} \quad (3.17)$$

$$r_\mu \to \infty, \qquad g_\mu \approx 0$$

$$r_o = \frac{1}{g_o} = \frac{VAF}{I_C} = \frac{VAF}{g_m V_{th}}$$

In small-signal AC analysis charge-storage effects are modeled by nonlinear capacitances. The diffusion charges are modeled by two diffusion capacitors, C_{DE} and C_{DC}:

$$C_{DE} = \frac{dQ_{DE}}{dV_{BE}} = TF\frac{\partial I_{CT}}{\partial V_{BE}} = TF \cdot g_{mF}$$

$$C_{DC} = \frac{dQ_{DC}}{dV_{BC}} = TR\frac{\partial I_{CT}}{\partial V_{BC}} = TR \cdot g_{mR}$$

(3.18)

where g_{mF} and g_{mR} are the forward and reverse transconductances of the BJT (Eqs. 3.15). The junction capacitances are defined by Eqs. 3.14. In the small-signal BJT model (Figure 3.9) the two types of capacitances for the BE and BC regions are consolidated in C_π and C_μ, corresponding to Q_{BE} and Q_{BC}, respectively.

$$C_\pi = C_{DE} + C_{JE}$$

$$C_\mu = C_{DC} + C_{JC}$$

(3.19)

An important characteristic of a BJT is the cutoff frequency, f_T, where the current gain drops to unity; f_T can be expressed as a function of the small-signal parameters:

$$f_T = \frac{g_m}{2\pi(C_\pi + C_\mu)}$$

(3.20)

3.3.3 Model Parameters

The general form of the BJT model statement is

.MODEL *MODname* **NPN/PNP** <**IS**=*IS* <**BF**=*BF* ...>>

In every model one of the keywords **NPN** or **PNP**, indicating the transistor type, must be specified.

Table 3.2 summarizes the model parameters introduced in this section together with the default values assigned by SPICE2.

EXAMPLE 3.2

The following are two BJT specifications.

```
QX12 14 15 21 QMOD IC = 0.6,5.0
.MODEL QMOD NPN BF=200 RB=100 CJC=5P TF=10N
```

Table 3.2 BJT Model Parameters

Name	Parameter	Units	Default	Example	Scale Factor
IS	Saturation current	A	10^{-16}	1E-16	*area*
BF	Forward current gain	—	100	80	—
BR	Reverse current gain	—	1	3	—
NF	Forward emission coefficient	—	1	2	—
NR	Reverse emission coefficient	—	1	1.5	—
VAF	Forward Early voltage	V	∞	100	—
VAR	Reverse Early voltage	V	∞	250	—
RC	Collector resistance	Ω	0	200	1/*area*
RE	Emitter resistance	Ω	0	2	1/*area*
RB	Base resistance	Ω	0	100	1/*area*
TF	Forward transit time	s	0	1N	—
TR	Reverse transit time	s	0	100N	—
CJE	BE zero-bias junction capacitance	F	0	2P	*area*
VJE	BE built-in potential	V	0.75	0.6	—
MJE	BE grading coefficient	—	0.33	0.33	—
CJC	BC zero-bias junction capacitance	F	0	2P	*area*
VJC	BC built-in potential	V	0.75	0.6	—
MJC	BC grading coefficient	—	0.33	0.5	—
CJS	CS zero-bias junction capacitance	F	0	2P	*area*
VJS	CS built-in potential	V	0.75	0.6	—
MJS	CS grading coefficient	—	0	0.5	—

QX12 is initialized at $v_{BE0} = 0.6$ V and $v_{CE0} = 5.0$ V in a transient analysis (see Chap. 6); its switching time is governed by the BC junction capacitance and the forward transit time. Defaults are used for the remaining parameters.

```
QFF1 1 3 0 QQ
QFF2 2 4 0 QQ OFF
.MODEL QQ PNP IS=1P BF=50 CJE=1P CJC=2P
```

QFF1 and QFF2 are the two transistors of a flip-flop. QFF2 is specified OFF in order to speed up the solution of the DC bias point. For more detail on flip-flop initialization, see Example 4.8.

For some standard parts SPICE parameters are available from semiconductor manufacturers because of the widespread use of SPICE simulation in circuit design. Generally, however, model parameters must be derived, and how the SPICE parameters can be derived for physical transistors becomes an important question. This topic is addressed in the following pages. The characteristics of IC transistors are derived from a test structure built on the same wafer. The measurement techniques leading to the SPICE parameters of bipolar transistors are presented in detail by Getreu (1976). The same measurements can be used to characterize discrete transistors if lab equipment and the transistor of interest are available. A different approach is necessary for discrete transistors when the only information available is a data sheet. The extraction of the main parameters from a data sheet is outlined in the following example.

EXAMPLE 3.3

Derive the SPICE DC model parameters for the MPS2222 npn transistor, similar to a 2N2222. Use the *Motorola Semiconductors Data Book* (Motorola Inc. 1988) for electrical characteristics. Simulate the $I_C = f(V_{CE}, I_B)$ characteristics with SPICE.

Solution
The information needed for extracting the model parameters is found in the MPS2222 data sheet, Figure 3.10, under *electrical characteristics*. Several categories of characteristics are included, namely the *OFF, ON, small-signal*, and *switching* characteristics. In addition to these data, provided as *minimum* or *maximum* values, many transistor data sheets contain graphs of several electrical quantities. When graphs are available, they should be used as primary sources of characteristic data, because the graphs represent a typical device.

The first parameter to be extracted is the saturation current, *IS*. From the graph entitled *ON Voltages*, we obtain the BE voltage in saturation, V_{BEsat}, as a function of I_C for a set ratio I_C/I_B, equal to 10 for this transistor. Choose $I_C = 150$ mA and the resulting $V_{BEsat} = 0.85$ V. Obtain from Eqs. 3.11 a relation between I_C, I_B, and V_{BEsat} by substituting the current I_{CE} from the I_B equation into the I_C equation; Eqs. 3.10 are then used with $NF = 1$ to obtain the following expression for *IS*:

$$IS = I_B \left(\frac{I_C}{I_B} + \frac{1}{1 - \alpha_R} \right) e^{-V_{BEsat}/V_{th}} = 0.015 \cdot 12 e^{-0.85/0.0258} = 1.14 \cdot 10^{-15} \text{ A}$$

$$(3.21)$$

In the above calculation it has been assumed that $\alpha_R = 0.5$, or $BR = 1$, which is the default value in SPICE (see Table 3.2). This approximation is of no consequence unless the transistor is operated in the reverse region most of the time.

Next, we derive a value for *BF*, the current gain factor. This can be obtained from the plot of the *DC current gain*, h_{FE}, as a function of I_C. Because *BF* is a constant and because h_{FE} dependence on I_C is not supported in the first-order model, an average value should be chosen for h_{FE} over the I_C range that the transistor is expected to operate. Choose

$$h_{FE1} = 150 \text{ at } I_C = 1 \text{ mA}$$
$$h_{FE2} = 200 \text{ at } I_C = 10 \text{ mA}$$
$$h_{FE3} = 240 \text{ at } I_C = 100 \text{ mA}$$

The average of these three values results in $BF = 190$.

MAXIMUM RATINGS

Rating	Symbol	MPS2222	MPS2222A	Unit
Collector-Emitter Voltage	V_{CEO}	30	40	Vdc
Collector-Base Voltage	V_{CBO}	60	75	Vdc
Emitter-Base Voltage	V_{EBO}	5.0	6.0	Vdc
Collector Current — Continuous	I_C	600		mAdc
Total Device Dissipation @ T_A = 25°C Derate above 25°C	P_D	625 5.0		mW mW/°C
Total Device Dissipation @ T_C = 25°C Derate above 25°C	P_D	1.5 12		Watts mW/°C
Operating and Storage Junction Temperature Range	T_J, T_{stg}	−55 to +150		°C

THERMAL CHARACTERISTICS

Characteristic	Symbol	Max	Unit
Thermal Resistance, Junction to Case	$R_{\theta JC}$	83.3	°C/W
Thermal Resistance, Junction to Ambient	$R_{\theta JA}$	200	°C/W

MPS2222, A*

CASE 29-04, STYLE 1
TO-92 (TO-226AA)

GENERAL PURPOSE TRANSISTORS

NPN SILICON

ELECTRICAL CHARACTERISTICS (T_A = 25°C unless otherwise noted.)

Characteristic		Symbol	Min	Max	Unit
OFF CHARACTERISTICS					
Collector-Emitter Breakdown Voltage (I_C = 10 mAdc, I_B = 0)	MPS2222 MPS2222A	$V_{(BR)CEO}$	30 40	— —	Vdc
Collector-Base Breakdown Voltage (I_C = 10 μAdc, I_E = 0)	MPS2222 MPS2222A	$V_{(BR)CBO}$	60 75	— —	Vdc
Emitter-Base Breakdown Voltage (I_E = 10 μAdc, I_C = 0)	MPS2222 MPS2222A	$V_{(BR)EBO}$	5.0 6.0	— —	Vdc
Collector Cutoff Current (V_{CE} = 60 Vdc, $V_{EB(off)}$ = 3.0 Vdc)	MPS2222A	I_{CEX}	—	10	nAdc
Collector Cutoff Current (V_{CB} = 50 Vdc, I_E = 0) (V_{CB} = 60 Vdc, I_E = 0) (V_{CB} = 50 Vdc, I_E = 0, T_A = 125°C) (V_{CB} = 50 Vdc, I_E = 0, T_A = 125°C)	 MPS2222 MPS2222A MPS2222 MPS2222A	I_{CBO}	 — — — —	 0.01 0.01 10 10	μAdc
Emitter Cutoff Current (V_{EB} = 3.0 Vdc, I_C = 0)	MPS2222A	I_{EBO}	—	10	nAdc
Base Cutoff Current (V_{CE} = 60 Vdc, $V_{EB(off)}$ = 3.0 Vdc)	MPS2222A	I_{BL}	—	20	nAdc
ON CHARACTERISTICS					
DC Current Gain (I_C = 0.1 mAdc, V_{CE} = 10 Vdc) (I_C = 1.0 mAdc, V_{CE} = 10 Vdc) (I_C = 10 mAdc, V_{CE} = 10 Vdc) (I_C = 10 mAdc, V_{CE} = 10 Vdc, T_A = −55°C) MPS2222A only (I_C = 150 mAdc, V_{CE} = 10 Vdc)(1) (I_C = 150 mAdc, V_{CE} = 1.0 Vdc)(1) (I_C = 500 mAdc, V_{CE} = 10 Vdc)(1) MPS2222 MPS2222A		h_{FE}	35 50 75 35 100 50 30 40	— — — — 300 — — —	—

Collector-Emitter Saturation Voltage(1)
 (I_C = 150 mAdc, I_B = 15 mAdc)   MPS2222
   MPS2222A
 	
 (I_C = 500 mAdc, I_B = 50 mAdc)   MPS2222
   MPS2222A
		$V_{CE(sat)}$	—
			 —
			
			 —
			 —
				0.4
				 0.3
				
				 1.6
				 1.0
					Vdc

*Also available as a PN2222,A.

Figure 3.10 MPS2222 data sheet.

ELECTRICAL CHARACTERISTICS (continued) (T_A = 25°C unless otherwise noted.)

Characteristic		Symbol	Min	Max	Unit
Base-Emitter Saturation Voltage(1)		$V_{BE(sat)}$			Vdc
(I_C = 150 mAdc, I_B = 15 mAdc)	MPS2222		—	1.3	
	MPS2222A		0.6	1.2	
(I_C = 500 mAdc, I_B = 50 mAdc)	MPS2222		—	2.6	
	MPS2222A		—	2.0	

SMALL-SIGNAL CHARACTERISTICS

Characteristic		Symbol	Min	Max	Unit
Current-Gain — Bandwidth Product(2)		f_T			MHz
(I_C = 20 mAdc, V_{CE} = 20 Vdc, f = 100 MHz)	MPS2222		250	—	
	MPS2222A		300	—	
Output Capacitance		C_{obo}	—	8.0	pF
(V_{CB} = 10 Vdc, I_E = 0, f = 1.0 MHz)					
Input Capacitance		C_{ibo}			pF
(V_{EB} = 0.5 Vdc, I_C = 0, f = 1.0 MHz)	MPS2222		—	30	
	MPS2222A		—	25	
Input Impedance		h_{ie}			kΩ
(I_C = 1.0 mAdc, V_{CE} = 10 Vdc, f = 1.0 kHz)	MPS2222A		2.0	8.0	
(I_C = 10 mAdc, V_{CE} = 10 Vdc, f = 1.0 kHz)	MPS2222A		0.25	1.25	
Voltage Feedback Ratio		h_{re}			X 10^{-4}
(I_C = 1.0 mAdc, V_{CE} = 10 Vdc, f = 1.0 kHz)	MPS2222A		—	8.0	
(I_C = 10 mAdc, V_{CE} = 10 Vdc, f = 1.0 kHz)	MPS2222A		—	4.0	
Small-Signal Current Gain		h_{fe}			—
(I_C = 1.0 mAdc, V_{CE} = 10 Vdc, f = 1.0 kHz)	MPS2222A		50	300	
(I_C = 10 mAdc, V_{CE} = 10 Vdc, f = 1.0 kHz)	MPS2222A		75	375	
Output Admittance		h_{oe}			μmhos
(I_C = 1.0 mAdc, V_{CE} = 10 Vdc, f = 1.0 kHz)	MPS2222A		5.0	35	
(I_C = 10 mAdc, V_{CE} = 10 Vdc, f = 1.0 kHz)	MPS2222A		25	200	
Collector Base Time Constant		$rb'C_c$	—	150	ps
(I_E = 20 mAdc, V_{CB} = 20 Vdc, f = 31.8 MHz)	MPS2222A				
Noise Figure		NF	—	4.0	dB
(I_C = 100 μAdc, V_{CE} = 10 Vdc, R_S = 1.0 kΩ, f = 1.0 kHz)	MPS2222A				

SWITCHING CHARACTERISTICS   MPS2222A only

		Symbol	Min	Max	Unit
Delay Time	(V_{CC} = 30 Vdc, $V_{BE(off)}$ = 0.5 Vdc,	t_d	—	10	ns
Rise Time	I_C = 150 mAdc, I_{B1} = 15 mAdc) (Figure 1)	t_r	—	25	ns
Storage Time	(V_{CC} = 30 Vdc, I_C = 150 mAdc,	t_s	—	225	ns
Fall Time	I_{B1} = I_{B2} = 15 mAdc) (Figure 2)	t_f	—	60	ns

(1) Pulse Test: Pulse Width ≤ 300 μs, Duty Cycle ≤ 2.0%.
(2) f_T is defined as the frequency at which $|h_{fe}|$ extrapolates to unity.

SWITCHING TIME EQUIVALENT TEST CIRCUITS

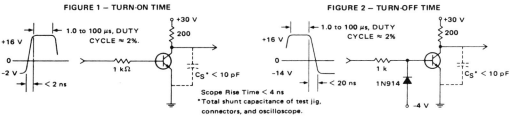

FIGURE 1 – TURN-ON TIME

FIGURE 2 – TURN-OFF TIME

Figure 3.10   *(continued)*

MPS2222, A

FIGURE 3 – DC CURRENT GAIN

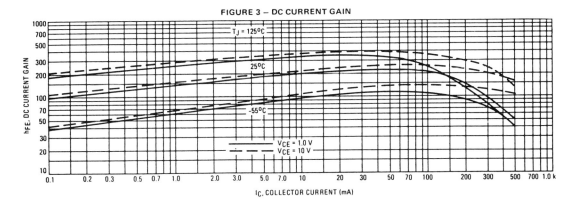

FIGURE 4 – COLLECTOR SATURATION REGION

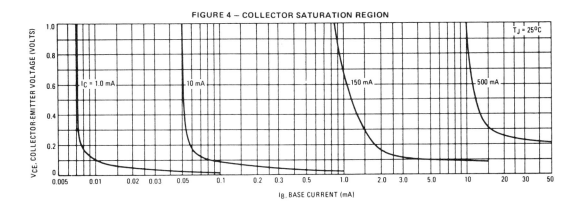

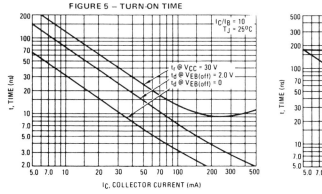

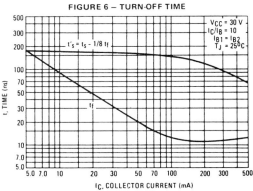

Figure 3.10 *(continued)*

FIGURE 7 – FREQUENCY EFFECTS

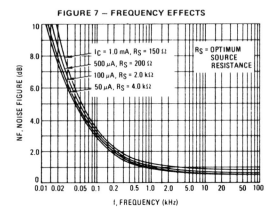

FIGURE 8 – SOURCE RESISTANCE EFFECTS

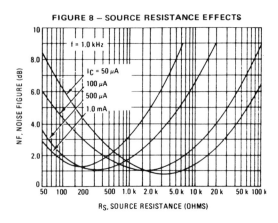

FIGURE 9 – CAPACITANCES

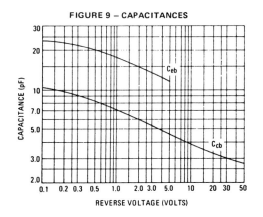

FIGURE 10 – CURRENT-GAIN BANDWIDTH PRODUCT

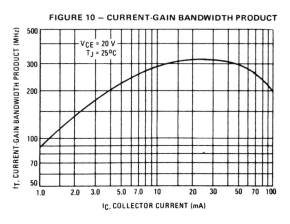

FIGURE 11 – "ON" VOLTAGES

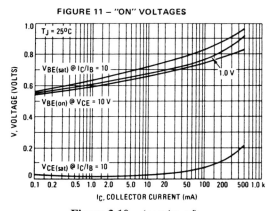

FIGURE 12 – TEMPERATURE COEFFICIENTS

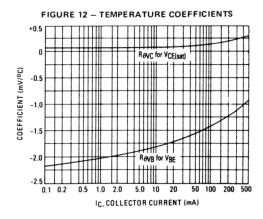

Figure 3.10 *(continued)*

The next parameter to be evaluated is the Early voltage for forward operation, *VAF*. This parameter can be calculated from the *small-signal characteristics*, which are given as *h* parameters, or hybrid parameters. The hybrid model is based on a two-port representation of the transistor with i_{in} and v_o as the independent variables. For a transistor in a CE configuration, v_{in} and i_o relate to i_{in} and v_o through the hybrid parameters h_{ie}, h_{re}, h_{fe}, and h_{oe}. A transistor can therefore be described by the following hybrid equations:

$$v_{in} = h_{ie}i_{in} + h_{re}v_o$$
$$i_o = h_{fe}i_{in} + h_{oe}v_o$$

(3.22)

h_{oe} is the data sheet parameter used for evaluating *VAF*. Both a minimum and a maximum value are listed for two values of I_C, 1 mA and 10 mA. The small-signal value of h_{oe} is the slope of the I_C-V_{CE} curve at the measured I_C and V_{CE}. Ideally the extrapolations of the two tangents to the I_C-V_{CE} curves intersect at $V_{CE} \approx -VAF$, as shown in Figure 3.7. A reasonable estimate for *VAF* is obtained by first choosing a value between *Min* and *Max* for h_{oe} at the higher current, if this current is not outside the range of the application. Then *VAF* is computed from geometric considerations, as in Figure 3.7.

$$VAF \approx \frac{I_C}{h_{oe}} - V_{CE} = \frac{10^{-2}\ \text{A}}{10^2 \cdot 10^{-6}\ \text{mho}} - 10\ \text{V} \approx 100\ \text{V}$$

The data for h_{oe} are measured at $V_{CE} = 10$ V.

We have now obtained the main BJT SPICE parameters for simulating the DC behavior of transistor MPS2222. With the information presented so far about SPICE, we can describe a measurement setup for displaying the I_C-V_{CE} characteristics of the BJT. The setup is shown in Figure 3.11. The corresponding SPICE input follows:

```
I-V CHARACTERISTICS OF 2N2222
*
Q1 2 1 0 Q2N2222
IB 0 1 .075M
VC 2 0 10
*
* PARAMETERS DERIVED BY HAND
*
*     W/ TYPICAL VALUES FROM GRAPHS
*.MODEL Q2N2222 NPN IS=1.14F BF=190 VAF=100
*
*     W/ MAX VALUES, NO GRAPHS
.MODEL Q2N2222 NPN IS=5.25E-14 BF=190 VAF=100 RC=3.4 RB=37
.OP
.DC VC 0 10 0.5 IB 0.4M 4M 0.9M
.PRINT DC I(VC)
.END
```

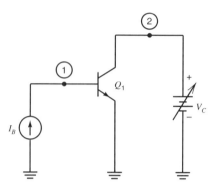

Figure 3.11   Measurement setup for I_C-V_{CE} characteristics.

The only statement that has not been defined so far is

```
.DC VC 0 10 0.5 IB 0.4M 4M 0.9M
```

which defines the range over which the V_C supply and the base current source I_B are swept; this statement is described in detail in the following chapter. The plot of the $I_C = f(V_{CE}, I_B)$ characteristics simulated by SPICE with the above parameters is shown in Figure 3.12. Note that the different curves are equally spaced in the first-order model based on a constant value of h_{FE}, equal to BF. No $I_C = f(V_{CE}, I_B)$ characteristic is available from the data sheet for comparison with the above simulated curves; the validity of the model can be verified for a few operating points given in the data sheet.

A slightly different approach must be followed if no graphs are available and all parameters must be derived from the *electrical characteristics* data. The two maximum values, V_{CEsatM} and V_{BEsatM}, are obtained from the saturation characteristics. These values are too high for following the extraction procedure outlined above.

The difference in V_{CEsatM} for the two values of I_C can be attributed to the ohmic collector resistance, **RC**, the value of which is

$$RC = \frac{V_{CEsatM1} - V_{CEsatM2}}{I_{C1} - I_{C2}} = \frac{1.2 \text{ V}}{0.35 \text{ A}} = 3.4 \ \Omega \tag{3.23}$$

Similarly, the difference in V_{BEsat} can be attributed in large part to the voltage drop across the parasitic base resistance, **RB**, resulting in a resistor value

$$RB = \frac{V_{BEsatM1} - V_{BEsatM2}}{I_{B1} - I_{B2}} = \frac{1.3 \text{ V}}{0.035 \text{ A}} = 37 \ \Omega \tag{3.24}$$

The approximation of attributing the V_{BEsatM} difference to an ohmic voltage drop is supported by the fact that it takes approximately only $V_{BE} = V_{th} = 26$ mV to increase I_B from 15 mA to 50 mA.

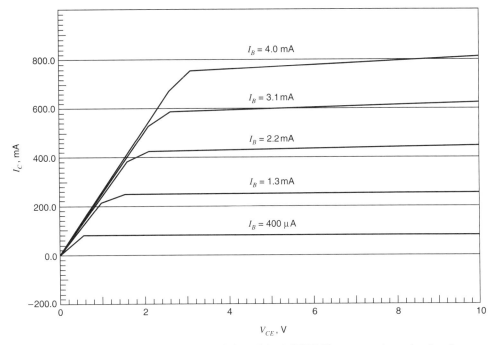

Figure 3.12 $I_C = f(V_{CE}, I_B)$ characteristics of the MPS2222 npn transistor simulated by SPICE.

Now **IS** can be derived from Eqs. 3.10 and 3.11 and the corrected values of V_{BEsatM} and V_{CEsatM}:

$$V_{BEsatM} = V_{BEsatM1} - RB \cdot I_{B1} = 0.74 \text{ V}$$
$$V_{CEsatM} = V_{CEsatM1} - RC \cdot I_{C1} = 0.69 \text{ V}$$
$$IS \approx I_{C1}e^{-V_{BEsatM}/V_{th}} = 0.15 \cdot 3.5 \cdot 10^{-13} \text{ A} = 5.25 \cdot 10^{-14} \text{ A}$$

In the absence of the *DC current gain* plots, *BF* can be estimated from the *ON characteristics* table, which lists the minimum values of h_{FE} for several values of I_C. Usually both a minimum and a maximum h_{FE} value are given for a single I_C value, the highest. Choose h_{FE1} in the range between *Min* and *Max*, closer to the minimum. Calculate h_{FE2} and h_{FE3} at two other values of I_C such that they represent the same multiple of the corresponding minimum values as h_{FE1}. *BF* results as the average of the three midrange values of h_{FE}: *BF* = 110. The extraction of **VAF** is the same as the above.

The charge-storage characteristics can be derived from the plots of *capacitances* versus reverse voltage and *switching characteristics*. The junction capacitances, **CJC** and **CJE**, are derived from the former characteristics, and the transit time, **TF**, is estimated from the latter. The three characteristic parameters of a junction capacitance,

CJX, MJX, and *VJX,* should be evaluated from a plot of log C_J versus $f(V_J)$, which is a straight line; see Eqs. 3.14.

The above parameter extraction approach can be automated by writing a small program for repeated use. Second-order effects can be added, such as low- and high-current behavior. The package *Parts*, from MicroSim Corp., computes SPICE parameters for all the supported semiconductor devices from data book characteristics. For transistor MPS2222, *Parts* finds the following DC parameters using typical data:

```
* Q2N2222 MODEL CREATED USING PARTS VERSION 4.03 ON 08/02/91 AT 13:59
*          W/ VALUES FROM GRAPHS OR AVERAGED
*
.MODEL Q2N2222 NPN(IS=15.01F XTI=3 EG=1.11 VAF=90.7 BF=223.7 NE=2.348
+              ISE=70.78P IKF=3.837 XTB=1.5 BR=1 NC=2 ISC=0 IKR=0 RC=0 CJC=2P
+              MJC=.3333 VJC=.75 FC=.5 CJE=5P MJE=.3333 VJE=.75 TR=10N TF=1N
+              ITF=0 VTF=0 XTF=0)
```

A number of parameters that are not in Table 3.2 are present in the above `.MODEL` statement because *Parts* uses the complete Gummel-Poon model in the extraction. Another parameter extraction package, *MODPEX*, from Symmetry Design Systems (1992), can scan a data sheet and generate SPICE parameters.

Exercise
Verify that the parameters from *Parts* result in $I_C = f(V_{CE}, I_B)$ characteristics similar to those in Figure 3.12.

3.4 JUNCTION FIELD EFFECT TRANSISTORS (JFETS)

The general form of a junction field effect transistor (JFET) statement is

> *Jname nd ng ns MODname* <*area*> <**OFF**> <**IC**= v_{DS0}, v_{GS0}>

The letter **J** must be the first character in *Jname*; **J** identifies a JFET and can be followed by up to seven characters in SPICE2. *nd, ng,* and *ns* are the drain, gate, and source nodes, respectively.

MODname is the name of the model that defines the parameter values for this transistor. Two JFET models are supported, n-channel (**NJF**) and p-channel (**PJF**). The schematic representations of the two types of JFETs are shown in Figure 3.13. *area* is a scale factor equal to the number of identical transistors connected in parallel; *area* defaults to 1. The keyword **OFF** initializes the transistor in the cut-off region for the initial iterations of the DC bias solution. By default, JFETs are initialized at the threshold voltage, with $V_{GS} = VTO$ and $V_{DS} = 0.0$ for the DC solution. The keyword **IC** defines the values of the terminal voltages, v_{DS0} and v_{GS0}, at time $t = 0$ in a time-domain analysis. v_{DS0} and v_{GS0} are used as initial values only when the **UIC** option is present in the `.TRAN` statement.

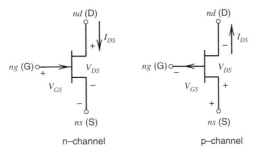

Figure 3.13 n- and p-channel JFETs.

The quadratic Shichman-Hodges model (Shichman and Hodges 1968) is used in SPICE to solve the BCEs. The drain-source current, I_{DS}, is defined by the following three equations for the three regions of operations, cut-off, saturation, and linear, respectively:

$$
I_{DS} = \begin{cases} 0 & \text{for } V_{GS} \leq VTO \\ BETA\,(V_{GS} - VTO)^2(1 + LAMBDA \cdot V_{DS}) & \text{for } 0 < V_{GS} - VTO \leq V_{DS} \\ BETA\, \cdot V_{DS}(2(V_{GS} - VTO) - V_{DS})(1 + LAMBDA \cdot V_{DS}) & \\ & \text{for } 0 < V_{DS} < V_{GS} - VTO \end{cases}
$$

(3.25)

VTO, BETA, and *LAMBDA* are the threshold, or pinch-off, voltage, the transconductance factor, and output conductance factor in saturation, respectively. Note that *LAMBDA* is measured in V^{-1} and is equivalent to the inverse of the Early voltage for the BJT, introduced in Sec. 3.3.1. There is an additional current component due to the pn junction current, characterized by the saturation current *IS*. The gate pn junctions are reverse biased, and therefore the pn junction current is negligible.

The above equations are valid for $V_{DS} > 0$. If V_{DS} changes sign, the behavior of the JFET is symmetrical, the drain and the source swapping roles; V_{GS} is replaced by V_{GD}, and V_{DS} is replaced by its absolute value in the above equations. The current I_{DS} flows in the opposite direction.

The dynamic behavior of a JFET is modeled by two charges associated with the gate-drain, GD, and gate-source, GS, junctions, respectively. These charges are accumulated on the depletion capacitances of the two reverse-biased junctions and are described by Eq. 3.3.

$$
C_{GS} = \frac{CGS}{(1 - V_{GS}/PB)^{0.5}}
$$
$$
C_{GD} = \frac{CGD}{(1 - V_{GD}/PB)^{0.5}}
$$

(3.26)

CGS, CGD, and PB are the zero-bias gate-source capacitance, zero-bias gate-drain capacitance, and built-in gate junction potential, respectively. These capacitances are used in the small-signal analysis. In the large-signal time-domain analysis the gate charges are computed from Eq. 3.4 using Eqs. 3.26. The large-signal JFET model is shown in Figure 3.14 and the small-signal model in Figure 3.15. The transconductance g_m and the drain-source resistance r_{ds} of the small-signal model are defined as follows:

$$g_m = \frac{dI_{DS}}{dV_{GS}}$$

$$g_{ds} = \frac{1}{r_{ds}} = \frac{dI_{DS}}{dV_{DS}}$$

(3.27)

$V_{GS'}$ and $V_{D'S'}$ must be used in Eqs. 3.25, 3.26, and 3.27 when RD and RS are nonzero. The expression of I_{DS} appropriate to the region of operation (Eqs. 3.25) must be used when deriving g_m and r_{ds}. Note that in normal operation the GD and GS diodes are reverse biased and the corresponding small-signal conductances can be neglected.

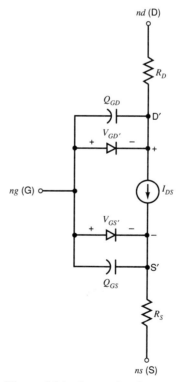

Figure 3.14   Large-signal
n-channel JFET model.

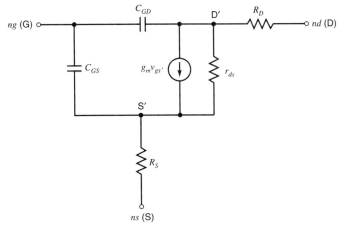

Figure 3.15   Small-signal JFET model.

The general form of the JFET model statement is

`.MODEL` *MODname* `NJF/PJF` $<$`VTO=`*VTO* $<$`BETA=`*BETA* $\dots$ $>>$

In every model statement one of the keywords `NJF` or `PJF`, indicting the transistor type, must be specified.

Table 3.3 summarizes the model parameters introduced so far, with the corresponding default values assigned by SPICE2.

The threshold, or pinch-off, voltage, *VTO*, is sign-sensitive. JFETs operate in depletion mode; that is, they are *normally on*. Therefore *VTO* is negative for n-channel devices (`NJF`) and should be positive for p-channel devices (`PJF`). Because of a bug in all recent versions of SPICE2, however, the sign of *VTO* for a depletion p-channel JFET should be entered as negative; in other words, *VTO* for a p-channel JFET is defined with the same sign as for an n-channel JFET. This discrepancy is present in SPICE3 as well as PSpice.

Table 3.3   JFET Model Parameters

Name	Parameter	Units	Default	Example	Scale Factor
VTO	Threshold (pinch-off) voltage	V	-2.0	-2.5	—
BETA	Transconductance parameter	AV^{-2}	10^{-4}	1.0E-3	*area*
LAMBDA	Channel length modulation parameter	V^{-1}	0	1.0E-4	—
RD	Drain ohmic resistance	Ω	0	100	1/*area*
RS	Source ohmic resistance	Ω	0	100	1/*area*
CGS	Zero-bias GS junction capacitance	F	0	5P	*area*
CGD	Zero-bias GD junction capacitance	F	0	1P	*area*
PB	Gate junction potential	V	1	0.6	—
IS	Gate junction saturation current	A	10^{-14}	1.0E- 16	*area*

EXAMPLE 3.4

```
J1 20 1 21 MODJ
.MODEL MODJ NJF VTO=-3 RD=20 RS=20
```

J1 is a normally-on n-channel JFET with parasitic series drain and source resistances. A normally-on p-channel JFET that conducts the same current as J1 in similar bias conditions is described by the following **MODEL** statement:

```
.MODEL MODJ PJF VTO=-3 RD=20 RS=20
```

EXAMPLE 3.5

Derive from the Motorola semiconductor data book the SPICE DC model parameters for the 2N4221 n-channel JFET.

Solution
As in Example 3.3, we choose to derive the SPICE parameters from graphical data if such graphs are available. The threshold, or pinch-off, voltage, **VTO**, is referred to as the *gate source cutoff voltage*, $V_{GS(off)}$, in the data book. For the 2N4221 this value is readily available from the *common-source transfer characteristics* plot; the value is $VTO = V_{GS(off)} = -3.5$ V.

The transconductance parameter BETA can be obtained from the zero-gate-voltage drain current, I_{DSS}. The measurement in the data book is taken at $V_{DS} = 15$ V, and therefore the transistor is saturated, according to Eqs. 3.25:

$$V_{GS} - VTO = 3.5 \text{ V} < V_{DS}$$

From Eqs. 3.25, *BETA* is computed as a function of I_{DSS}.

$$BETA = \frac{I_{DSS}}{(V_{GS} - VTO)^2} = \frac{4 \text{ mA}}{(3.5 \text{ V})^2} = 0.327 \text{ mA V}^{-2}$$

I_{DSS} has been set at 4 mA, which is the average between the minimum and maximum values provided by the data sheet.

The last parameter of importance for DC is **LAMBDA**, the channel modulation parameter, which measures the output conductance in saturation. This value is obtained from equating the expression of g_{ds}, the small-signal DS conductance, defined by Eqs. 3.27, to Y_{os}, the common-source output admittance in the data sheet:

$$g_{ds} = BETA \cdot LAMBDA(V_{GS} - VTO)^2 = Y_{os} \tag{3.28}$$

and

$$LAMBDA \approx \frac{Y_{os}}{I_{DSS}} = \frac{10^{-5} \text{ mho}}{4 \cdot 10^{-3} \text{ A}} = 2.5 \cdot 10^{-3} \text{ V}^{-1}$$

These values are used in Example 4.4, in the following chapter, for computing the $I_D = f(V_{DS}, V_{GS})$ characteristics, which can be checked against the characteristics in the data book.

SPICE3 and PSpice additionally support a metal-semiconductor FET, or MESFET. MESFETs can be represented by the same models as those shown in Figures 3.14 and 3.15 for JFETs. The details of the MESFET models can be found in Sec. 3.6.

3.5 METAL-OXIDE-SEMICONDUCTOR FIELD EFFECT TRANSISTORS (MOSFETs)

The general form of a metal-oxide-semiconductor field effect transistor (MOSFET) statement is

> **M**name nd ng ns nb MODname <<**L=**>L> <<**W=**>W> <**AD=**AD> <**AS=**AS>
> +   <**PD=**PD> <**PS=**PS> <**NRD=**NRD> <**NRS=**NRS> <**OFF**>
> +   <**IC=**v_DS0, v_GS0, v_BS0>

The letter **M** must be the first character in *Mname*; **M** identifies a MOSFET and can be followed by up to seven characters in SPICE2. *nd, ng, ns*, and *nb* are four numbers that specify the drain, gate, source, and bulk nodes, respectively. *MODname* is the name of the model that defines the parameters for this transistor. Two MOSFET models are supported, n-channel (**NMOS**) and p-channel (**PMOS**). The schematic representations for the two types of MOSFETs are shown in Figure 3.16. Note that a MOSFET can be represented either as a four-terminal or three-terminal device; the bulk terminal is often

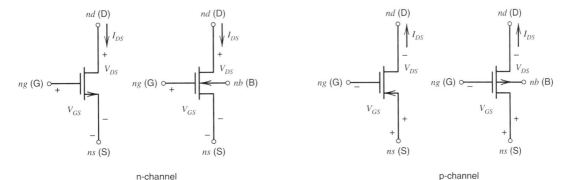

Figure 3.16   n- and p-channel MOSFET elements.

omitted, because all n-channel transistors have the bulk connected to the most nega-tive voltage and all p-channel transistors have the bulk connected to the most positive voltage.

Up to eight geometry parameters can be specified for each MOSFET.

L and W are the length and width of the conducting channel beneath the gate, be-tween source and drain, in meters. The key letters **L** and **W** are optional; if they are omitted, the two values that follow *MODname* are interpreted as length and width, in that order. Default values can be defined as *DEFL* and *DEFW* in an **.OPTIONS** con-trol statement (see also Section 9.5.5). The SPICE2 built-in defaults for L and W are 1 m; other SPICE versions may differ in the values assigned to these defaults. It is recommended to set either L and W or *DEFL* and *DEFW*.

AD and *AS* are the areas, in square meters, of the drain and source diffusions of the MOSFET. *AD* and *AS* multiply the bulk junction capacitance per square meter, *CJ*, in the computation of the drain-bulk (DB) and source-bulk (SB) junction capacitances. **CJ** is a model parameter defined by model *MODname*. Default values can be set at *DEFAD* and *DEFAS* in an **.OPTIONS** statement; alternatively, in SPICE2 the defaults of the two areas are each 100 μm^2.

Note that the following parameters are used only for the very accurate modeling of MOS ICs, such as RAMs, where the operations of the memory cell, sense amplifiers, and decoders need to be predicted reliably. This detail of geometry specification is not necessary for first-order analysis.

PD and *PS* are the perimeters of the drain and source diffusions in meters. *PD* and *PS* multiply the sidewall bulk junction capacitance per meter, *CJSW*, in the computation of the DB and SB junction sidewall capacitances, respectively. In SPICE2 the defaults for *PD* and *PS* are zero.

NRD and *NRS* are the equivalent number of squares of the drain and source dif-fusions. *NRD* and *NRS* multiply the sheet resistance, *RSH*, in the computation of the parasitic drain and source series resistances, respectively. The defaults for *NRD* and *NRS* are 1.

The keyword **OFF** initializes the transistor in the cutoff region for the initial itera-tions of the DC bias solution. By default MOSFETs are initialized cutoff at the limit of turn-on, with $V_{GS} = VTO$, the threshold voltage, $V_{DS} = 0.0$, and $V_{BS} = -1$, for the DC solution.

The keyword **IC** defines the values of the terminal voltages, v_{DS0}, v_{GS0}, and v_{BS0} at time $t = 0$ in a time-domain analysis. v_{DS0}, v_{GS0}, and v_{BS0} are used as initial values only when the **UIC** option is present in the **.TRAN** statement.

3.5.1   DC Model

The most basic MOSFET model used in SPICE to describe the static BCEs of a MOS-FET, the **LEVEL** = 1 model, is the quadratic Shichman-Hodges model (Shichman and Hodges 1968). More complex SPICE MOSFET models, which incorporate second-order effects, are described in Appendix A and references (Antognetti and Massobrio 1988; Vladimirescu and Liu 1981).

The drain-source current, I_{DS}, of an n-channel device is defined by the following three equations for the three regions of operations cutoff, saturation, and linear, respectively:

$$I_{DS} = \begin{cases} 0 & \text{for } V_{GS} \leq V_{TH} \\ \dfrac{KP}{2}\dfrac{W}{L_{eff}}(V_{GS} - V_{TH})^2(1 + LAMBDA \cdot V_{DS}) & \text{for } 0 < V_{GS} - V_{TH} \leq V_{DS} \\ \dfrac{KP}{2}\dfrac{W}{L_{eff}}V_{DS}(2(V_{GS} - V_{TH}) - V_{DS})(1 + LAMBDA \cdot V_{DS}) \\ & \text{for } 0 < V_{DS} < V_{GS} - V_{TH} \end{cases}$$

$$(3.29)$$

where $L_{eff} = L - 2 \cdot LD$ is the effective channel length corrected for the lateral diffusion, LD, of the drain and source, and

$$V_{TH} = VTO + GAMMA\left(\sqrt{PHI - V_{BS}} - \sqrt{PHI}\right) \qquad (3.30)$$

is the threshold voltage in the presence of back-gate bias, $V_{BS} < 0.0$.

VTO, **KP**, **GAMMA**, **PHI**, and **LAMBDA** are the electric parameters of a MOSFET model, representing the threshold voltage, transconductance factor, bulk threshold parameter, surface potential, and output conductance factor in saturation, respectively. This model, similar to the JFET model, is generally applicable to FETs with minor changes to account for the specifics of each device category. For a MOSFET the transconductance factor KP depends both on the device geometry, W and L, and the process characteristics, surface mobility, and thin-oxide thickness.

There is an additional current component, due to the pn-junction current, characterized by the saturation current, **IS**, or, equivalently, by its density, **JS**. The drain and source pn junctions are reverse biased, and therefore the pn-junction currents are negligible in a first-order analysis.

The above equations are valid for $V_{DS} > 0$. If V_{DS} changes sign, the behavior of the MOSFET is symmetrical, the drain and the source swapping roles: V_{GS} is replaced by V_{GD}, V_{BS} is replaced by V_{BD}, and V_{DS} is replaced by its absolute value in the above equations. The current I_{DS} flows in the opposite direction.

For a p-channel MOSFET the same current equations apply; the absolute values are used for the terminal voltages and the current flows in the opposite direction.

3.5.2 Dynamic and Small-Signal Models

The dynamic behavior of a MOSFET is governed by the charge associated with the gate-oxide-semiconductor interface and by the charges associated with the drain and source diffusions.

Three distinct charges can be identified on the plates of the MOS capacitor: a gate, a channel, and a bulk charge. These charges are voltage-dependent. Voltage-dependent capacitances are always computed for the **LEVEL** = 2 and **LEVEL** = 3 models; variable gate capacitances are computed for a **LEVEL** = 1 model only if the value of the thin-oxide thickness parameter, *TOX*, is specified. Details on the gate charge and capitance formulations can be found in Appendix A.

For a first-order model it can be assumed that the three MOS charges are associated with three constant capacitors, represented by *CGDO*, the GD overlap capacitance per unit channel width; *CGSO*, the GS overlap capacitance per unit channel width; and *CGBO*, the GB overlap capacitance per channel length. The thin-oxide capacitance per unit area is defined by

$$C_{ox} = \frac{\epsilon_{ox}\epsilon_0}{TOX} \tag{3.31}$$

For a first-order analysis the constant gate capacitances are specified as overlap capacitances approximated by

$$
\begin{aligned}
CGDO &= CGSO = \tfrac{1}{2}C_{ox}L \\
CGBO &= 0
\end{aligned}
\tag{3.32}
$$

where ϵ_{ox} and ϵ_0 are the permitivities of SiO_2 and free space, respectively, and *TOX* is the thin-oxide thickness. The actual gate capacitances, C_{GD}, C_{GS}, and C_{GB}, are computed in SPICE by multiplying *CGDO* by *W*, *CGSO* by *W*, and *CGBO* by *L*, respectively.

The above approximation is appropriate for digital circuits. For analog circuits a more careful evaluation of *CGDO* is necessary. This capacitance has an important effect on the bandwidth of an MOS amplifier. It is recommended to set it to zero for a transistor biased in saturation and to the value given in Eqs. 3.32 for a transistor in the linear region. This distinction is especially important in small-signal frequency analysis. It is preferable to let SPICE use the voltage-dependent capacitances shown in Figure 3.17 by specifying *TOX* in the **.MODEL** statement. In the three regions of operation the three capacitances are

$$
\begin{array}{llll}
C_{GD} = C_{GS} = 0, & C_{GB} = C_{ox}WL & & \text{for } V_{GS} \leq V_{TH} \\
C_{GD} = 0, & C_{GS} = \tfrac{2}{3}C_{ox}WL, & C_{GB} = 0 & \text{for } 0 < V_{GS} - V_{TH} \leq V_{DS} \\
C_{GD} = C_{GS} = \tfrac{1}{2}C_{ox}WL, & C_{GB} = 0 & & \text{for } 0 \leq V_{DS} < V_{GS} - V_{TH}
\end{array}
\tag{3.33}
$$

The definition of the overlap capacitances needs to be changed when voltage-dependent capacitances are used. *CGDO, CGSO,* and *CGBO* should be used to describe the actual overlap of the drain, source, and bulk by the gate, beyond the channel. The three gate capacitances used by SPICE are computed by adding the capacitances in Eq. 3.33 to the respective overlap capacitances:

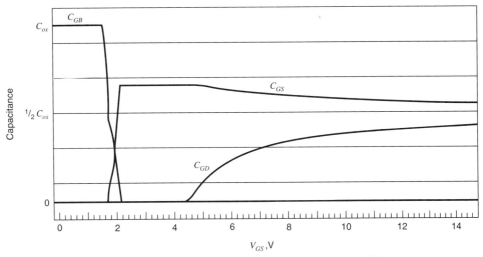

Figure 3.17 MOSFET gate capacitances per unit channel area versus V_{GS}.

$$CGDO = CGSO = \tfrac{1}{2}C_{ox}LD$$

$$CGBO = C_{fox}W_{ov}$$

(3.34)

where LD is the length of the gate extension over the drain and source diffusions, C_{fox} is the field-oxide (isolation) thickness, and W_{ov} is the gate width extension beyond the channel. Unless a very accurate simulation is needed, in which case one of the more accurate models presented in Appendix A is recommended, one can omit the overlap capacitances when voltage-dependent gate capacitances are used in **LEVEL** = 1.

The depletion charges are accumulated on the depletion capacitances of the two reverse-biased junctions and are described by Eqs. 3.3:

$$C_{BD} = \frac{CBD}{(1 - V_{BD}/PB)^{MJ}}$$

$$C_{BS} = \frac{CBS}{(1 - V_{BS}/PB)^{MJ}}$$

(3.35)

where *CBD, CBS, PB,* and *MJ* are the zero-bias bulk-drain capacitance, the zero-bias bulk-source capacitance, the built-in bulk junction potential, and the junction grading coefficient.

The design of LSI and VLSI circuits requires the most accurate representation of the actual physical realization of the circuit. The DB and SB junction capacitances have a great impact on the operation speed of an IC; SPICE2 provides the means for an accurate specification of the junction capacitance for each device geometry and diffusion profile. Both drain and source junction can be characterized by a bottom junction capacitance per unit area, *CJ*, and a sidewall junction capacitance per unit length,

CJSW. The reason for this differentiation is the smaller grading coefficient of the side-wall of the diffusion, which can make the sidewall contribution the dominant junction capacitance.

Scaled by the device geometry parameters, *AD, AS, PD*, and *PS*, introduced above, the DB and SB junction capacitances can be expressed as

$$C_{BD} = \frac{AD \cdot CJ}{(1 - V_{BD}/PB)^{MJ}} + \frac{PD \cdot CJSW}{(1 - V_{BD}/PB)^{MJSW}}$$

$$C_{BS} = \frac{AS \cdot CJ}{(1 - V_{BS}/PB)^{MJ}} + \frac{PS \cdot CJSW}{(1 - V_{BS}/PB)^{MJSW}}$$

(3.36)

Note that whenever both the total zero-bias junction capacitance, **CBD** or **CBS**, and the geometry-oriented specification are available, the total capacitances take precedence.

The capacitances introduced so far are used in the small-signal analysis. In the large-signal time-domain analysis the gate charges and bulk junction charges are computed from Eq. 3.4 using Eqs. 3.32 through 3.36; the large-signal MOSFET model is shown in Figure 3.18.

The small-signal MOSFET model is shown in Figure 3.19. The nonlinear current generator I_{DS} is replaced by the resistance r_{ds} and transconductances g_m and g_{mbs} de-

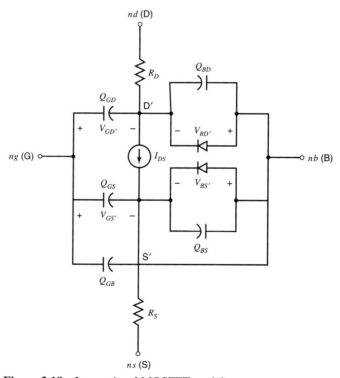

Figure 3.18 Large-signal MOSFET model.

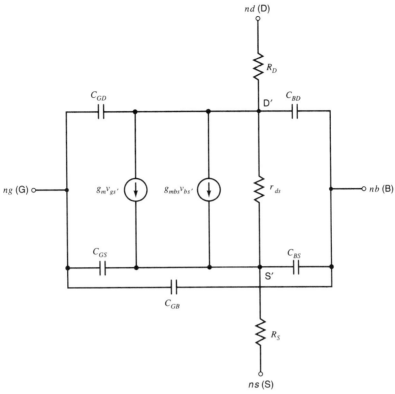

Figure 3.19 Small-signal MOSFET model.

fined by the following equations:

$$g_{ds} = \frac{1}{r_{ds}} = \frac{dI_{DS}}{dV_{DS}}$$

$$g_m = \frac{dI_{DS}}{dV_{GS}} \qquad (3.37)$$

$$g_{mbs} = \frac{dI_{DS}}{dV_{BS}}$$

The expression of I_{DS} appropriate to the region of operation (Eqs. 3.29) must be used when deriving g_m, g_{ds}, and g_{mbs}. In all MOSFET equations terminal voltages must be referred to nodes D' and S' when the parasitic terminal resistances RD and RS are nonzero.

The diodes representing the drain and source junctions are modeled by the conductances g_{bd} and g_{bs} in the small-signal analysis. The values of g_{bd} and g_{bs} are computed using Eq. 3.5 for the conductance of a diode. Usually, these diodes are turned off and g_{bd} and g_{bs} have very small values.

3.5.3   Model Parameters

The general form of the MOSFET model statement is

> .MODEL *MODname* **NMOS** / **PMOS** <VTO=*VTO* <KP=*KP* ... >>

In every model statement one of the keywords **NMOS** or **PMOS**, indicating the transistor type, must be specified.

Table 3.4 summarizes the model parameters introduced so far along with the default values assigned by SPICE2. The model parameter is multiplied by the geometry parameter listed in the Scale Factor column on the device statement. The SPICE parameters of MOSFETs are often derived from measurements of I-V characteristics of test structures on IC wafers. The sequence is similar to the one described in Example 3.5 for a JFET, except that the characteristics are obtained from a curve tracer.

EXAMPLE 3.6

The following statements are examples of MOSFET definitions in SPICE.

```
M1 1 2 0 0 MOD1
.MODEL MOD1 NMOS VTO=1.5
```

M1 is an NMOS transistor with no geometry data specified. The channel length, L, and width, W, default to 1 meter; the only model parameter that is defined is the threshold voltage, **VTO**, and defaults are used for **KP** and the other parameters. **LEVEL** = 1 MOSFET equations are used if this parameter is absent from the .**MODEL** statement. Such limited specification may be useful only for a DC analysis, where only the ratio of W/L affects the solution, and not the individual values of W and L. The absence of any charge-storage elements from the above model may cause the simulator to abort a transient analysis; see Examples 10.8 and 10.10 for details.

```
MN 2 1 0 0 MODN L=10U W=20U
MP 2 1 3 3 MODP L=10U W=40U
.MODEL MODN NMOS VTO=1 KP=30U LAMBDA=.001 CGDO=3.45N CGSO=3.45N
.MODEL MODP PMOS VTO=-1 KP=15U LAMBDA=.001 CGDO=3.45N CGSO=3.45N
```

These statements describe a CMOS inverter. The drains are connected together at node 2, the gates are connected at node 1, the source and bulk of the NMOS are connected to ground, node 0, and the source and bulk of the PMOS are connected to the supply, node 3. Note that the *VTO* specification of the PMOS incorporates the sign.

```
MDRIV 2 1 0 0 ENH L=10U W=20U
MLOAD 3 2 2 0 DEP L=20U W=10U
.MODEL ENH NMOS VTO=1 LAMBDA=.005 TOX=.1U
.MODEL DEP NMOS VTO=-3 GAMMA=.5 LAMBDA=.01 TOX=.1U
```

Table 3.4 MOSFET Model Parameters

Name	Parameter	Units	Default	Example	Scale Factor
VTO	Threshold voltage	V	0	1.0	—
KP	Transconductance parameter	AV^{-2}	2.0×10^{-5}	1.0E-3	—
GAMMA	Bulk threshold parameter	$V^{1/2}$	0	0.5	—
PHI	Surface potential	V	0.6	0.7	—
LAMBDA	Channel length modulation parameter	V^{-1}	0	1.0E-4	—
RD	Drain ohmic resistance	Ω	0	10	—
RS	Source ohmic resistance	Ω	0	10	—
RSH	D and S diffusion sheet resistance	Ω/sq	0	10	*NRD* *NRS*
CBD	Zero-bias BD junction capacitance	F	0	5P	—
CBS	Zero-bias BS junction capacitance	F	0	1P	—
CJ	Zero-bias bulk junction bottom capacitance	Fm^{-2}	0	2.0E-4	*AD* *AS*
MJ	Bulk junction grading coefficient	—	0.5	0.5	
CJSW	Zero-bias bulk junction sidewall capacitance	Fm^{-1}	0	1.0E- 9	*PD* *PS*
MJSW	Bulk junction grading coefficient	—	0.33	0.25	
PB	Bulk junction potential	V	1	0.6	
IS	Bulk junction saturation current	A	10^{-14}	1.0E-16	
CGDO	GD overlap capacitance per unit channel width	Fm^{-1}	0	4.0E-11	*W*
CGSO	GS overlap capacitance per unit channel width	Fm^{-1}	0	4.0E-11	*W*
CGBO	GB overlap capacitance per unit channel length	Fm^{-1}	0	2.0E-10	*L*
TOX	Thin-oxide thickness	m	∞	0.1U	
LD	Lateral diffusion	m	0	0.2U	

The statements on the previous page describe an enhancement-depletion NMOS inverter. The depletion transistor is *normally on* and its threshold voltage is negative.

Exercise
Build SPICE decks for the CMOS inverter and enhancement-depletion inverter defined above and trace the I/O transfer characteristic using the **.DC** statement, introduced in Example 3.3.

3.6 METAL-SEMICONDUCTOR FIELD EFFECT TRANSISTORS (MESFETS)

In SPICE3 the general form of a metal-semiconductor field effect transistor (MESFET) statement is

Z*name nd ng ns MODname* <*area*> <**OFF**> <**IC**=v_{DS0}, v_{GS0}>

The letter **Z** must be the first character in *Zname*; **Z** identifies a MESFET only in SPICE3. In PSpice the identification character is **B**. *nd*, *ng*, and *ns* are the drain, gate, and source nodes, respectively. *MODname* is the name of the model that defines the parameters for this transistor. Two MESFET models are supported, n-channel (**NMF**) and p-channel (**PMF**). The schematic representations of the two types of MESFETs are identical to those of the corresponding types of JFETs, shown in Figure 3.13. The scale factor *area* is equal to the number of identical transistors connected in parallel, and defaults to 1. The keyword **OFF** initializes the transistor in the cutoff region for the initial iterations of the DC bias solution. By default MESFETs are initialized conducting, with $V_{GS} = VTO$, the pinch-off voltage, and $V_{DS} = 0.0$ for the DC solution. The keyword **IC** sets the terminal voltages, v_{DS0} and v_{GS0}, at time $t = 0$ in a time domain analysis. v_{DS0} and v_{GS0} are used as initial values only when the **UIC** option is present in the **.TRAN** statement.

The SPICE3 BCEs for this device are given by the Raytheon model (Statz, Newman, Smith, Pucel, and Haus, 1987). The drain-source current, I_{DS}, is given by the following three equations for the three regions of operations, cutoff, saturation, and linear, respectively:

$$I_{DS} = \begin{cases} 0 & \text{for } V_{GS} \leq VTO \\ \beta(V_{GS} - VTO)^2 \left[1 - \left(1 - ALPHA\frac{V_{DS}}{3}\right)^3\right] \cdot (1 + LAMBDA \cdot V_{DS}) & \\ & \text{for } 0 < V_{DS} \leq 3/ALPHA \\ \beta(V_{GS} - VTO)^2(1 + LAMBDA \cdot V_{DS}) & \text{for } V_{DS} > 3/ALPHA \end{cases}$$

$$\beta = \frac{BETA}{1 + B(V_{GS} - VTO)} \tag{3.38}$$

VTO, BETA, ALPHA, B, and *LAMBDA* are the threshold voltage, transconductance factor, saturation voltage parameter, doping tail extending parameter, and output conductance factor in saturation, respectively.

The above equations are valid for $V_{DS} > 0$. If V_{DS} changes sign, the behavior of the MESFET is symmetrical, the drain and the source swapping roles: V_{GS} is replaced by V_{GD}, and V_{DS} is replaced by its absolute value in the above equations. The current I_{DS} flows in the opposite direction.

The dynamic behavior of a MESFET is modeled by two charges associated with the GD and GS junctions. These charges are accumulated on the depletion capacitances of the two reverse-biased junctions and are described by Eq. 3.3.

$$C_{GS} = \frac{CGS}{(1 - V_{GS}/PB)^{0.5}}$$
$$C_{GD} = \frac{CGD}{(1 - V_{GD}/PB)^{0.5}} \tag{3.39}$$

CGS, CGD, and *PB* are the zero-bias gate-source capacitance, zero-bias gate-drain capacitance, and the built-in gate junction potential, respectively. The MESFET implementation of these capacitances uses voltage-dependent factors that control the continuity of the equations around $V_{DS} = 0$.

The large-signal and small-signal MESFET equivalent models are similar to the corresponding JFET models shown in Figures 3.14 and 3.15.

The general form of the MESFET model statement is

.**MODEL** *MODname* **NMF**/**PMF** <**VTO**=*VTO* <**BETA**=*BETA* ...>>

In every model statement, one of the keywords **NMF** or **PMF** must be specified for the transistor type.

Table 3.5 summarizes the model parameters introduced so far along with the default values assigned by SPICE3. The threshold voltage, *VTO*, is sign-sensitive. MESFETs operate in depletion mode, i.e., they are normally on; therefore *VTO* is negative for n-channel devices, **NMF**.

EXAMPLE 3.7

```
Z1 1 2 3 MODZ
.MODEL MODZ NMF VTO=-3 ALPHA=1 RD=20 RS=20
```

Z1 is a normally on n-channel MESFET with parasitic series drain and source resistances. A normally on p-channel MESFET that conducts the same current as Z1 in similar bias conditions is described by the following .**MODEL** statement:

```
.MODEL MODZ PMF VTO=3 ALPHA=1 RD=20 RS=20
```

PSpice supports two MESFET models, both with the *MODtype* keyword **GASFET**. In addition to the Raytheon model (Statz, Newman, Smith, Pucel, and Haus, 1987) available in SPICE3, PSpice also supports the Curtice model (Curtice 1980).

Table 3.5   MESFET Model Parameters

Name	Parameter	Units	Default	Example	Scale Factor
VTO	Threshold (pinch-off)voltage	V	-2.0	-2.5	—
BETA	Transconductance parameter	AV^{-2}	10^{-4}	1.0E-3	*area*
B	Doping tail extending parameter	V^{-1}	0.3	0.3	*area*
ALPHA	Saturation voltage parameter	V^{-1}	2	2	*area*
LAMBDA	Channel length modulation parameter	V^{-1}	0	1.0E-4	—
RD	Drain ohmic resistance	Ω	0	100	1/*area*
RS	Source ohmic resistance	Ω	0	100	1/*area*
CGS	Zero-bias GS junction capacitance	F	0	5P	*area*
CGD	Zero-bias GD junction capacitance	F	0	1P	*area*
PB	Gate junction potential	V	1	0.6	—

3.7   SUMMARY

This chapter has described the semiconductor devices implemented in the most common SPICE programs. The SPICE analytical models and syntax for the diode, the bipolar junction transistor, and the three kinds of field effect transistors, JFET, MOSFET, and MESFET, have been presented in detail. Examples have demonstrated the meanings and the derivations of the model parameters.

Each of the semiconductor devices is defined by an element statement and a set of parameters contained in a **.MODEL** statement. The same **.MODEL** statement, that is, the same set of parameters, can be common to more than one device.

The diode is defined by the following line:

$$D name \; n+ \; n- \; MODname \; <area> <\mathbf{OFF}> <\mathbf{IC}=v_{D0} >$$

The model parameters describing a diode are listed in Table 3.1 and can be specified in statements of model type **D**.

The BJT specification is

$$Q name \; nc \; nb \; ne \; <ns> \; MODname \; <area> < \mathbf{OFF}> <\mathbf{IC} = v_{BE0}, v_{CE0} >$$

Two types of BJTs are supported in SPICE, **NPN** and **PNP**; the model parameters are summarized in Table 3.2.

The format for JFETs is

$$J name \; nd \; ng \; ns \; MODname \; <area> <\mathbf{OFF}> <\mathbf{IC}=v_{DS0}, v_{GS0} >$$

Two types of JFETs are available in SPICE, **NJF** and **PJF**; the model parameters can be found in Table 3.3.

A MOSFET is defined by the following line:

$$M name \; nd \; ng \; ns \; nb \; MODname \; <<\mathbf{L}=>L> <<\mathbf{W}=>W> <\mathbf{AD}= AD> <\mathbf{AS}= AS>$$
$$+ \qquad <\mathbf{PD}=PD> <\mathbf{PS}= PS> <\mathbf{NRD}=NRD> <\mathbf{NRS}=NRS> <\mathbf{OFF}>$$
$$+ \qquad <\mathbf{IC}=v_{DS0}, v_{GS0}, v_{BS0}>$$

The two types of MOSFETs supported in SPICE are the **NMOS** and **PMOS** devices; the model parameters are listed in Table 3.4.

MESFETs are not supported in SPICE2 but are available in SPICE3, PSpice, and most commercial SPICE programs; the syntax differs among SPICE versions. SPICE3 uses the following format:

$$Z name \; nd \; ng \; ns \; MODname \; <area> <\mathbf{OFF}> <\mathbf{IC}=v_{DS0}, v_{GS0} >$$

The same syntax is used also in PSpice with the sole difference that the identification character is **B**. The two types of MESFET devices are **NMF** and **PMF**. The model parameters are summarized in Table 3.5.

The BJT and the MOSFET are described by very complex equations having many parameters. The description in this chapter has not covered second-order effects. Complete equations of the semiconductor models implemented in SPICE can be found in the book by Antognetti and Massobrio (1988) or in Appendix A.

REFERENCES

Antognetti, P., and G. Massobrio. 1988. *Semiconductor Device Modeling with SPICE*. New York: McGraw-Hill.

Curtice, W. R. 1980. A MESFET model for use in the design of GaAs integrated circuits. *IEEE Transactions on Microwave Theory and Techniques* MTT-28, pp. 448–456.

Getreu, I. 1976. *Modeling the Bipolar Transistor*. Beaverton, OR: Tektronix Inc.

Gray, P. R., and R. G. Meyer. 1993. *Analysis and Design of Analog Integrated Circuits*. 3d ed. New York: John Wiley & Sons.

Grove, A. S. 1967. *Physics and Technology of Semiconductor Devices*. New York: John Wiley & Sons.

Motorola Inc. 1988. *Motorola Semiconductors Data Book*. Phoenix, AZ: Author.

Muller, R. S., and T. I. Kamins. 1977. *Device Electronics for Integrated Circuits*. New York: John Wiley & Sons.

Shichman, H., and D. A. Hodges. 1968. Modeling and simulation of insulated-gate field-effect transistor switching circuits. *IEEE Journal of Solid-State Circuits* SC-3 (September), pp. 285–289.

Statz, H., P. Newman, I. W. Smith, R. A. Pucel, and H.A. Haus, 1987. GaAs FET device and circuit simulation in SPICE. *IEEE Transactions on Electron Devices* (February), pp. 160–169.

Symmetry Design Systems. 1992. *MODPEX*. Los Altos, CA: Author.

Sze, S. M. 1981. *Physics of Semiconductor Devices*. New York: John Wiley & Sons.

Vladimirescu, A., and S. Liu. 1981. The simulation of MOS integrated circuits using SPICE2. Univ. of California, Berkeley, ERL Memo UCB/ERL M80/7 (March).

Four

DC ANALYSIS

4.1 ANALYSIS OVERVIEW

This chapter and the following two chapters describe the different *analysis types* performed by SPICE in its three *simulation modes*, DC, small-signal AC, and large-signal transient. Each simulation mode supports more than one analysis type. The specifics of each analysis are explained in the chapters on the corresponding simulation modes.

All statements introduced in this chapter and the following two chapters, analysis specifications and output requests, are control statements as defined in Chap. 1 and start with a period in the first column.

4.1.1 Simulation Modes and Analysis Types

The first simulation mode, DC, always computes and lists the voltages at every node in the circuit. The DC node voltages are computed prior to an AC or transient (TRAN) simulation. In the TRAN mode the DC solution can be specifically prohibited. DC supports the following four analysis types:

OP	DC voltages and operating point information for nonlinear elements
TF	Small-signal midfrequency transfer function
DC	Transfer curves
SENS	Sensitivity analysis

These analyses are presented in the following four sections of this chapter. The last section describes node voltage initialization. SPICE finds the DC solution in most cases without any additional information; in the situations where SPICE fails to find the DC solution, initialization options are available.

The AC analysis described in Chap. 5 computes the frequency response of linear circuits and of the small-signal equivalents of nonlinear circuits linearized near the DC bias point. Two additional analysis types can be performed in the frequency domain:

NOISE Small-signal noise response analysis

DISTO Small-signal distortion analysis of diode and BJT circuits

In the TRAN mode SPICE computes the large-signal time-domain response of the circuit. An initial transient solution, which is identical to the DC bias, precedes by default a TRAN simulation. The only additional analysis in the time-domain is

FOUR Fourier analysis

The time-domain analysis is presented in Chap. 6.

4.1.2 Result Processing and Output Variables

The results of the different analyses must be requested in **.PRINT** or **.PLOT** statements, introduced in Sec. 1.3.3. These statements identify circuit variables, voltages, and currents, to be computed and stored for specific analyses. Analyses are omitted in SPICE2 if no results are requested. SPICE3 performs the specified analyses even in the absence of a **.PRINT** or **.PLOT** statement; the results are stored in a rawfile and can be displayed using the postprocessing utility Nutmeg. Similarly, PSpice runs the analysis and stores the results only if a **.PRINT**, **.PLOT**, or the proprietary **.PROBE** line is present in the input file; **.PROBE** saves the output results in a binary or a text file, which is used by the graphic display program Probe.

.PRINT provides tabular outputs, whereas **.PLOT** generates line-printer plots of the desired variables, as seen in Figure 1.7. The general format of the output request statement is

.PRINT/PLOT *Analysis_TYPE OUT_var1* <*OUT_var2...*> <*plot_limits*>

where *Analysis_TYPE* can be **DC**, **AC**, **NOISE**, **DISTO**, or **TRAN** and is followed by up to eight output variables (*OUT_var*), which are voltages or currents. If more than eight output variables are desired, additional **PRINT/PLOT** statements must be used. There is no limit on the number of output variables.

Output variables can be node voltages, branch voltages, and currents through voltage sources. A voltage output variable has the general form **V**(*node1*<,*node2*>). If only *node1* is present, that node voltage is stored; if two nodes are specified, the output variable is the branch voltage across elements connected between *node1* and *node2*.

A current output variable is of the form **I**(*Vname*) where *Vname* is an independent voltage source defined in the input circuit. The current measured by *Vname* flows through the source from the positive to the negative source node. PSpice provides the convenience of identifying the current flowing through any circuit element by the expression I<*pin*>(*Element_name*) where *Element_name* corresponds to an element present in the circuit file and *pin* must be used only for multiterminal devices, such as transistors. I(R3), I(L1), and IC(Q7) are accepted current variables in PSpice representing the currents flowing through the resistor R1, the inductor L1, and the collector of the BJT Q7.

In AC analysis the **V** and the **I** are followed by one or more characters specifying the desired format of the complex variable. In the **NOISE** and **DISTO** analyses, output variables are limited to the specific functions detailed in Sec. 5.3 and Sec. 5.4, respectively.

4.1.3   Analysis Parameters: Temperature

All element values specified in a SPICE deck are assumed to have been measured at a nominal temperature, *TNOM*, equal to 27° Celsius (300 K). The simulation of the circuit operation is performed at the nominal temperature of 27° C. The nominal temperature can be set to a different value using the **.OPTIONS** statement, described in Chap. 9.

In practical design situations the operation of the circuit must be verified over a range of temperatures. In SPICE the circuit can be simulated at other temperatures defined in a global statement, **.TEMP**, with the following syntax:

> **.TEMP** *temp1* <*temp2...*>

The simulation is performed at temperatures *temp1, temp2,...* when a **.TEMP** line is present in the SPICE input file. The temperature values must be specified in degrees Celsius. Note that if the value of the nominal temperature is not present on the **.TEMP** line, the circuit is not simulated at *TNOM*.

The effects of temperature on the values of different elements is computed by SPICE, and the updated values are used to simulate the circuit. Resistor values are adjusted for temperature variations by the following quadratic equation:

$$value(TEMP) = value(TNOM)[1 + tc1(TEMP - TNOM)$$
$$+ tc2(TEMP - TNOM)^2] \qquad (4.1)$$

where *TEMP* is the circuit temperature, *TNOM* is the nominal temperature, and *tc1* and *tc2* are the first- and second-order temperature coefficients. The behavior of semiconductor devices is affected significantly by temperature; for example, temperature appears explicitly in the exponential terms of the BJT and diode current equations (see Chap. 3), as well as in the expressions of the saturation currents (I_S), built-in potentials (ϕ_J), gain factor (β_F), and pn-junction capacitance (C_J). The detailed temperature dependence of the model parameters of semiconductor devices is described in Appendix A.

When a circuit is analyzed at a temperature different from *TNOM*, SPICE2 lists the TEMPERATURE-ADJUSTED VALUES for each element or model affected by temperature. Note that SPICE3 does not support the `.TEMP` statement; the ambient temperature must be defined on an `.OPTIONS` line.

Exercise
Add the statement

```
.TEMP 100
```

to the one-transistor input file used in Example 1.3, run SPICE2, and note the differences in the model parameters and DC operating point. Which behavior of the circuit is most affected by temperature variation?

4.2   OPERATING (BIAS) POINT

The DC mode solves for the stable operating point of the circuit with only DC supplies applied. Capacitors are open circuits and inductors are shorts in DC. The DC solution consists of two sets of results; first, the DC bias solution, or the voltages at all nodes; and second, the operating point information, or the current, the terminal voltages, and the element values of the small-signal linear equivalent, computed only for the nonlinear devices in the circuit. SPICE computes and prints the bias solution prior to any other analysis. The operating point, however, is not printed unless requested by an `.OP` statement. The only time this information is printed without the presence of `.OP` is when no analysis request is present in the input file. The voltages at all nodes, the total power consumption, and the current through each supply are printed as part of the SMALL-SIGNAL BIAS SOLUTION (SSBS). Currents, terminal voltages, and small-signal equivalent conductances of all nonlinear devices are listed in the OPERATING POINT INFORMATION (OPI) section of the output.

The information provided by SPICE about the DC operation of a circuit is best explained by two examples, a linear and a nonlinear circuit.

EXAMPLE 4.1

Replace resistors R_1 and R_3 in the bridge-T circuit of Figure 1.1, by two capacitors, $C_1 = C_2 = 1 \ \mu F$. Verify the DC solution with SPICE. The new circuit is shown in Figure 4.1.

Solution
The DC solution for the resistive circuit was computed in Example 1.1. A capacitor is equivalent to an open circuit in DC; therefore we expect the following DC solution: $V(1) = V(3) = 12$ V; $V(2) = 0$ V.

The modified SPICE input file and the results of the analysis are shown in Figure 4.2. The information in the SSBS is a complete characterization of the circuit. The

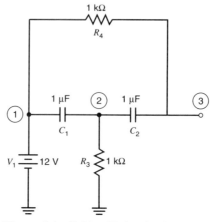

Figure 4.1   Bridge-T circuit with capacitors.

```
BRIDGE-T CIRCUIT

 ****        CIRCUIT DESCRIPTION
*
V1 1 0 12 AC 1
C1 1 2 1u
C2 2 3 1u
R3 2 0 1k
R4 1 3 1k
*
.END

 BRIDGE-T CIRCUIT
 ****      SMALL SIGNAL BIAS SOLUTION      TEMPERATURE = 27.000 DEG C

NODE    VOLTAGE     NODE   VOLTAGE    NODE VOLTAGE   NODE VOLTAGE

(   1)  12.0000    (    2)   0.0000  (   3)   12.0000

   VOLTAGE SOURCE CURRENTS
   NAME          CURRENT

   V1            0.000E+00

   TOTAL POWER DISSIPATION  0.00E+00 WATTS
```

Figure 4.2   DC solution of bridge-T circuit with capacitors.

only data not computed by SPICE are the branch currents, but the node voltages and element values are sufficient for the derivation of any current. If a specific current is desired from SPICE, a dummy voltage source must be added in series with the element of interest. As seen in the two DC solutions of the bridge-T circuit, Figures 1.3 and 4.2, the currents through voltage sources are listed in the SSBS.

EXAMPLE 4.2

Consider next the one-transistor amplifier of the first chapter, reproduced here in Figure 4.3. Relate the SPICE2 parameters listed under the OPI with the BJT model presented in Sec. 3.3.2. The SPICE2 results obtained in Chap. 1 are repeated in Figure 4.4 for convenience.

Solution
The values of the small-signal equivalent model components of Q_1 shown in Figure 3.9 are part of the OPI section. GM and RPI are computed according to Eqs. 3.15 and 3.17:

$$GM = \frac{IC}{V_{th}} = \frac{2.1 \cdot 10^{-3} \text{A}}{0.0258 \text{ V}} = 8.14 \cdot 10^{-2} \text{ mho}$$

$$RPI = \frac{BETADC}{GM} = 1.23 \text{ k}\Omega$$

RX is equal to *RB*, the series parasitic base resistance, plus a second-order resistance (see Appendix A for more detail), and RO is the collector-emitter output resistance. In the absence of the Early voltage, *VAF*, introduced in Chap. 3, RO is infinite; internally

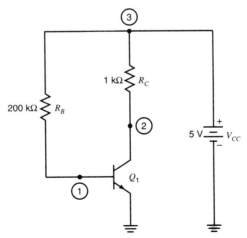

Figure 4.3 One-transistor circuit.

```
*******12/20/88 ********  SPICE 2G.6  3/15/83 ********17:11:05*****

ONE-TRANSISTOR CIRCUIT (FIG. 4.3)

****       INPUT LISTING              TEMPERATURE =   27.000 DEG C

*************************************************************************

*
Q1 2 1 0 QMOD
RC 2 3 1K
RB 1 3 200K
VCC 3 0 5
*
.MODEL QMOD NPN
*
.OP
*
.WIDTH OUT=80
.END
*******12/20/88 ********  SPICE 2G.6  3/15/83 ********17:11:05*****

ONE-TRANSISTOR CIRCUIT (FIG. 4.3)

****       SMALL SIGNAL BIAS SOLUTION    TEMPERATURE =   27.000 DEG C

*************************************************************************

 NODE    VOLTAGE    NODE    VOLTAGE    NODE     VOLTAGE

 ( 1)     0.7934    ( 2)    2.8967    ( 3)     5.0000

     VOLTAGE SOURCE CURRENTS

     NAME CURRENT

     VCC  -2.124D-03

     TOTAL POWER DISSIPATION   1.06D-02 WATTS
*******12/20/88 ********  SPICE 2G.6  3/15/83 ********17:11:05*****

ONE-TRANSISTOR CIRCUIT (FIG. 4.3)

****       OPERATING POINT INFORMATION    TEMPERATURE =   27.000 DEG C

*************************************************************************
```

Figure 4.4   .OP results for one-transistor circuit.

```
**** BIPOLAR JUNCTION TRANSISTORS

                Q1
MODEL         QMOD
 IB           2.10E-05
 IC           2.10E-03
 VBE             0.793
 VBC            -2.103
 VCE             2.897
 BETADC       100.000
 GM           8.13E-02
 RPI          1.23E+03
 RX           0.00E+00
 RO           1.00E+12
 CPI          0.00E+00
 CMU          0.00E+00
 CBX          0.00E+00
 CCS          0.00E+00
 BETAAC       100.000
 FT           1.29E+18
```

Figure 4.4   (*continued*)

SPICE2 clamps the maximum resistance to 1/*GMIN*, which defaults to 10^{12} Ω and can be set as an option parameter (see Chaps. 9 and 10). The small-signal capacitances, CPI and CMU, and the corresponding cutoff frequency, FT, are also part of the bias information and are computed according to Eqs. 3.19 and 3.20. Since no values are defined for *CJE, CJC, TF*, or *TR*, in the **.MODEL** statement the cutoff frequency, f_T, is infinite; this explains the very high value of FT in Figure 4.4. The remaining capacitances, CBX and CCS, are only relevant for second-order effects.

For more than one transistor or other nonlinear element the OPI is computed for each such element.

Another example of DC operating point information is included for the depletion-load NMOS inverter shown in Figure 4.5.

EXAMPLE 4.3

Two important characteristic values of a logic gate are the high and low output voltages, V_{OH} and V_{OL}, respectively. For a correctly designed enhancement-depletion (E-D) inverter, V_{OH}, which corresponds to a low V_{IN}, is equal to V_{DD}, the supply voltage. V_{OL}, however, which corresponds to high V_{IN}, depends on the model parameters of the two

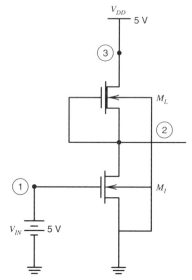

Figure 4.5   Enhancement-depletion
NMOS inverter circuit.

transistors, M_I and M_L, and the geometry ratio, K_R, of the inverter (Hodges and Jackson 1983):

$$K_R = \frac{(W/L)_I}{(W/L)_L} \tag{4.2}$$

Find the low output voltage, V_{OL}, of the E-D inverter of Figure 4.5 using the following MOS model parameters: $VTO_E = 1$ V; $VTO_D = -3$ V; $KP = 20 \ \mu A/V^2$; $GAMMA = 0.5 \ V^{1/2}$; and $PHI = 2\phi_F = 0.6$ V; where VTO_E and VTO_D are the enhancement and depletion transistor threshold voltage, respectively. Assume that $V_{IN} = V_{DD} = 5$ V, $(W/L)_I = 4$, and $(W/L)_L = 1$. Verify the result with SPICE.

Solution
First calculate the threshold voltage of the depletion device taking into account the body effect, as in Eq. 3.30:

$$V_{THD} = VTO_D + GAMMA\left(\sqrt{PHI + V_{OL}} - \sqrt{PHI}\right)$$
$$= -3 \text{ V} + 0.14 \text{ V} = -2.86 \text{ V} \tag{4.3}$$

In the above calculation V_{OL} has only been estimated at 0.5 V in order to account for the body effect of the depletion device. V_{OL} is calculated by equating the currents of the enhancement transistor, M_I, and depletion transistor, M_L, according to Eqs. 3.29:

$$I_{DI} = I_{DL}$$

$$\tfrac{1}{2}(W/L)_I KP[2(V_{IN} - VTO_E)V_{OL} - V_{OL}^2] = \tfrac{1}{2}(W/L)_L KP(0 - V_{THD})^2 \qquad (4.4)$$

M_I is assumed to operate in the linear region, because

$$V_{DSI} = V_{OL} < V_{GSI} - VTO_E$$

since V_{OL} has been approximated at 0.5 V.

Substitution of the guess for V_{OL} in Eq. 4.3 yields a simple quadratic equation in V_{OL} for Eqs. 4.4; the new solution is $V_{OL} = 0.27$ V. The correction in V_{THD} has been overestimated, but the result is very close to the SPICE solution. The current flowing through the inverter is

$$I_{DI} = I_{DL} = \tfrac{1}{2}(W/L)_L KP \cdot V_{THD}^2 = \tfrac{1}{2} \cdot 20(-2.92)^2 \ \mu A = 85.2 \ \mu A$$

These results of the SPICE2 analysis are shown in Figure 4.6. The circuit description is listed first, followed by the MOSFET MODEL PARAMETERS of transistors MI and ML, models EMOS and DMOS. Note that only the parameter values specified in the input file are listed, and not the default values for other parameters, such as the value of *PHI*, 0.6 V. For all the default values of MOSFET parameters see Table 3.4. The SSBS and OPI list the node voltages and small-signal values of the transistors.

```
******** 12/29/88 ******** PSpice 3.02 (Mar 1987) ******** 00:11:01 ********

E-D NMOS INVERTER

 ****     CIRCUIT DESCRIPTION

***************************************************************************

*
MI 2 1 0 0 EMOS W=40U L=10U
ML 3 2 2 0 DMOS W=10U L=10U
VDD 3 0 5
VIN 1 0 5
*
.MODEL EMOS NMOS VTO=1 KP=20U
.MODEL DMOS NMOS VTO=-3 KP=20U GAMMA=.5
*
.OP
*
.END
```

Figure 4.6 SPICE2 input and operating point of MOS inverter.

```
****         MOSFET MODEL PARAMETERS


                    EMOS            DMOS
      TYPE          NMOS            NMOS
      LEVEL         1.000           1.000
      VTO           1.000          -3.000
      KP            2.00E-05        2.00E-05
      GAMMA         0.0             0.500


****      SMALL SIGNAL BIAS SOLUTION         TEMPERATURE =  27.000 DEG C

NODE     VOLTAGE     NODE     VOLTAGE       NODE     VOLTAGE    NODE VOLTAGE

(    1)    5.0000  (    2)      0.2758   (    3)     5.0000


      VOLTAGE SOURCE CURRENTS

      NAME          CURRENT

       VDD         -8.523E-005

       VIN          0.000E+000


   TOTAL POWER DISSIPATION 4.26E-004 WATTS

    ****      OPERATING POINT INFORMATION  TEMPERATURE =  27.000 DEG C

**** MOSFETS

                 MI              ML
      MODEL      EMOS            DMOS
      ID         8.52E-05        8.52E-05
      VGS        5.00E+00        0.00E+00
      VDS        2.76E-01        4.72E+00
      VBS        0.00E+00       -2.76E-01
      VTH        1.00E+00       -2.92E+00
      VDSAT      4.00E+00        2.92E+00
      GM         2.21E-05        5.84E-05
      GDS        2.98E-04        0.00E+00
      GMB        0.00E+00        1.56E-05
      CBD        0.00E+00        0.00E+00
      CBS        0.00E+00        0.00E+00
      CGSOV      0.00E+00        0.00E+00
      CGDOV      0.00E+00        0.00E+00
      CGBOV      0.00E+00        0.00E+00
      CGS        0.00E+00        0.00E+00
      CGD        0.00E+00        0.00E+00
      CGB        0.00E+00        0.00E+00
```

Figure 4.6 (*continued*)

124

The output voltage at the drain of MI agrees to two decimal points with the hand calculation. The bias-point information contains the drain current, I_D, the terminal voltages, V_{DS}, V_{GS}, and V_{BS}, and VTH, the back-gate bias-corrected threshold voltage. Note that VTH for the load transistor ML is corrected to -2.92 V according to Eq. 4.3. VTH also contains corrections due to small-size geometry for higher-level MOSFET models, as presented in more detail in the works by Antognetti and Massobrio (1988) and Vladimirescu and Liu (1981) and Appendix A.

The small-signal characteristics for a MOSFET correspond to the small-signal model of Figure 3.19 and are GDS, GM, GMBS, CBS, CBD, CGS, CGD, and CGB, as well as the overlap capacitances.

4.3   DC TRANSFER CURVES

This analysis computes the DC states of a circuit while a voltage or current source is swept over a given interval. The following statement defines the source and the range of swept values:

.DC V/I*name1 start1 stop1 step1* <V/I*name2 start2 stop2 step2*>

The voltage or current source names *Vname1, Vname2* or *Iname1, Iname2* must be defined in another independent source statement. SPICE allows a second source to be varied as an outer variable. In addition to the voltage and current of an independent source, SPICE2 and other SPICE versions allow the user to vary the value of a resistor, *Rname*, or the temperature, *TEMP*. Any node voltage or current through a voltage source can be defined as an output variable. The value of the output variable is evaluated by sweeping the variables in the following order:

```
v/i2_value = start2
for (v/i2_value ≤ stop2, v/i2_value = v/i2_value + step2)
        v/i1_value = start1
        for (v/i1_value ≤ stop1, v/i1_value = v/i1_value + step1)
            OUT_var = f (v/i1_value,v/i2_value)
```

The value *v/i1_value* of source *V/Iname1* is swept first over the interval from *start1* to *stop1* for each value, *v/i2_value*, of *V/Iname2*. Source *V/Iname1* is also called the inner variable and source *V/Iname2*, the outer variable. The variation of two sources does not result in a comprehensive line-printer plot; the program generates one plot for each value of the second source, the outer variable, rather than gather all curves in a single plot. The graphic display tools Nutmeg and Probe can overcome this problem.

The output variables of interest must be requested as either a tabular print or a plot

.PRINT DC *OUT_var1* <*OUT_var2*...>

.PLOT DC *OUT_var1* <*OUT_var2*...> <*plot_lim1,plot_lim2*>

In SPICE2 if no DC output variable is defined in either print or plot format, the DC sweep analysis is omitted.

Example

```
.DC VIN 0.5 0.8 0.25
.PRINT DC V(6) V(10,11)
.PLOT DC I(VICQ9)
```

The above three statements are part of a SPICE input file requiring the computation of DC voltage and current transfer curves. The first line specifies that voltage source VIN is to be swept from 0.5 V to 0.8 V in steps of 0.25 V. The second statement produces a tabular output listing of the values of the voltage at node 6 and the voltage difference between nodes 10 and 11. The third statement generates a line-printer plot of the current flowing through the voltage source VICQ9, a dummy voltage source connected at the collector of a transistor, Q_9, to measure the values of I_C.

Two very common applications for using DC transfer curves are described in the following examples.

EXAMPLE 4.4

Use PSpice and Probe to represent graphically the $I_{DS} = f(V_{DS}, V_{GS})$ characteristics of the junction field effect transistor 2N4221; use the *Motorola Semiconductors Data Book* (Motorola Inc. 1988) for the electrical characteristics. A different SPICE simulator and plotting package can be used.

Solution

The measurement setup is shown in Figure 4.7, and from the Motorola data book we can derive the following model parameters: $VTO = V_{GS(off)} = -3.5$ V; $BETA = 4.1 \cdot 10^{-4}$ A$/$V^2; $LAMBDA = 0.002$ V^{-1}, $RD = 200$. The derivation of the SPICE model parameters from data book characteristics is described in Example 3.5.

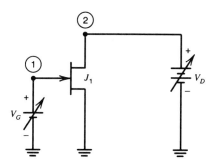

Figure 4.7 Measurement setup for JFET $I_{DS} = f(V_{DS}, V_{GS})$ characteristics with SPICE deck.

The SPICE input description is listed below. Note that no **.PRINT** or **.PLOT** lines are necessary for graphical results in PSpice; all that is needed is a **.PROBE** statement. Eqs. 3.25 are used to compute the current I_{DS} for values of V_{DS} from 0 to 25 V in steps of 1 V at four values of V_{GS}, -3 V, -2 V, -1 V, and 0 V. The range of values and increments for the two bias sources, V_D and V_G, are defined by the **.DC** statement.

```
I-V CHARACTERISTICS OF JFET 2N4221
*
J1 2 1 0 MODJ
VD 2 0 25
VG 1 0 -2
*
.MODEL MODJ NJF VTO=-3.5 BETA=4.1E-4 LAMBDA=0.002 RD=200
*
*.OP
.DC VD 0 25 1 VG -3 0 1
* ONLY FOR PSPICE
.PROBE
* OTHER SPICE
*.PRINT DC I(VD)
.END
```

The output characteristics of the transistor computed by SPICE are shown in Figure 4.8. Note that in spite of using a very simple model with just basic parameters, the

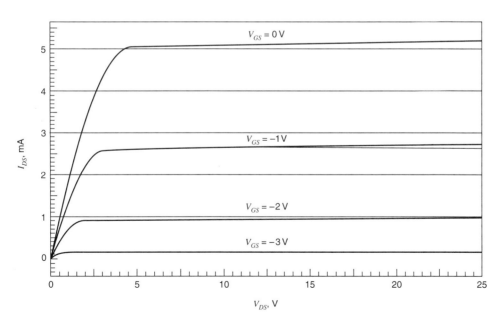

Figure 4.8 Simulated $I_{DS} = f(V_{DS}, V_{GS})$ characteristics of a JFET.

computed I-V characteristics are close to those in the data book. More accurate model parameters can be obtained using parameter extraction programs, such as *Parts* from MicroSim.

Another common use of DC transfer curves is for the design of logic gates, which require a thorough characterization of the noise margins.

EXAMPLE 4.5

Find the noise margins NM_H and NM_L of the NMOS inverter in Figure 4.5; the gate of transistor M_I is swept from 0 to 5 V. Use the parameters of Example 4.3.

Solution
The following two statements must be added to the input circuit description in Figure 4.6:

```
.DC VIN 0 5 0.25
.PLOT DC V(2)
```

The resulting SPICE2 plot is shown in Figure 4.9. The four voltages V_{OH}, V_{OL}, V_{IH}, and V_{IL}, defining the high and low noise margins, NM_H and NM_L, respectively, are marked on the plots. The input voltages V_{IH} and V_{IL} are defined by the points where the slope of the voltage transfer characteristic is unity (Hodges and Jackson 1983). The four voltages are: $V_{OH} = 5$ V, $V_{OL} = 0.28$ V, $V_{IH} = 1.5$ V, and $V_{IL} = 2.5$ V. These result in

$$NM_H = V_{OH} - V_{IH} = 3.5 \text{ V}$$

and

$$NM_L = V_{IL} - V_{OL} = 2.22 \text{ V}$$

The above values guarantee a proper operation of the E-D NMOS inverter when connected with other logic gates implemented in the same technology that is, in NMOS. The effect of changes in the geometry of the transistors, that is, in W and L, can be easily observed in a repetition of the above analysis.

This example also shows the usefulness of a line-printer plot for an accurate reading of the output voltages for given input voltages.

```
E-D NMOS INVERTER

****      DC TRANSFER CURVES                TEMPERATURE =   27.000 DEG C

   VIN        V(2)
   (*)----------   0.0000E+00   2.0000E+00   4.0000E+00   6.0000E+00   8.0000E+00
                - - - - - - - - - - - - - - - - - - - - - - - - - - - - - - -
       0.000E+00  5.000E+00 .            .            .         *  .            .
       2.500E-01  5.000E+00 .            .            .         *  .            .
       5.000E-01  5.000E+00 .            .            .         *  .            .
       7.500E-01  5.000E+00 .            .            .         *  .            .
       1.000E+00  5.000E+00 .            .            .         *  .            .
       1.250E+00  4.943E+00 .            .            .        *   .            .
  VIH  1.500E+00  4.763E+00 . - - - - - - - - - - - - - - - *  45°  .          .
       1.750E+00  4.432E+00 .            .            .    *       .            .
       2.000E+00  3.866E+00 .            .            . *.         .            .
       2.250E+00  2.546E+00 .            .         *  .            .            .
  VIL  2.500E+00  9.240E-01 . - - - - / 45° - .         .            .            .
       2.750E+00  7.101E-01 .      /*     .            .            .            .
       3.000E+00  5.923E-01 .    *        .            .            .            .
       3.250E+00  5.124E-01 .  *          .            .            .            .
       3.500E+00  4.541E-01 .  *          .            .            .            .
       3.750E+00  4.087E-01 .  *          .            .            .            .
       4.000E+00  3.721E-01 . *           .            .            .            .
       4.250E+00  3.419E-01 . *           .            .            .            .
       4.500E+00  3.165E-01 . *           .            .            .            .
       4.750E+00  2.947E-01 . *           .            .            .            .
       5.000E+00  2.758E-01 . *           .            .            .            .
                - - - - - - - - - - - - - - - - - - - - - - - - - - - - - - -
                        V_OL                         V_OH
```

Figure 4.9 DC transfer characteristic of NMOS inverter.

4.4 SMALL-SIGNAL TRANSFER FUNCTION

At the completion of the DC bias solution the linearized network of a nonlinear input circuit is available. For example, for a BJT SPICE has computed the values of all the elements of the *small-signal BJT model* shown in Figure 3.9 and listed in Figure 4.4. The two-port characteristics of the linearized circuit can be obtained by using the

.**TF** *OUT_var* **V/I***name*

control statement. The output variable can assume any of the forms described for output variables on **PRINT** and **PLOT DC** statements. *V/Iname* identifies an independent voltage or current source connected at the input of the two ports defined by the above statement.

SPICE2 computes the gain and the input and output resistances of the two-port circuit defined by the .**TF** statement. The gain can be a voltage or current gain, dV_o/dV_i or dI_o/dI_i, a transconductance, dI_o/dV_i, or a transresistance, dV_o/dI_i.

An important assumption of .TF is that the midfrequency behavior of the circuit is to be computed. This assumption is valid for frequencies at which charge-storage effects can be neglected; that is, coupling capacitors are shorted and high-frequency capacitors are open. Coupling capacitors are used to decouple amplifier stages from each other and from signal generators for proper bias. Their values range from 1 nF to 1 μF. High-frequency capacitors limit the bandwidth at high frequencies and are exemplified by transistor internal capacitances, such as C_π and C_μ for a BJT and C_{GS} and C_{GD} for a MOSFET. Inductors at midfrequency are assumed to be shorted or open depending on which end of the frequency range they affect.

SPICE treats *all capacitors* as open and *all inductors* as shorted in the DC analysis. Therefore, the validity of the .TF analysis is limited to circuits that contain only high-frequency capacitors and low-frequency inductors. A pure resistive network is assumed in the transfer function solution.

The inclusion of a small-signal transfer function in a large-signal DC analysis needs a few more explanations. In general, the voltage or current at any node is

$$v_N = V_N + v_n = V_N + V_n \sin(\omega t + \phi)$$

where V_N is the DC bias value, V_n is the amplitude, and ϕ is the phase shift of the AC signal. In the large-signal time-domain analysis V_n can have any value; in the frequency-domain analysis the assumption that the signal is small limits V_n to V_{th}, the thermal voltage. With this assumption the AC response can be computed on a linear network, which is a valid approximation of the nonlinear circuit as long as it does not deviate significantly from its bias point. The additional midfrequency assumption simplifies the AC component of the signal to a real rather than complex value.

EXAMPLE 4.6

Find the two-port midfrequency characteristics of the one-transistor amplifier of Figure 4.3. Verify the results using the .TF analysis of SPICE2.

Solution

The midfrequency small-signal equivalent circuit of the one-transistor amplifier is shown in Figure 4.10. The values of the linearized network can be computed using Eqs. 3.15 through 3.20 or taken directly from the OPI of transistor Q_1, listed in Figure 4.4.

In order that a .TF analysis can be performed, an input signal source must be added to the circuit in Figure 4.3. A source in SPICE has also a DC component, which can perturb the DC bias of the circuit; the addition of a voltage input source at the base of Q_1, node 1, would change the operating point of the transistor. The easiest approach to adding a signal source without perturbing the DC state is to use a current source I_i that has a DC value of zero at the input. This source represents the AC input signal i_i.

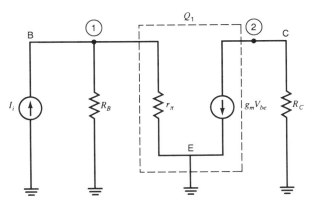

Figure 4.10   Small-signal midfrequency equivalent of
one-transistor amplifier.

The transfer function characteristics of the one-transistor circuit for a voltage output, V_2, at the collector of Q_1, and a current input, I_i, are the transresistance, a_r, the input resistance, R_i, and output resistance, R_o:

$$a_r = \frac{v_o}{i_i} = \frac{V_2}{I_i} = -\frac{R_B}{R_B + r_\pi}\beta_F R_C \tag{4.5}$$

$$R_i = \frac{v_i}{i_i} = R_B \parallel r_\pi \tag{4.6}$$

$$R_o = \frac{v_o}{i_o} = R_C \parallel r_o \approx R_C \tag{4.7}$$

The voltage gain can be obtained from the transresistance transfer function:

$$a_v = \frac{v_o}{v_i} = \frac{a_r}{R_i} \tag{4.8}$$

The following two statements must be added to the input description of Figure 4.4 for performing the transfer function analysis in SPICE2:

```
II 0 1
.TF V(2) II
```

The modified input circuit and the SPICE2 results of the analysis are shown in Figure 4.11. Note in the SSBS that the zero-valued input current source, II, does not disturb the operating point of the circuit. The results confirm the above hand calculations.

```
ONE-TRANSISTOR CIRCUIT (FIG. 4.3)
****      INPUT LISTING                    TEMPERATURE = 27.000 DEG C

*
Q1 2 1 0 QMOD
RC 2 3 1K
RB 1 3 200K
VCC 3 0 5
*
II 0 1
*
.MODEL QMOD NPN
*
.TF V(2) II
*
.WIDTH OUT=80
.END


****      SMALL SIGNAL BIAS SOLUTION        TEMPERATURE = 27.000 DEG C

NODE     VOLTAGE     NODE     VOLTAGE     NODE     VOLTAGE

(  1)     0.7934    (  2)     2.8967    (  3)     5.0000

     VOLTAGE SOURCE CURRENTS

     NAME         CURRENT

     VCC          -2.124D-03

     TOTAL POWER DISSIPATION 1.06D-02 WATTS

****     SMALL-SIGNAL CHARACTERISTICS
     V(2)/II                              = -9.939D+04
     INPUT RESISTANCE AT II               = 1.222D+03
     OUTPUT RESISTANCE AT V(2)            = 1.000D+03
```

Figure 4.11   **.TF** analysis results for a one-transistor circuit.

Exercise

Confirm the value of the voltage gain, a_v, derived above in Eq. 4.8, using a voltage transfer function. A voltage transfer function can be obtained by replacing resistor R_B with a Thevenin equivalent at node 1 and specifying the added voltage source as input in the **.TF** statement.

4.5   SENSITIVITY ANALYSIS

Sensitivity analysis offers insight into the effect of the values of circuit elements and variations of model parameters on selected output variables and hence on circuit performance. The sensitivity analysis request has the following form:

.**SENS** *OUT_var1* <*OUT_var2...*>

where output variables *OUT_var1*, *OUT_var2*,... are defined in the same manner as for the .**PRINT** and .**PLOT** statements. The sensitivities with respect to every element in the circuit and all DC model parameters of diodes and BJTs are computed for each output variable defined in the .**SENS** statement. No sensitivities with respect to model parameters of JFET or MOSFET transistors are available. SPICE3 does not support this type of analysis.

There are two sensitivity numbers listed for each parameter value: the absolute sensitivity, $\partial V_i / \partial p_j$, and the relative sensitivity, $(\partial V_i / \partial p_j)(p_j / 100)$. These values reflect the sensitivities of DC voltages and currents with respect to perturbations in circuit element values. No sensitivity analysis is available in SPICE2 for AC or time-domain response.

EXAMPLE 4.7

Use SPICE2 to compute the sensitivity of the current provided by the current mirror (Gray and Meyer 1993) shown in Figure 4.12 with respect to the circuit parameters. Assume that $\beta_F = 100$ and assign the default values to the remaining BJT model parameters.

Solution
The current supplied by this current source is

$$I_{C2} = \frac{I_{REF}}{1 + 2/\beta_F} = \frac{1\,A}{1.02} = 0.98\,mA$$

$$I_{REF} = \frac{V_{CC} - V_{BE(on)}}{REF} = \frac{5 - 0.7\,V}{4.3 \cdot 10^3\,\Omega} = 1\,mA$$

The input specification, operating point, and sensitivity results of the SPICE2 simulation are listed in Figure 4.13. First, note in the OPI section that the value of I_{C2} is very close to the above estimate. Second, consider the sensitivity results in the DC SENSITIVITY ANALYSIS section. The same data are computed for each output variable on the .SENS statement. In this example a single output variable has been requested, and all results appear under the header DC SENSITIVITIES OF OUTPUT I(VMEAS). The four-column tabular output lists the ELEMENT NAME, ELEMENT VALUE, ELEMENT SENSITIVITY, the absolute sensitivity in amperes or volts per

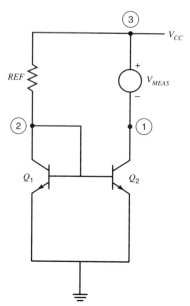

Figure 4.12   Current mirror
current source.

unit of the respective element, and the NORMALIZED SENSITIVITY in amperes or
volts per 1% variation in the value of the respective element.

The most informative data are the normalized sensitivities. For this small circuit it
is easy to spot that a 1% change in the value of any of the following elements causes
roughly a 10-μA variation in I_{C2}: the reference resistor, *REF*, the supply, V_{CC}, and *IS*,
the saturation current of transistors Q_1 and Q_2. An increase of *REF* and *IS* causes I_{C2} to

```
CURRENT MIRROR CURRENT SOURCE
*
REF 3 2 4.3k
Q1 2 2 0 QMOD
Q2 1 2 0 QMOD
VMEAS 3 1
VCC 3 0 5
*
.MODEL QMOD NPN BF=100 VA=50
*
.OP
.SENS I(VMEAS)
*
.WIDTH OUT=80
.END
```

Figure 4.13   SPICE2 sensitivity results for the current mirror.

```
****        SMALL SIGNAL BIAS SOLUTION        TEMPERATURE = 27.000 DEG C

NODE        VOLTAGE        NODE    VOLTAGE        NODE    VOLTAGE        NODE VOLTAGE

(    1)        5.0000  (    2)        0.7733  (    3)        5.0000

        VOLTAGE SOURCE CURRENTS

        NAME CURRENT

        VMEAS 1.045E-003
          VCC -2.028E-003

    TOTAL POWER DISSIPATION 1.01E-002 WATTS

****        OPERATING POINT INFORMATION   TEMPERATURE = 27.000 DEG C

**** BIPOLAR JUNCTION TRANSISTORS
NAME            Q1          Q2
MODEL        QMOD        QMOD
IB        9.64E-06    9.64E-06
IC        9.64E-04    1.05E-03
VBE       7.73E-01    7.73E-01
VBC       0.00E+00   -4.23E+00
VCE       7.73E-01    5.00E+00

    ****      DC SENSITIVITY ANALYSIS          TEMPERATURE = 27.000 DEG C
DC SENSITIVITIES OF OUTPUT I(VMEAS)
            ELEMENT         ELEMENT        ELEMENT       NORMALIZED
            NAME            VALUE          SENSITIVITY   SENSITIVITY
                                          (AMPS/UNIT)   (AMPS/PERCENT)


            REF             4.300E+03      -2.415E-07    -1.038E-05
            VMEAS           0.000E+00      -1.927E-05     0.000E+00
            VCC             5.000E+00       2.649E-04     1.325E-05
Q1
            RB              0.000E+00       0.000E+00     0.000E+00
            RC              0.000E+00       0.000E+00     0.000E+00
            RE              0.000E+00       0.000E+00     0.000E+00
            BF              1.000E+02       1.018E-07     1.018E-07
            ISE             0.000E+00       0.000E+00     0.000E+00
            BR              1.000E+00       0.000E+00     0.000E+00
            ISC             0.000E+00       0.000E+00     0.000E+00
            IS              1.000E-16      -1.028E+13    -1.028E-05
            NE              1.500E+00       0.000E+00     0.000E+00
            NC              2.000E+00       0.000E+00     0.000E+00
            IKF             0.000E+00       0.000E+00     0.000E+00
            IKR             0.000E+00       0.000E+00     0.000E+00
            VAF             5.000E+01       0.000E+00     0.000E+00
            VAR             0.000E+00       0.000E+00     0.000E+00
```

Figure 4.13   (*continued*)

Q2

RB	0.000E+00	0.000E+00	0.000E+00
RC	0.000E+00	0.000E+00	0.000E+00
RE	0.000E+00	0.000E+00	0.000E+00
BF	1.000E+02	1.018E-07	1.018E-07
ISE	0.000E+00	0.000E+00	0.000E+00
BR	1.000E+00	-2.056E-16	-2.056E-18
ISC	0.000E+00	0.000E+00	0.000E+00
IS	1.000E-16	1.035E+13	1.035E-05
NE	1.500E+00	0.000E+00	0.000E+00
NC	2.000E+00	0.000E+00	0.000E+00
IKF	0.000E+00	0.000E+00	0.000E+00
IKR	0.000E+00	0.000E+00	0.000E+00
VAF	5.000E+01	-1.629E-06	-8.147E-07
VAR	0.000E+00	0.000E+00	0.000E+00

Figure 4.13   (*continued*)

decrease, whereas an increase in V_{CC} brings about an increase in I_{C2}. The effect of a perturbation in *BF* is far less important.

The number of requested output variables in sensitivity analysis should be kept small, because a large amount of information is generated by this analysis.

4.6   NODE VOLTAGE INITIALIZATION

All nonlinear electrical simulation programs compute the solution iteratively. The iterative process starts with an initial guess of the voltages. Initially SPICE assumes that all node voltages are zero. In most cases the user does not need to specify any information about initial voltages. There are exceptions when SPICE cannot find the solution in the default number of 100 iterations. For these situations a node voltage initialization statement is available with the following general format:

.NODESET V(*node1*)=*value1* <V(*node2*)=*value2*...>

The effect of this statement is to assign *value1* to the voltage of node *node1*, *value2* to node *node2*, and so on, in the first few solution iterations.

The final solution may differ from the values specified by .NODESET. SPICE2 uses the initial values only as a guidance until it finds a first solution; the search for the DC voltages continues, however, with the initialization constraint removed until the final solution is reached. The final solution is probably in agreement with the .NODESET values, if they are correct, but it must not be identical with them.

A good example of the use of the .NODESET statement is the DC analysis of bistable circuits. The following example shows the effect .NODESET has on the final solution.

EXAMPLE 4.8

Find the DC solution of the flip-flop circuit shown in Figure 4.14. Use the same MOS-FET model parameters as in Example 4.3.

Solution
The input specification and the bias point obtained from SPICE2 are shown in Figure 4.15. The flip-flop or bistable circuit, of Figure 4.14 has two stable operating points: the first is with M_{I1} ON and M_{I2} OFF, and the second is with M_{I1} OFF and M_{I2} ON. The solution found by SPICE2 has both inverters biased identically, with both M_{I1} and M_{I2} conducting. This is a metastable state, which in reality would not last. In reality the two inverters are not physically identical, and upon connecting the supply, V_{DD}, one inverter would assume the ON state, and the other would be OFF.

The physical imbalance can be reproduced in SPICE2 by initializing the drain voltage of M_{I1} to 5 V ($\text{V}(2) = 5$) and the drain voltage of M_{I2} to 0.25 V ($\text{V}(1) = 0.25$), as listed in the modified input in Figure 4.16. The latter value is roughly equal to V_{OL} estimated in Example 4.3 for the same inverter. The solution obtained by SPICE2 in the presence of the **.NODESET** statement and shown also in Figure 4.16 is according to expectations. Note that the voltage at node 1, $\text{V}(1) = 0.2758$, which is equal to the solution found in Example 4.3, is a corrected value of the initial guess. The same result can be obtained by adding the keyword **OFF** to the M_{I1} line; this is equivalent to initializing transistor M_{I1} in the OFF state.

Another approach to node voltage initialization, **.IC**, is presented in Section 6.3 in conjunction with the time-domain analysis. **.IC** can be used to find the DC bias

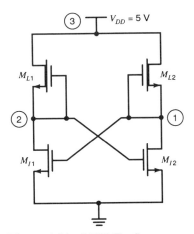

Figure 4.14 MOS flip-flop.

```
NMOS FLIP-FLOP
*
.WIDTH OUT=80
.OPTION NOPAGE
*
MI1 2 1 0 0 EMOS W=40U L=10U
ML1 3 2 2 0 DMOS W=10U L=10U
MI2 1 2 0 0 EMOS W=40U L=10U
ML2 3 1 1 0 DMOS W=10U L=10U
VDD 3 0 5
*
.MODEL EMOS NMOS VTO=1 KP=20U
.MODEL DMOS NMOS VTO=-3 KP=20U GAMMA=.5
*
.OP
*
.END
```

```
    ****     SMALL SIGNAL BIAS SOLUTION          TEMPERATURE = 27.000 DEG C

   NODE   VOLTAGE      NODE   VOLTAGE       NODE    VOLTAGE   NODE   VOLTAGE

(   1)     2.2701 (   2)     2.2701   (    3)     5.0000
```

Figure 4.15   SPICE2 bias solution of a MDS flip-flop without .NODESET.

```
NMOS FLIP-FLOP
*
* CIRCUIT DESCRIPTION COMES HERE
*
.NODESET V(1)=0.25 V(2)=5
.OP
*
.END
```

```
    ****     SMALL SIGNAL BIAS SOLUTION        TEMPERATURE = 27.000 DEG C

  NODE    VOLTAGE      NODE   VOLTAGE     NODE    VOLTAGE    NODE   VOLTAGE

(   1)     0.2758  (    2)     5.0000 (    3)     5.0000
```

Figure 4.16   SPICE2 bias solution of a MOS flip-flop with .NODESET.

solution; the major difference from `.NODESET` is that node voltages are forced to the values specified by the user in the `.IC` statement and are not corrected after an initial pass.

4.7 SUMMARY

This chapter has presented an overview of the SPICE analysis modes and has described in detail the analysis types of the DC mode. Analysis parameters, output variables, and result processing are also outlined as part of the analysis overview. Examples have shown how to apply the various DC analyses to specific circuit problems.

The DC analysis types—operating point, transfer curves, small-signal transfer function, and sensitivity—are specified by the following control lines:

`.OP`

`.DC` V/I*name1 start1 stop1 step1* <V/I*name2 start2 stop2 step2*>

`.TF` *OUT_var* V/I*name*

`.SENS` *OUT_var1* <*OUT_var2...*>

With the exception of the DC operating point information, results are stored in the output file only for specified circuit variables, *OUT_var*, which can be voltages or currents:

V(*node1*<,*node2*>)

I(*Vname*)

The output variables for a `.DC` transfer curve analysis can be saved either in tabular or line-printer plot format using the `.PRINT` or `.PLOT` control statement, respectively. The general format of the output request is

`.PRINT/PLOT` *Analysis_TYPE OUT_var1* <*OUT_var2...*> <*plot_limits*>

In the DC mode, *Analysis_TYPE* can be only `DC`.

The ambient temperature for the circuit analysis is defined by

`.TEMP` *temp1* <*temp2...*>

All circuit element values and model parameter values are defined at the nominal temperature, *TNOM*, which defaults to 27°C.

Node voltages can be initialized for a DC computation using the following statement:

`.NODESET` V(*node1*)=*value1* <V(*node2*)=*value2...*>

Note that the final values of the voltages at the nodes *node1, node2,...* may differ from the initialization values.

REFERENCES

Antognetti, P., and G. Massobrio. 1988. *Semiconductor Device Modeling with SPICE.* New York: McGraw-Hill.

Gray, P. R., and R. G. Meyer. 1993. *Analysis and Design of Analog Integrated Circuits.* 3d ed. New York: John Wiley & Sons.

Hodges, D. A., and H. G. Jackson. 1983. *Analysis and Design of Digital Integrated Circuits* New York: McGraw-Hill.

Motorola Inc. 1988. *Motorola Semiconductors Data Book.* Phoenix, AZ: Author.

Vladimirescu, A., and S. Liu. 1981. The simulation of MOS integrated circuits using SPICE2. Memo UCB/ERL M80/7 (March).

Five

AC ANALYSIS

5.1 INTRODUCTION

In AC mode SPICE computes the frequency response of linear circuits. Small input signals, with amplitudes less than the thermal voltage, $V_{th} = kT/q$, are assumed for nonlinear circuits that are linearized around the DC operating point. In AC the node admittances are complex, frequency-dependent entities of the form

$$\mathbf{Y} = G + j\omega C + \frac{1}{j\omega L} \tag{5.1}$$

where $\omega = 2\pi f$ is the angular frequency measured in radians per second and f is the frequency in Hertz. In the frequency domain the voltages and currents of the circuit are also complex numbers, called phasors:

$$\mathbf{V} = V_R + jV_I = |\mathbf{V}|e^{j\phi}$$
$$|\mathbf{V}| = \sqrt{V_R^2 + V_I^2}$$
$$\phi = \arctan\left(\frac{V_I}{V_R}\right) \tag{5.2}$$

Phasors consist of a real part, V_R, and an imaginary part, V_I, and can also be expressed as magnitude, $|\mathbf{V}|$, and phase, ϕ. The periodicity factor, $\sin \omega t$, is implicitly assumed for all variables in an AC analysis.

141

SPICE2 supports several small-signal analysis types in the frequency domain. These are **.AC**, for a frequency sweep, presented in Sec. 5.2, **.NOISE**, for input and output noise computation, presented in Sec. 5.3, and **.DISTO**, for analysis of distortion due to semiconductor device nonlinearities, presented in Sec. 5.4. SPICE3 also offers a pole-zero analysis, **.PZ**, which is described in Sec. 5.5.

Prior to an AC analysis SPICE always computes the DC operating point, which becomes the reference for linearizing nonlinear circuit elements.

5.2   AC FREQUENCY SWEEP

This analysis computes the values of node voltages in the circuit over a specified frequency interval. The following statement specifies the frequency interval and scale:

.AC *Interval numpts fstart fstop*

where *Interval* is one of the three keywords that indicate whether the frequency varies by decade (**DEC**), by octave (**OCT**), or linearly (**LIN**), between *fstart*, the starting frequency, and *fstop*, the final frequency. The variable *numpts* specifies the number of frequency points used per interval; for the linear interval *numpts* is the total number of frequency values between *fstart* and *fstop*. This analysis provides meaningful results if there is at least one independent source with a specified AC value in the input circuit.

EXAMPLE 5.1

Describe the differences in the AC analysis for the three types of intervals.

Solution
The following three **.AC** statements cover the same frequency range but cause circuit evaluations at different frequency points:

```
.AC DEC 10 1K 1MEG
.AC OCT 4 1K 1MEG
.AC LIN 1000 1K 1MEG
```

The first statement divides the frequency interval between 1 kHz and 1 MHz into three subintervals, with the endpoint of each subinterval being $f_2 = 10 f_1$, where f_1 is the starting frequency of the subinterval. The 10 frequency values in each subinterval are selected on a logarithmic scale. The circuit is evaluated at 30 frequencies.

The second statement divides the frequency range into subintervals defined by the following relation between endpoints: $f_2 = 2 f_1$. Ten subintervals are needed, since $fstop = 1000 fstart$ and $2^{10} = 1024$. The points in each interval are selected on a logarithmic scale, four per octave. A total of 41 circuit evaluations are performed, with the last analysis at 1.024 MHz.

The third statement divides the frequency range in 1000 equal parts, and the frequency varies linearly between *fstart* and *fstop*. One thousand evaluations are necessary in this analysis.

The decade is the most commonly used frequency interval, because it is consistent with a Bode plot of the circuit response.

The results of an AC analysis can be viewed in either tabular or line-printer format by adding one or more of the following statements:

.PRINT AC *AC_OUT_var1 <AC_OUT_var2 ... >*

.PLOT AC *AC_OUT_var1 <AC_OUT_var2 ... > <plot_lim1, plot_lim2>*

Output variables for the AC analysis, *AC_OUT_var1, ...* contain additional information besides the type, **V** or **I**, and the node numbers. The extra characters contained in the output variable's name differentiate among various representations of complex numbers, as in Eqs. 5.2. The accepted names for *AC_OUT_var* are the following:

VR or **IR**	Real part of complex value
VI or **II**	Imaginary part of complex value
VM or **IM**	Magnitude of complex number, $|V|$ or $|I|$
VP or **IP**	Phase of complex number
VDB or **IDB**	Decibel value of magnitude, $20 \log_{10}(|V|)$ or $20 \log_{10}(|I|)$

As in DC analysis a current output variable is specified as *I(Vname)* where *Vname* can be any voltage source in the circuit description.

At least one **.PRINT AC** or **.PLOT AC** statement is necessary in order for SPICE2 to perform the analysis. SPICE3 does not require an output statement, and PSpice needs either a **PRINT/PLOT** line or a **.PROBE** line.

EXAMPLE 5.2

Derive the transfer function V_3/V_1 of the bridge-T circuit shown in Figure 4.1, sketch its Bode plot, and verify with an **.AC** analysis performed with SPICE.

Solution

The transfer function can be derived from the KVL and the BCE relations. For $R = R_1 = R_2$ and $C = C_1 = C_2$ it is equal to:

$$\frac{\mathbf{V}_o}{\mathbf{V}_i} = \frac{\mathbf{V}_3}{\mathbf{V}_1} = \frac{R^2C^2s^2 + 2RCs + 1}{R^2C^2s^2 + 3RCs + 1} \tag{5.3}$$

where s is the complex frequency; $s = \sigma + j\omega$. The quadratic equations in the numerator and denominator must be solved for the zeros and poles, respectively, in order to be able to represent the Bode plot (Dorf 1989). The two zeros are equal and are $z_1 = z_2 = -10^3$ rad/s, corresponding to 159 Hz. The two poles surround the double zero on the negative real frequency axis and are

$$p_1 = \frac{-3 + \sqrt{5}}{2}\frac{1}{RC} = -3.82 \cdot 10^2 \text{ rad}/\text{s} \rightarrow -61\text{Hz}$$

$$p_2 = \frac{-3 - \sqrt{5}}{2}\frac{1}{RC} = -2.62 \cdot 10^3 \text{ rad}/\text{s} \rightarrow -417\text{Hz}$$

The locations of the poles and zeros point to a dip in the frequency characteristic centered around 159 Hz. Because it attenuates signals of a given frequency, this circuit is also called a notch filter.

The SPICE input for the circuit is listed below. Note that a decadic interval is specified and the transfer function requested is from 10 Hz to 10 kHz. When this example is run on PSpice, the **.PROBE** line should be added in order to save all the phasors of the circuit. The **.PLOT AC** statement requests a line-printer plot, similar to the one of Figure 1.7, of the magnitude in decibels, VDB(3), and the phase, VP(3), of the voltage at node 3 to be saved in the output file.

```
BRIDGE-T CIRCUIT
*
V1 1 0 12 AC 1
C1 1 2 1u
C2 2 3 1u
R3 2 0 1k
R4 1 3 1k
*
.AC DEC 10 10 10k
.PLOT AC VDB(3) VP(3)
.WIDTH OUT=80
* PSPICE ONLY
*.PROBE
*
.END
```

The Bode plot of VDB(3) and VP(3) is reproduced in Figure 5.1. The graphical representation validates the above hand calculations. In addition to the frequency sweep, a pole-zero analysis can be performed in SPICE3; see also Sec. 5.5.

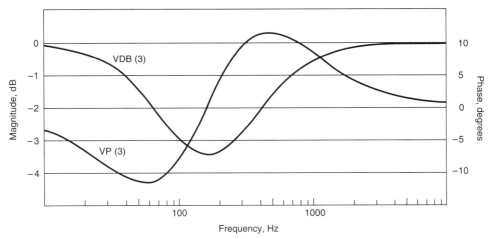

Figure 5.1 Magnitude and phase of the bridge-T transfer function.

EXAMPLE 5.3

Find the frequency variation of the current gain of the one-transistor amplifier of Figure 4.3 by running an **.AC** analysis. Identify important frequency points. Add the values of the BE and BC junction capacitances to the model parameters: $CJE = 1\text{pF}$, $CJC = 2\text{pF}$.

Solution
The transfer function of interest is

$$\mathbf{A_i} = \frac{\mathbf{I}_o}{\mathbf{I}_i} = \frac{\mathbf{I}_c}{\mathbf{I}_b} = \frac{g_m}{g_\pi + j\omega(C_\pi + C_\mu)} = \beta(j\omega) \qquad (5.4)$$

Several statements must be added to the SPICE2 input for the one-transistor amplifier in Figure 4.3. The equivalent small-signal circuit for the frequency sweep of the current gain is shown in Figure 5.2 and the modified SPICE2 input in Figure 5.3.

First, the input signal source I_i must be connected to the base of Q_1, node 1:

```
II 1 0 AC 1
```

This source has a zero DC current and therefore does not disturb the DC bias point. The AC amplitude of 1A at the input scales the resulting complex voltages and currents to represent the transfer functions with respect to the input. A value different from 1, A

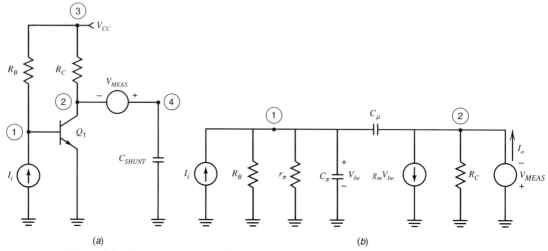

Figure 5.2   Current-gain amplifier: (*a*) amplifier circuit; (*b*) small-signal equivalent.

```
*******01/14/89 ********  SPICE 2G.6    3/15/83 ********17:36:16*****

ONE-TRANSISTOR CIRCUIT (FIG. 5.2)

****     INPUT LISTING              TEMPERATURE =   27.000 DEG C

***********************************************************************
*
Q1 2 1 0 QMOD
RC 2 3 1K
RB 1 3 200K
VCC 3 0 5
II 0 1 AC 1
VMEAS 4 2
CSHUNT 4 0 .1U
*
.MODEL QMOD NPN CJE=1P CJC=2P
*
.OP
.AC DEC 10 0.1MEG 10G
*
.PLOT AC IDB(VMEAS) IP(VMEAS)
.WIDTH OUT=80
.END
```

Figure 5.3   SPICE2 one-transistor current-gain circuit with bias information.

```
****        BJT MODEL PARAMETERS                TEMPERATURE =   27.000 DEG C


             QMOD
TYPE         NPN
IS           1.00D-16
BF           100.000
NF             1.000
BR             1.000
NR             1.000
CJE          1.00D-12
CJC          2.00D-12


****        SMALL SIGNAL BIAS SOLUTION         TEMPERATURE =   27.000 DEG C

  NODE    VOLTAGE       NODE    VOLTAGE       NODE    VOLTAGE       NODE    VOLTAGE

(   1)     0.7934     (   2)    2.8967      (   3)     5.0000     (   4)    2.8967


     VOLTAGE SOURCE CURRENTS

     NAME         CURRENT

     VCC       -  2.124D-03
     VMEAS        0.000D+00


     TOTAL POWER DISSIPATION 1.06D-02 WATTS

****        OPERATING POINT INFORMATION        TEMPERATURE =   27.000 DEG C

**** BIPOLAR JUNCTION TRANSISTORS


             Q1
MODEL        QMOD
IB           2.10E-05
IC           2.10E-03
VBE             0.793
VBC            -2.103
VCE             2.897
BETADC       100.000
GM           8.13E-02
RPI          1.23E+03
RX           0.00E+00
RO           1.00E+12
CPI          1.72E-12
CMU          1.29E-12
CBX          0.00E+00
CCS          0.00E+00
BETAAC       100.000
FT           4.30E+09
```

Figure 5.3 (*continued*)

or V, for the AC input amplitude proves useful when the output of a circuit needs to be calibrated at 1 V, or 0 dB.

The zero-DC offset current source connected above and in Example 4.6 is very useful for adding AC input signals or for purposely introducing an open circuit in DC without SPICE2 flagging the error LESS THAN TWO ELEMENTS CONNECTED AT NODE x.

Second, the AC component of the output current I_o must be separated from the DC component and measured. A decoupling capacitor, C_{SHUNT}, connected to the collector in parallel with R_C is equivalent to a short at high frequencies and an open circuit at DC. A dummy voltage source, V_{MEAS} must be added in series with C_{SHUNT} to measure the output current, I_o.

Third, the zero-bias junction capacitance values must be added to the .MODEL statement of the npn transistor. These capacitances together with the diffusion capacitances, which are omitted in this example, limit the gain bandwidth of the transistor.

Fourth, an AC control statement is needed. The frequency range of interest can be estimated roughly by locating the pole of the current gain at

$$\omega_\beta = \frac{1}{r_\pi(C_\pi + C_\mu)} = \frac{1}{1.23 \cdot 10^3 \cdot 3 \cdot 10^{-12}} \text{ rad/s} = 0.27 \cdot 10^9 \text{ rad/s} \qquad (5.5)$$

The value of r_π has been taken from the bias point computation in Figure 4.4, and the zero-bias junction capacitances values have been used for C_π and C_μ. This pole corresponds to a -3-dB frequency of 43 MHz. The unity-gain frequency f_T, defined in Eq. 3.20, is two orders of magnitude higher:

$$f_T = \beta_F f_\beta = 100 \cdot \frac{0.27 \cdot 10^9}{2\pi} \text{ Hz} = 4.3 \text{ GHz} \qquad (5.6)$$

The frequency interval of interest is therefore between 10 MHz and 10 GHz:

```
.AC DEC 10 10MEG 10G
```

Last, the output control statement must be added:

```
.PLOT AC IDB(VMEAS) IP(VMEAS)
```

It produces a Bode plot of the magnitude in decibels and phase in degrees of the current transfer function, Eq. 5.4.

The input circuit with the bias point information is listed in Figure 5.3, and the Bode plot is shown in Figure 5.4.

In the OPERATING POINT INFORMATION (*OPI*), section note that the values CPI and CMU have been modified using the actual junction voltages VBE and VBC according to Eqs. 3.14, 3.18, and 3.19, C_π has increased, because V_{BE} is positive, and C_μ has decreased, because V_{BC} is negative. Also listed in this section is the value of the unity-gain frequency, f_T. This value is very close to the one computed in Eq. 5.6.

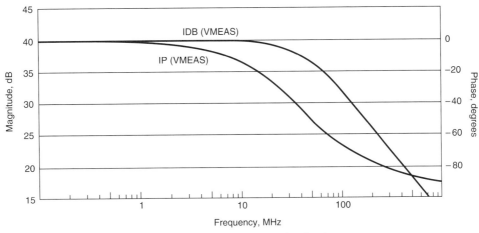

Figure 5.4 Bode plot for the one-transistor current-gain circuit.

The next step is to check the values of f_β and f_T in the AC plot. The pole position, f_β, is located where

$$|\beta(j\omega_\beta)| = \beta_F/\sqrt{2} \qquad (5.7)$$

which translated into decibels is equal to -3 dB below the midfrequency magnitude of β_F. The phase of the current gain at f_β is $-45°$. From the AC plot the midfrequency magnitude of the current gain is 39.93 dB, which is approximately equal to the value of $\beta_F = 100$. The magnitude drops to 37.28 dB and the phase to $-40°$ at 39.81 MHz. (These accurate values are obtained with a `.PRINT AC` statement.)

The unity-gain frequency is located where the magnitude curve crosses the 0-dB mark. This frequency is between 3.98 GHz and 5 GHz. Both f_β and f_T derived from the AC plot confirm the above hand calculations.

5.3 NOISE ANALYSIS

There is a lower limit to the amplitude of a signal that can be processed by electronic circuits. This limit is imposed by the *noise* that is generated in electronic components. Noise characterization of a circuit can be performed by adding to each SPICE component a noise generator.

Noise generation has a random character and can be due to a number of phenomena. The most common is the *thermal noise* generated in resistors. Semiconductor devices produce *shot noise, flicker noise*, and *burst noise* (Gray and Meyer 1993). A common origin of the noise phenomenon is the conduction of electric current by individual carriers, electrons and holes, in semiconductor circuits. The various types of noise have different frequency behaviors; some types cover the entire frequency spectrum uniformly and are also known as *white noise*, whereas other types are greater at one end of the spectrum than at the other.

SPICE models noise in resistors and all semiconductor devices. Capacitors, inductors, and controlled sources are noise-free. The noise current or noise voltage generators associated with different elements are characterized by a mean-square value, $\overline{i^2}$ or $\overline{v^2}$, respectively. A mean-square value is used for noise sources because the phenomena underlying the charge-flow mechanism are random. The noise generators of the different elements in a circuit are uncorrelated. The total effect of all the noise sources at the output of the circuit is obtained by adding all the mean-square values of the noisy elements reflected at the output:

$$\overline{v_{out}^2} = \sum_{i=1}^{n} \overline{v_i^2} \tag{5.8}$$

Noise-source values are proportional to the frequency bandwidth, Δf, of the measurement. A useful measure is the spectral density, $\overline{v^2}/\Delta f$ or $\overline{i^2}/\Delta f$, of the noise source. The spectral density is measured in V^2/Hz or A^2/Hz.

The best known noise behavior of an electronic component is the generation of thermal noise in a resistor. The theoretical mean square of the noise voltage source in series with the noise-free resistor (see Figure 5.5) is given by

$$\overline{v_R^2} = 4kTR\Delta f \tag{5.9}$$

where k, equal to $8.62 \cdot 10^{-5}$ eV/K, is Boltzman's constant; T is the absolute temperature, measured in degrees Kelvin; R is the resistance; and Δf is the frequency bandwidth

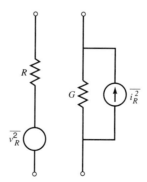

Figure 5.5   Noise source of a resistor.

of the measurement. Another way of representing the noise contribution is to connect a current generator in parallel with mean-square value

$$\overline{i_R^2} = 4kTG\Delta f \tag{5.10}$$

with $G = 1/R$, as shown in Figure 5.5. The latter approach is used in SPICE because of the ease of adding the contribution of current generators in nodal equations.

The major source of noise in semiconductor devices is associated with the flow of DC current and is known as *shot noise*. The small-signal equivalent model of a BJT with the shot noise sources of the base and collector currents, $\overline{i_b^2}$, and $\overline{i_c^2}$, is shown in Figure 5.6. The mean-square value of each source is proportional to the corresponding DC current, I_B or I_C:

$$\overline{i_b^2} = 2qI_B\Delta f \tag{5.11}$$

$$\overline{i_c^2} = 2qI_C\Delta f \tag{5.12}$$

The mean-square values of the noise sources are small compared to the thermal voltage, V_{th}, and therefore the analysis can be performed on the linear equivalent of a nonlinear circuit. The noise analysis is performed by SPICE in conjunction with an `.AC` request; both the `.AC` and the `.NOISE` control statements must be present in the input file. SPICE computes the output noise voltage at a specified output and an equivalent input noise voltage or current, depending on whether the circuit input is defined by a voltage or current source, respectively. The equivalent input noise is obtained by dividing the output noise by the transfer function of the circuit and represents the measure of all noise sources concentrated in a single noise source at the input. Additionally, a report on each noise source's contribution can be generated by SPICE2 at specified frequencies. This report can produce a large amount of printout.

The general form of the `.NOISE` control statement is

`.NOISE V`(*n1*<,*n2*>) `V/I`*name nums*

which defines the two-port connections of the circuit for the noise computation; **V**(*n1*<, *n2*>) defines the output port as a voltage between nodes *n1* and *n2*. If only one node

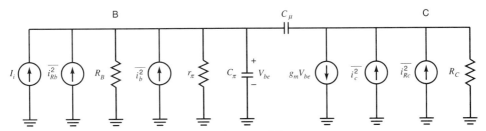

Figure 5.6   Noise sources of the one-transistor amplifier.

is specified, the output is between it and ground. The input of the two-port circuit is identified by an input source, **V/I**name, which can be a voltage or current source and must be present in the circuit description. SPICE2 can list the individual contribution of each noise generator at given frequencies; a summary of each noise source value is listed in the result file once every *nums* frequency points in the interval *fstart* to *fstop*, the frequency interval specified in the **.AC** statement. The number of noise-source summaries for a **DEC** interval with *numpts* frequency points per decade and a number of frequency decades in the interval *fstart* to *fstop* equal to *decades* is equal to *decades · numpts/nums* + 1. A zero or the absence of a value for *nums* disables the individual noise-source report.

The results of a noise analysis can be requested in tabular form with a **.PRINT** statement or as a line-printer plot with a **.PLOT** statement. The general form of the output request is

> **.PRINT NOISE ONOISE**<(M/DB)> <**INOISE**<(M/DB)>>
>
> **.PLOT NOISE ONOISE**<(M/DB)> <**INOISE**<(M/DB)>>

ONOISE represents the total noise voltage, **V**(*n1*<, *n2*>), at the output nodes defined in the **.NOISE** statement, and **INOISE** is the equivalent input noise, voltage or current, at **V/I**name, also defined by the **.NOISE** line. At least one resulting noise value must appear on a **.PRINT** or **.PLOT NOISE** statement. The optional qualifiers differentiate between magnitude, **M**, which is the default, and decibels, **DB**. The output noise and the input noise are computed at all frequencies between *fstart* and *fstop*, as specified in the **.AC** statement.

EXAMPLE 5.4

```
.AC DEC 10 1K 100MEG
.NOISE V(11) VIN1 10
```

The above two statements define the frequency interval, from *fstart* = 1 kHz to *fstop* = 100 MHz; define the two-port circuit, of which the input is VIN1 and the output is node 11; and request six noise-source summaries, one for each decade. No output is generated in the absence of a **.PRINT** or **.PLOT** statement for **AC** or **NOISE**. If, in addition, the value 10 is missing from the **.NOISE** statement, then SPICE2 does not generate any output related to the two statements.

The addition of the statement

```
.PLOT NOISE ONOISE
```

to the above two statements causes SPICE2 to produce a line-printer plot of the total *root-mean-square (rms)* value of the output voltage noise at node 11.

EXAMPLE 5.5

Compute the contribution of each noise source to the output voltage noise, the total output noise, and the equivalent input noise for the one-transistor amplifier in Figure 5.2 without the V_{MEAS}, C_{SHUNT} net. Check your results with SPICE2.

Solution

The noise sources are shown in the small-signal equivalent circuit of the one-transistor amplifier in Figure 5.6. The values of all noise sources can be computed using the definitions of thermal and shot noise, Eqs. 5.9 to 5.12:

$$\frac{\overline{v_{Rb}^2}}{\Delta f} = 4kTR_B = 1.6 \cdot 10^{-20} \cdot 2 \cdot 10^5 \text{ V}^2/\text{Hz} = 3.2 \cdot 10^{-15} \text{ V}^2/\text{Hz}$$

$$\frac{\overline{v_{Rc}^2}}{\Delta f} = 4kTR_C = 1.6 \cdot 10^{-20} \cdot 10^3 \text{ V}^2/\text{Hz} = 1.6 \cdot 10^{-17} \text{ V}^2/\text{Hz}$$

$$\frac{\overline{i_b^2}}{\Delta f} = 2qI_B = 2 \cdot 1.6 \cdot 10^{-19} \cdot 2.1 \cdot 10^{-5} \text{ A}^2/\text{Hz} = 6.72 \cdot 10^{-24} \text{ A}^2/\text{Hz}$$

$$\frac{\overline{i_c^2}}{\Delta f} = 2qI_C = 2 \cdot 1.6 \cdot 10^{-19} \cdot 2.1 \cdot 10^{-3} \text{ A}^2/\text{Hz} = 6.72 \cdot 10^{-22} \text{ A}^2/\text{Hz}$$

The contribution of each of the above sources to the output noise voltage is calculated next. The contribution of the noise sources connected at the base of the transistor can be obtained by multiplying the mean-square values of i_{Rb}^2 and i_b^2 by the square of the transfer function V_o/I_i. The value of the transfer function at mid-frequency, as defined in Sec. 4.4, is

$$|A_{vi}| = \frac{V_o}{I_i} = \beta_F R_C = 10^2 \cdot 10^3 \ \Omega = 10^5 \ \Omega$$

The contribution of the noise sources connected to the collector is obtained by multiplying i_{Rc}^2 and i_c^2 by the square of the output resistance. All contributions to the output noise voltage are spectral densities of the mean-square values.

First, the contributions of the two BJT noise currents are evaluated:

$$\frac{\overline{v_{o1}^2}}{\Delta f} = \frac{\overline{i_c^2}R_C^2}{\Delta f} = 6.72 \cdot 10^{-16} \text{ V}^2/\text{Hz}$$

$$\frac{\overline{v_{o2}^2}}{\Delta f} = \frac{\overline{i_b^2}A_{vi}^2}{\Delta f} = 6.72 \cdot 10^{-14} \text{ V}^2/\text{Hz}$$

The noise contributed by the base current at the output is significant because the current amplification available in BJTs is high. For this reason low-noise amplifiers often use FETs in the input stage.

Second, the noise seen at the output due to R_C and R_B is:

$$\frac{\overline{v_{o3}^2}}{\Delta f} = \frac{\overline{v_{Rc}^2}}{\Delta f} = 1.6 \cdot 10^{-17} \text{ V}^2/\text{Hz}$$

$$\frac{\overline{v_{o4}^2}}{\Delta f} = \frac{\overline{i_{Rb}^2}A_{vi}^2}{\Delta f} = 8 \cdot 10^{-26} \cdot 10^5 \text{ V}^2/\text{Hz} = 8 \cdot 10^{-21} \text{ V}^2/\text{Hz}$$

The total mean-square output noise voltage $\overline{v_o^2}$ is the sum of the mean-square values of all contributions, resulting in

$$\frac{\overline{v_o^2}}{\Delta f} = \frac{1}{\Delta f}\sum_{n=1}^{4} \overline{v_{on}^2} = 6.79 \cdot 10^{-14} \text{ V}^2/\text{Hz}$$

or $2.6 \cdot 10^{-7} \text{ V}/\sqrt{\text{Hz}}$, expressed as an rms value.

The following two statements must be added to the SPICE2 input used in the AC sweep (Example 5.3), in order to have a noise analysis performed:

```
.NOISE V(2) II 10
.PLOT NOISE ONOISE INOISE
```

The element lines VMEAS and CSHUNT must be deleted since an output voltage must be sampled. The .**NOISE** statement defines node 2 as the noise output and current source II as the noise input. One summary report of each noise source is computed for each frequency decade. A frequency sweep of the total output noise voltage and equivalent input noise current is also requested.

The input circuit, the summary report printed by SPICE2 of the noise analysis at 100 kHz, and the frequency variation of the rms values of $v_o/\Delta f$ and $i_{ieq}/\Delta f$ computed from 100 kHz to 10 GHz are listed in Figure 5.7.

Below the NOISE ANALYSIS header are the mean-square values of all individual noise sources computed at the output. The FREQUENCY precedes each such report. The RESISTOR SQUARED NOISE VOLTAGES are in agreement with the hand calculations. All noise sources associated with a BJT and their values are listed under TRANSISTOR SQUARED NOISE VOLTAGES. The parasitic terminal resistors, **RB**, **RC**, and **RE**, generate thermal noise, which is equal to zero in this case because no parasitic resistances have been specified in the .**MODEL** statement. The shot noise contributions from IB and IC are in agreement with the hand calculations. The last noise source of a BJT, FN, is the *flicker noise* component, which is described in more detail in the reference text by Gray and Meyer (1993).

Each NOISE ANALYSIS summary ends with the mean-square and rms values of TOTAL OUTPUT NOISE VOLTAGE; TRANSFER FUNCTION VALUE: V(2)/II

```
*******02/03/89 ********   SPICE 2G.6    3/15/83 ********11:40:54*****

ONE-TRANSISTOR CIRCUIT (FIG. 5.2)

****     INPUT LISTING                  TEMPERATURE =   27.000 DEG C

**********************************************************************
*
Q1 2 1 0 QMOD
RC 2 3 1K
RB 1 3 200K
VCC 3 0 5
*
II 0 1 AC 1
*
.MODEL QMOD NPN CJE=1P CJC=2P
*
.OP
.AC DEC 10 0.1MEG 10G
.NOISE V(2) II 10
*
.PLOT NOISE ONOISE INOISE
.WIDTH OUT=80
.END
*******02/03/89 ********   SPICE 2G.6    3/15/83 ********11:40:54*****

ONE-TRANSISTOR CIRCUIT (FIG. 5.2)

****     NOISE ANALYSIS                  TEMPERATURE =   27.000 DEG C

**********************************************************************

    FREQUENCY = 1.000D+05 HZ

**** RESISTOR SQUARED NOISE VOLTAGES (SQ V/HZ)

          RC        RB
TOTAL   1.646D-17 8.130D-16

**** TRANSISTOR SQUARED NOISE VOLTAGES (SQ V/HZ)

          Q1
RB      0.000D+00
RC      0.000D+00
RE      0.000D+00
IB      6.612D-14
IC      6.693D-16
FN      0.000D+00
TOTAL   6.679D-14

**** TOTAL OUTPUT NOISE VOLTAGE       =  6.762D-14 SQ V/HZ
                                      =  2.600D-07 V/RT HZ

    TRANSFER FUNCTION VALUE:
      V(2)/II                         =  9.904D+04
      EQUIVALENT INPUT NOISE AT II    =  2.625D-12 /RT HZ
```

(continued on next page)

Figure 5.7 Results of one-transistor amplifier noise analysis.

```
*******02/03/89 ********   SPICE 2G.6    3/15/83 ********11:40:54*****
ONE-TRANSISTOR CIRCUIT (FIG. 5.2)
****      INPUT LISTING                        TEMPERATURE =   27.000 DEG C
*****************************************************************************
 LEGEND:
*: ONOISE
+: INOISE
     FREQ      ONOISE
*)------------- 1.000D-10    1.000D-09    1.000D-08    1.000D-07  1.000D-06
              - - - - - - - - - - - - - - - - - - - - - - - - - - - - - -
+)------------- 1.000D-12    3.162D-12    1.000D-11    3.162D-11  1.000D-10
              - - - - - - - - - - - - - - - - - - - - - - - - - - - - - -
 1.000D+05 2.600D-07 .        + .           .          .       *      .
 1.259D+05 2.595D-07 .        + .           .          .       *      .
 1.585D+05 2.587D-07 .        + .           .          .       *      .
 1.995D+05 2.574D-07 .        + .           .          .       *      .
 2.512D+05 2.554D-07 .        + .           .          .       *      .
 3.162D+05 2.523D-07 .        + .           .          .       *      .
 3.981D+05 2.476D-07 .        + .           .          .       *      .
 5.012D+05 2.407D-07 .        + .           .          .      *       .
 6.310D+05 2.309D-07 .        + .           .          .      *       .
 7.943D+05 2.175D-07 .        + .           .          .      *       .
 1.000D+06 2.003D-07 .        + .           .          .     *        .
 1.259D+06 1.799D-07 .        + .           .          .     *        .
 1.585D+06 1.574D-07 .        + .           .          .    *         .
 1.995D+06 1.343D-07 .        + .           .          .   *          .
 2.512D+06 1.123D-07 .        + .           .          .  *           .
 3.162D+06 9.245D-08 .        + .           .          . *            .
 3.981D+06 7.520D-08 .        + .           .          *.             .
 5.012D+06 6.067D-08 .        + .           .        * .              .
 6.310D+06 4.868D-08 .        + .           .       * .               .
 7.943D+06 3.892D-08 .        + .           .      * .                .
 1.000D+07 3.105D-08 .        + .           .     * .                 .
 1.259D+07 2.473D-08 .        + .           .    * .                  .
 1.585D+07 1.968D-08 .        + .           .   * .                   .
 1.995D+07 1.566D-08 .        + .           .  * .                    .
 2.512D+07 1.245D-08 .        + .          .* .                       .
 3.162D+07 9.905D-09 .        + .          * .                        .
 3.981D+07 7.883D-09 .        + .        *. .                         .
 5.012D+07 6.278D-09 .        + .       *  .                          .
 6.310D+07 5.007D-09 .        + .      *   .                          .
 7.943D+07 4.002D-09 .        + .    *     .                          .
 1.000D+08 3.209D-09 .        + .  *       .                          .
 1.259D+08 2.587D-09 .        + .*         .                          .
 1.585D+08 2.102D-09 .       +.  *         .                          .
 1.995D+08 1.727D-09 .       +.*           .                          .
 2.512D+08 1.441D-09 .        + *          .                          .
 3.162D+08 1.226D-09 .        +*           .                          .
 3.981D+08 1.069D-09 .        *+           .                          .
 5.012D+08 9.564D-10 .       *   +         .                          .
 6.310D+08 8.774D-10 .      *.    +        .                          .
 7.943D+08 8.230D-10 .      *.      +      .                          .
 1.000D+09 7.855D-10 .      *.        +    .                          .
 1.259D+09 7.590D-10 .     * .         +   .                          .
 1.585D+09 7.388D-10 .     * .          +  .                          .
 1.995D+09 7.211D-10 .     * .        .    +                          .
 2.512D+09 7.028D-10 .     * .        .      +                        .
 3.162D+09 6.807D-10 .     * .        .       +                       .
 3.981D+09 6.521D-10 .    * .         .         +                     .
 5.012D+09 6.147D-10 .    * .         .          +.                   .
 6.310D+09 5.675D-10 .    * .         .           +                   .
 7.943D+09 5.111D-10 .   * .          .         .   +                 .
 1.000D+10 4.485D-10 .  * .           .         .     +               .
              - - - - - - - - - - - - - - - - - - - - - - - - - - - - - -
```

Figure 5.7   (*continued*)

at the corresponding frequency; and the EQUIVALENT INPUT NOISE AT II. One such summary is printed also for 1 MHz, 10 MHz, 100 MHz, 1 GHz, and 10 GHz; these summaries have been omitted from Figure 5.7.

In the frequency plot entitled AC ANALYSIS, the output noise voltage, ONOISE, is seen to fall off above 1 MHz, and the equivalent input noise, INOISE, increases starting at 158 MHz. The falloff of the output noise voltage is due to the 3 dB/octave roll-off of the transfer function, A_{vi}, and the increase in equivalent input noise, proportional to f^2, is due to the frequency dependence of the current gain, $\beta(j\omega)$.

Exercise
Explain the difference between the -3-dB frequency obtained in the AC frequency sweep, Example 5.3, and the -3-dB frequency obtained in the above analysis.

5.4 DISTORTION ANALYSIS

The signal applied to an active circuit is distorted because of a number of causes such as nonlinear elements and limiting. For small signals the cause of small distortions is the nonlinear I-V characteristic of semiconductor devices. SPICE2 and SPICE3 compute several harmonic distortion characteristics using AC small-signal analysis. This analysis is not available in PSpice.

The distortion measures derived below are computed for nearly linear circuits, operating at the bias point and at midfrequency, where capacitances and inductances can be neglected. The output signal, S_o, a voltage or a current, can be expressed as a power series of the input signal, S_i:

$$S_o = a_1 S_i + a_2 S_i^2 + a_3 S_i^3 + \cdots \tag{5.13}$$

where $a_1, a_2, a_3, \ldots$ are constants. a_1 is the transfer function at midfrequency computed by SPICE2 as part of the .**TF** or .**AC** analysis.

Harmonic distortion is generated in the circuit when one or more sinusoidal signals are applied at the input, such as

$$S_i = S_1 \cos \omega_1 t$$

When S_i is replaced by $S_1 \cos \omega_1 t$ in Eq. 5.13, signals of frequencies $2\omega_1$ and $3\omega_1$ (and higher) are generated, representing the *second* and *third harmonic distortion* terms.

When a second signal is also present in the input signal,

$$S_i = S_1 \cos \omega_1 t + S_2 \cos \omega_2 t \tag{5.14}$$

intermodulation distortion terms of frequencies $(\omega_1 + \omega_2)$ and $(\omega_1 - \omega_2)$ are generated as well. These terms are called *sum* and *difference second-order intermodulation* components. The third-order term in the power series of Eq. 5.13 produces third-order intermodulation terms at frequencies $(2\omega_2 \pm \omega_1)$ and $(2\omega_1 \pm \omega_2)$.

The following quantities are computed by SPICE2 as a measure of the different distortion components. *HD2* is the fractional second-harmonic distortion, of frequency $2\omega_1$ in the absence of the second signal, and is equal to

$$HD2 = \frac{\text{amplitude of second-harmonic distortion signal}}{\text{amplitude of fundamental}} = \frac{1}{2} \cdot \frac{a_2}{a_1} S_1 \qquad (5.15)$$

The expression of *HD2* as a function of the power series coefficients is derived by replacing S_i in Eq. 5.13 by $S_1 \cos \omega_1 t$ and ordering the terms of the fundamental and the harmonics.

The normalized third-harmonic distortion magnitude, *HD3*, is computed similarly, using the amplitude of the third-harmonic distortion signal.

The second-order sum and difference intermodulation components, *SIM2* and *DIM2*, respectively, are computed from the following equation and under the assumption that two signals are present at the input:

$$IM2 = \frac{\text{amplitude of second-order intermodulation component}}{\text{amplitude of fundamental}} \qquad (5.16)$$

The last distortion measure computed by SPICE2 is *DIM3*, the normalized third-order intermodulation component of frequency $(2\omega_1 - \omega_2)$.

Associated with each component of the small-signal equivalent model of a transistor is a distortion contribution at the output of the circuit. Each distortion contribution is computed as distortion power in a designated load resistor; the total harmonic distortion of a given order is obtained by summing up all individual contributions. The small distortions measures defined above are computed by SPICE2 in conjunction with an AC small-signal analysis; both an **.AC** and a **.DISTO** statement must be present in the input file.

The general form of the distortion control statement is

.DISTO *RLname* $<nums < f_2/f_1 <P_{ref} <S_2>>>>$

RLname is the name of the load resistor for computing the power contribution of the distortion. A resistor with the same name must be present in the input file. A summary of each distortion source in the circuit is listed in the result file once every *nums* frequency points in the interval from *fstart* to *fstop*, the frequency interval specified in the **.AC** statement. The number of summaries can be related to the type of frequency interval, specified in the **.AC** statement; for a **DEC** interval with *numpts* frequency points per decade, and *decades* decades in the interval from *fstart* to *fstop*, $decades \cdot numpts/nums + 1$ summaries are printed. A zero or the absence of a value for *nums* disables the report on individual distortion sources.

The expressions f_2/f_1 and S_2 define the frequency, f_2, and amplitude, S_2, of the second input signal, $S_2 \cos \omega_2 t$, in Eq. 5.14 for the evaluation of the intermodulation distortion terms. The frequency of the first signal, f_1, is the frequency being swept in the AC analysis. If f_2/f_1 is not specified, a value of $f_2/f_1 = 0.9$ is used by SPICE2 and S_2 defaults to 1.

The variable P_{ref} is the power used as reference in the computation of the distortion-power terms in resistor *RLname*. The power of the output signal, V_o, measured in the load resistor R_L is

$$P_{ref} = \frac{V_o^2}{2R_L} \tag{5.17}$$

By default SPICE2 uses 1 mW for P_{ref}. For a given value of R_L both the amplitude V_o and S_1 can be calculated, which serve as reference for deriving the distortion measures *HD2, HD3,* ... according to Eqs. 5.15 and 5.16. Example 5.7 provides more insight into the derivation of the distortion components of the one-transistor amplifier.

The frequency sweep of the distortion components can be requested in tabular form with a **.PRINT** statement or as a line-printer plot with a **.PLOT** statement. The general form of the output request is

```
.PRINT DISTO HD2<(x)> HD3<(x)> SIM2<(x)> DIM2<(x)>
+ DIM3<(x)>
.PLOT DISTO HD2<(x)> HD3<(x)> SIM2<(x)> DIM2<(x)>
+ DIM3<(x)>
```

HD2, **HD3**, **SIM2**, **DIM2**, and **DIM3** have the meanings defined above, and x stands for any of **R**, for real; **I** for imaginary; **M**, for magnitude, which is the default; **P**, for phase; and **DB**, for logarithmic representation. At least one distortion term must appear on a **.PRINT** or **.PLOT DISTO** statement. The distortion terms are computed at all frequencies between *fstart* and *fstop* as specified in the **.AC** statement.

EXAMPLE 5.6

```
.AC DEC 10 1K 100MEG
.DISTO ROUT 20 0.95 1M 0.5
```

The above two statements request the computation of the small distortion measures in the load resistor ROUT over a frequency interval from 1 kHz to 100 MHz. The output produced by SPICE2 will consist of three summaries of all distortion sources and the total distortion terms **HD2**, **HD3**, **SIM2**, **DIM2**, and **DIM3** at 1 kHz, 100 kHz, and 10 MHz. The value 20 in the **.DISTO** statement establishes the summary to be printed once every 20 frequency points resulting in the three summary frequencies mentioned above; according to the **.AC** statement specification there are 10 frequency points per decade.

The rest of the data in the **.DISTO** statement define the second signal and reference output power level, P_{ref}. The frequency of the second signal f_2 is set to $0.95 f_1$, and the amplitude is 0.5. P_{ref} is used to scale the distortion terms that the output voltage amplitude of the fundamental is equal to

$$V_o = \sqrt{2R_L P_{ref}}$$

If $P_{ref} = 1$ mW, the logarithmic values of the distortion components can be expressed in terms of *dBm*, a unit often used in telecommunications:

$$dBm \text{ distortion} = 20 \log HD$$

where 1 mW corresponds to 0 dBm. It should be noted that referencing distortion to the output power level uniquely determines the amplitude of the input signal producing that distortion. The value of P_{ref} can contradict the AC amplitude of the input source, which is usually 1V. The resulting distortion components reflect the actual input amplitude corresponding to P_{ref}. The computation of the distortion terms is presented in more detail in the following example.

The results of the **.AC** and **.DISTO** statements are detailed summaries of all distortion sources in the circuit. If only the total distortion is of interest, *nums* is set to zero in the **.DISTO** statement and a **.PRINT** or **.PLOT DISTO** statement is added, such as

```
.PRINT DISTO HD2 HD2(P) DIM2 DIM2(P)
```

which lists the magnitude and phase of the second harmonic distortion, **HD2**, and the second intermodulation difference distortion, **DIM2**, for all computed frequency points between 1 kHz and 100 MHz.

EXAMPLE 5.7

Compute all distortion measures introduced above for the one-transistor amplifier (Figure 4.3) and check the results using SPICE2.

Solution

Expressions for the different distortion terms can be derived based on the exponential I-V characteristic of the BJT. The one-transistor circuit is simplified, as shown in Figure 5.8, by removal of the bias base resistor, R_B, and replacement of it by a bias source of value $V_{BE} = 0.7934$, equal to the BE voltage obtained with the base resistor such that the quiescent collector current, $I_C = 2.1$ mA, is preserved. This simplification is necessary because distortion is a strong function of the input source resistance. A voltage signal source of amplitude V_{be1} is also connected at the input. The total voltage applied at the base of Q_1, v_{BE}, can be expressed as

$$v_{BE} = V_{BE} + v_{be} = V_{BE} + V_{be1} \cos \omega t \tag{5.18}$$

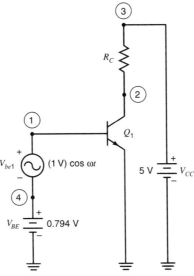

Figure 5.8 Simplified one-transistor circuit for distortion analysis.

The resulting total collector current, i_C, has a DC component, I_C, and an AC component, I_c:

$$i_C = I_C + I_c \cos \omega t \qquad (5.19)$$

The total collector current, i_C, is an exponential function of the total base-emitter voltage, v_{BE},

$$i_C = I_S \cdot e^{v_{BE}/V_{th}} = I_S \cdot e^{V_{BE}/V_{th}} \cdot e^{V_{be1}/V_{th}} = I_C e^{V_{be1}/V_{th}} \qquad (5.20)$$

where v_{BE} has been replaced by the sum of its components. Because of the assumption of small nonlinearities, the exponential in Eq. 5.20 can be expanded in a power series:

$$i_C = I_C \left[1 + \frac{V_i}{V_{th}} + \frac{1}{2} \left(\frac{V_i}{V_{th}} \right)^2 + \frac{1}{6} \left(\frac{V_i}{V_{th}} \right)^3 + \cdots \right] \qquad (5.21)$$

where V_i is the input signal, equal to V_{be1}. I_c follows from Eqs. 5.19 and 5.21:

$$I_c = i_C - I_C = \frac{I_C}{V_{th}} V_i + \frac{1}{2} \frac{I_C}{V_{th}^2} V_i^2 + \frac{1}{6} \frac{I_C}{V_{th}^3} V_i^3 + \cdots$$

The coefficients a_i of the power series in Eq. 5.13 can be set equal to the coefficients in the above power series, assuming that $S_i = V_i$ and $S_o = I_c$. The distortion terms follow naturally:

$$HD2 = \frac{1}{2} \cdot \frac{a_2}{a_1} S_i = \frac{1}{4} \cdot \frac{V_i}{V_{th}} \tag{5.22a}$$

or, using the relation $S_o = a_1 S_i$, $HD2$ can be rewritten as

$$HD2 = \frac{1}{2} \cdot \frac{a_2}{a_1^2} S_o = \frac{1}{4} \cdot \frac{I_c}{I_C} \tag{5.22b}$$

In order to evaluate the above expressions, V_i or I_c must be calculated. Both values result from the equation of the output power in R_C

$$P_o = \frac{V_o^2}{2R_C} = \frac{1}{2} I_c^2 R_C = P_{ref} = 1 \text{ mW}$$

$$V_o = \sqrt{2R_C P_{ref}} = 1.4 \text{ V}$$

$$I_c = \sqrt{\frac{2P_{ref}}{R_C}} = 1.4 \text{ mA}$$

The amplitude of the input signal V_i is

$$V_i = \frac{I_c}{a_1} = \frac{I_c}{g_m} = \frac{1.4 \text{ mA}}{0.081 \text{ mho}} = 17 \text{ mV}$$

The second-harmonic distortion is obtained by substituting for either V_i or I_c in Eq. 5.22a or Eq. 5.22b, respectively:

$$HD2 = \frac{1}{4} \frac{I_c}{I_C} = \frac{1}{4} \cdot \frac{1.4}{2.1} = 0.16$$

The remaining distortion terms can be derived similarly:

$$HD3 = \frac{1}{24} \cdot \left(\frac{V_i}{V_{th}} \right)^2 = \frac{1}{24} \cdot \left(\frac{I_c}{I_C} \right)^2 = 0.018 \tag{5.23}$$

$$SIM2 = DIM2 = \frac{1}{2} \cdot \frac{V_i}{V_{th}} = \frac{1}{2} \cdot \frac{I_c}{I_C} = 0.333 \tag{5.24}$$

$$DIM3 = \frac{1}{8} \cdot \left(\frac{V_i}{V_{th}}\right)^2 = \frac{1}{8} \cdot \left(\frac{I_c}{I_C}\right)^2 = 0.054 \qquad (5.25)$$

The input statements for the distortion analysis of the one-transistor amplifier of Figure 5.8 and the distortion measures at 1 kHz computed by SPICE2 are listed in Figure 5.9. Note that *fstart* and *fstop* in the **.AC** statement have been set equal since only

```
*******03/06/89 ********  SPICE 2G.6   3/15/83 ********11:10:26*****

ONE-TRANSISTOR CIRCUIT (FIG. 5.8)
****     INPUT LISTING              TEMPERATURE =   27.000 DEG C

********************************************************************
*
Q1 2 1 0 QMOD
RC 2 3 1K
*RB 1 3 200K
VCC 3 0 5
*
VBE 4 0 793.4M
VBE1 1 4 AC 1
*
.MODEL QMOD NPN
*+ CJE=1P CJC=2P
*
.OP
.AC LIN 1 1K 1K
.DISTO RC
*
.PRINT DISTO HD2 HD3 SIM2 DIM2 DIM3
.PRINT DISTO HD2(DB) HD3(DB) SIM2(DB) DIM2(DB) DIM3(DB)
.WIDTH OUT=80
.END
```

```
****     AC ANALYSIS                TEMPERATURE =   27.000 DEG C

    FREQ      HD2          HD3          SIM2         DIM2         DIM3

  1.000E+03   1.683E-01    1.889E-02    3.367E-01    3.367E-01    5.668E-02

**** AC ANALYSIS                    TEMPERATURE =   27.000 DEG C

    FREQ      HD2(DB)      HD3(DB)      SIM2(DB)     DIM2(DB)     DIM3(DB)

  1.000E+03   -1.548E+01   -3.447E+01   -9.455E+00   -9.455E+00   -2.493E+01
```

Figure 5.9 SPICE2 distortion analysis results.

the total distortion terms at midfrequency are of interest. The distortion terms computed by SPICE2 are in good agreement with the values calculated by hand according to Eqs. 5.22 through 5.25. The distortion terms in dBm are listed in the SPICE2 output as well.

5.5   POLE-ZERO ANALYSIS

The frequency sweep, **.AC**, introduced above yields the frequency response of circuits in the form of a graph. The locations of poles and zeros can in general be inferred from a Bode plot. In many applications, such as filters or feedback circuits, the succession of several poles and zeros makes reading their locations from a frequency plot difficult and makes it necessary to obtain the actual values. SPICE3 and high-end SPICE versions, such as SpicePLUS and HSPICE, provide a pole-zero analysis.

The pole-zero analysis computes the transfer function of the circuit represented as a two-port circuit:

$$\mathbf{H}(s) = \frac{\mathbf{V_o}(s)}{\mathbf{S_i}(s)} = a\frac{(s - z_1)(s - z_2)\cdots(s - z_n)}{(s - p_1)(s - p_2)\cdots(s - p_m)} \tag{5.26}$$

where $s = j\omega$, $\mathbf{V_o}$ is the output voltage, and $\mathbf{S_i}$ is the input signal, which can be either a current or a voltage. The output for this analysis is always a voltage.

The general form of the pole-zero statement in SPICE3 is

.PZ *ni*1 *ni*2 *no*1 *no*2 **CUR/VOL POL/ZER/PZ**

where *ni*1 and *ni*2 are the input nodes and *no*1 and *no*2 are the output nodes of the two-port representation. The field following the node specification defines the type of the input; **CUR** for current input and **VOL** for voltage input. One of the keywords must be present. The last field specifies whether the poles (**POL**), zeros (**ZER**), or both poles and zeros (**PZ**), should be computed. One of the three keywords must be present. In interactive mode the same command with the omission of the leading period must be typed at the SPICE3 shell prompt.

The results are stored in the output file if the following line is added to the input circuit:

```
.PRINT PZ ALL
```

The same command without the leading period can be issued in interactive mode.

A noteworthy difference between the frequency sweep and the pole-zero analysis is that in the former case one analysis computes the transfer function from input to any node in the circuit, whereas in the latter case a separate analysis must be performed each time the two-port representation is redefined.

The pole-zero analysis is very useful for relatively small circuits; for circuits containing more than 20 charge storage elements the results must be interpreted carefully.

The bridge-T circuit, which has exemplified the **.AC** frequency sweep, is used below for finding the poles and zeros and double- checking the Bode plots.

EXAMPLE 5.8

Use SPICE3 to compute the poles and zeros of the bridge-T filter; compare the results with the hand calculations in Example 5.2 and the Bode plot produced by SPICE.

Solution

The SPICE3 input and results for the pole-zero analysis are shown in Figure 5.10 on page 166. Note that the **.AC** statement has been replaced by a **.PZ** line defining the input of the two-port representation between nodes 1 and 0 and the output between nodes 3 and 0. Furthermore, the input is defined as a voltage. The output signal is always assumed to be a voltage.

The output contains the two real poles and zeros of the transfer function, which are identical to the hand calculations carried out in Example 5.2. Note that the pole-zero algorithm usually runs into difficulties when the transfer function is complex and has multiple poles or zeros.

5.6 SUMMARY

This chapter has described the analyses performed by SPICE in the AC mode. The control statements for each analysis have been introduced as well as the specifications of output variables and result-processing requests. Several examples have been used to show how to apply the various AC analyses to practical circuit problems. The implications of small-signal analysis for nonlinear circuits in the AC mode has been addressed in the examples.

The AC analysis types, frequency sweep, noise and distortion analysis, and pole-zero computations are specified by the following control lines:

.AC *Interval numpts fstart fstop*

.NOISE V($n1<,n2>$) **V/I**$name$ $nums$

.DISTO *RLname* $<nums <f_2/f_1 <P_{ref} <S_2>>>>$

.PZ $ni1$ $ni2$ $no1$ $no2$ **CUR/VOL POL/ZER/PZ**   (SPICE3)

Noise and distortion are frequency-domain analyses; therefore these statements must be used in conjunction with an **.AC** line.

With the exception of the **PZ** analysis, results are stored in the output file only for specified circuit variables, *AC_OUT_var*, which can be complex voltages or currents:

Vx($node1<,node2>$)

Ix($Vname$)

```
BRIDGE T FILTER
*
V1 1 0 12 AC 1
C1 1 2 1U
C2 2 3 1U
R3 2 0 1K
R4 1 3 1K
*
.OP
.PZ 1 0 3 0 VOL PZ
.PRINT PZ ALL
*
.END
```

Circuit: BRIDGE T FILTER

Circuit: BRIDGE T FILTER
Date: Fri Apr 19 14:55:55 1991

Operating point information:

```
      Node  Voltage
      ----  -------
      V(3)  1.000000e+01
      V(2)  0.000000e+00
      V(1)  1.000000e+01

      Source      Current
      ------      -------
      v1#branch  0.000000e+00
```

                          bridge T filter
                          pole-zero analysis Fri Apr 19 14:55:55 1991

Index   pole(1)   pole(2)
----
0   -2.618034e+03,   0.000000e+00   -3.819660e+02,   0.000000e+00

                          bridge T filter
                          pole-zero analysis Fri Apr 19 14:55:55 1991

Index   zero(1)   zero(2)
----
0   -1.000000e+03,   0.000000e+00   -1.000000e+03,   0.000000e+00

Figure 5.10 Input and results for pole-zero analysis of bridge-T circuit.

where x defines the output format of the complex variable; accepted formats are **R** and **I**, for real and imaginary part, respectively; **M** and **P**, for magnitude and phase, respectively; and **DB**, for the decibel value of the magnitude.

The output variables of a frequency-domain analysis can be saved either in tabular or line-printer-plot format using the **.PRINT** or **.PLOT** control statement, respectively. The general format of the output request is

> **.PRINT/PLOT AC** *AC_OUT_var1* <*AC_OUT_var2...*> <*plot_limits*>

REFERENCES

Dorf, R. C., 1989. *Introduction to Electric Circuits.* New York: John Wiley & Sons.
Gray, P. R., and R. G. Meyer. 1993. *Analysis and Design of Analog Integrated Circuits,* 3d ed. New York: John Wiley & Sons.

Six

TIME-DOMAIN ANALYSIS

6.1 ANALYSIS DESCRIPTION

The transient analysis of SPICE2 computes the time response of a circuit. This analysis mode takes into account all nonlinearities of the circuit. The input signals applied to the circuit can be any of the time-dependent functions described in Chap. 2: pulse, exponential, sinusoidal, piecewise linear, and single-frequency FM. In contrast, in AC analysis only sinusoidal signals with small amplitudes, for which circuits can be considered linear, are used.

Time-domain analysis computes, in addition to voltages and currents of time-invariant elements, the variation of charges, q, and fluxes, ϕ, associated with capacitors and inductors. These are described by the branch-constitutive equations (BCEs) for capacitors and inductors defined in Secs. 2.2.3 and 2.2.5:

$$i_C = \frac{dq}{dt} = C\frac{dv_C}{dt} \tag{6.1}$$

$$v_L = \frac{d\phi}{dt} = L\frac{di_L}{dt} \tag{6.2}$$

where i_C and v_C are the current and voltage of capacitor C and i_L and v_L are the current and voltage of inductor L. The BCE for resistors, Ohm's law, is time-invariant.

Two analysis types are supported in SPICE for the time-domain solution:

TRAN   Computes the voltage and current waveforms over a given time interval

FOUR   Computes the Fourier coefficients, or spectral components, of periodic signals

An additional utility for transient analysis, `.IC` (*initial conditions*), is used for specifying the initial voltages at selected or all nodes. An INITIAL TRANSIENT SOLUTION (*ITS*) precedes a time-domain analysis unless it is specifically disabled. It is a DC solution at $t = 0$.

6.2 TRANSIENT ANALYSIS

The following statement is required by SPICE to perform a transient analysis:

.**TRAN** *TSTEP TSTOP* <*TSTART* <*TMAX*>> <**UIC**>

The analysis is performed over the time interval from 0 to *TSTOP*, but results can be output starting from a user-defined time, *TSTART*, to *TSTOP*; *TSTART* is assumed to be 0 if it is not specified. *TSTEP* is the time interval used for printing or plotting the results requested by a `.PRINT` or a `.PLOT`. Note that SPICE2 and most programs derived from it use a different internal time step for solving the circuit equations, which is automatically adjusted by the program for accuracy. By default the internal time step is bound by the smaller of (*TSTOP-TSTART*)/50 or 2· *TSTEP*. Although in most cases the SPICE internal time-step selection algorithm is accurate enough, there are situations when for better accuracy a user may want to restrict the maximum time step. This can be achieved by specifying the value of the maximum allowed internal time step *TMAX*. The data on a `.TRAN` statement are order-sensitive, and a value for *TSTART* must always precede *TMAX*.

The time-domain solution is preceded by a DC solution, the ITS, which computes the initial values of voltages and currents necessary for the integration of the BCEs, Eqs. 6.1 and 6.2, or Eqs. 2.5 and 2.10. A user can avoid the initial transient, DC, solution by concluding the `.TRAN` statement with the keyword **UIC** (*use initial conditions*). In this case the initial value of every voltage and current is 0 except those voltages and currents initialized with the `IC` keyword in the element definition lines or the `.IC` statement. One scenario where **UIC** is useful is the computation of the steady-state solution without the transient response leading to it. For a correct solution the user must define the correct initial values for all charge-storage elements in the circuit.

EXAMPLE 6.1

Explain the meaning of the following transient analysis statements:

```
.TRAN 1n 100n
.TRAN 0.1u 100u 90u
.TRAN 100u 1m 0 10u
.TRAN 10n 1u UIC
.PRINT TRAN V(6) I(VCC)
.PLOT TRAN V(6) V(2,1) (0,5)
```

Solution

The first statement specifies that a time-domain analysis is to be performed from time 0 to 100 ns and that the results are to be output at a 1-ns interval. The second statement requests that the analysis be performed to 100 μs and that the results between 90μs and 100μs be output at a 0.1-μs interval. The third statement requests a long analysis to 1ms with results output every 100μs but limits the internal time step to 10μs. Finally, the fourth .TRAN statement requires SPICE to omit the initial transient, or DC, solution by concluding the statement with the UIC keyword.

The desired voltages and currents resulting from a transient analysis are identified in a .PRINT or .PLOT statement. The keyword TRAN must be present to identify the analysis type. The time interval and time step for the prints and plots are those specified on the .TRAN statement. At least one .PRINT or .PLOT statement must be present in the input file for the analysis to be performed. In this example the values of V(6) and the current through voltage source VCC are printed and V(6) and V(2,1) are plotted with a common voltage scale with values from 0 to 5 V at the output.

EXAMPLE 6.2

Compute the time-domain response of an RLC parallel circuit that at $t = 0$ is connected to a constant current source as shown in Figure 6.1. Verify the solution with SPICE.

Solution

The KCL applied to node 1 yields the following equation:

$$C\frac{dv_C}{dt} + i_L + \frac{v_C}{R} = I_S \tag{6.3}$$

The BCE of the inductor, Eq. 6.2, yields a substitution for i_L :

$$L\frac{di_L}{dt} = v_C \tag{6.4}$$

leading after differentiation with respect to time to the following second-order differential equation in v_C:

$$\frac{d^2v_C}{dt^2} + \frac{1}{RC}\frac{dv_C}{dt} + \frac{1}{LC}v_C = \frac{1}{C}\frac{dI_S}{dt} = 0 \tag{6.5}$$

The solution of this equation is of the form

$$v_C(t) = Ae^{pt} \tag{6.6}$$

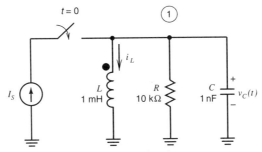

Figure 6.1   Parallel RLC circuit.

which put in the differential equation above leads to the *characteristic equation*:

$$p^2 + 2\alpha p + \omega_0^2 = 0 \tag{6.7}$$

where

$$\alpha = \frac{1}{2RC} = 5 \cdot 10^4 \, \text{s}^{-1}$$

$$\omega_0 = \frac{1}{\sqrt{LC}} = 10^6 \, \text{rad/s} \tag{6.8}$$

The solution of the quadratic equation is

$$p_{1,\,2} = -\alpha \pm \sqrt{\alpha^2 - \omega_0^2} = -\alpha \pm \sqrt{-(\omega_0^2 - \alpha^2)} = -\alpha \pm j\omega_d \tag{6.9}$$

which is put in Eq. 6.6 to obtain $v_C(t)$:

$$v_C(t) = e^{-\alpha t}(A_1 e^{j\omega_d t} + A_2 e^{-j\omega_d t})$$

$$= e^{-\alpha t}[(A_1 + A_2)\cos\omega_d t + j(A_1 - A_2)\sin\omega_d t] \tag{6.10}$$

α is the *damping factor*, and ω_d is the *damped radian frequency* (Nilsson 1990). Coefficients A_1 and A_2 are found from the initial conditions; at $t = 0$, $v_C(0) = 0$ and therefore $A_2 = -A_1$. A_1 is obtained by putting $v_C(0)$ in Eq. 6.3:

$$A_1 = \frac{1}{2}\frac{I_S}{j\omega_d C}$$

The time-dependent function $v_C(t)$ is

$$v_C(t) = \frac{1}{\omega_d C} I_S e^{-\alpha t} \sin \omega_d t \tag{6.11}$$

The angular frequency, ω, is close to ω_0 because α is negligible by comparison; therefore the period is

$$T = \frac{2\pi}{\omega_0} = 6.28\,\mu s \tag{6.12}$$

```
PARALLEL RLC CIRCUIT
*
IS 0 1 PWL 0 0 1N 1M 1 1M
L 1 0 1M
C 1 0 1N
R 1 0 10K
*
.TRAN 1U 100U
.PLOT TRAN V(1)
.END
```

The SPICE input file for this circuit entitled, PARALLEL RLC CIRCUIT, is shown above. A transient analysis for 100 μs is requested corresponding to approximately 16 periods according to the above calculations. The waveform computed by SPICE2 is shown in Figure 6.2 and has the damped sinusoidal shape predicted by Eq. 6.11. The complex algorithms in SPICE2 (see Sec. 9.4) can be verified to predict waveforms in agreement with the above hand-derived solution for this simple problem. According to Eq. 6.11, the amplitude of the oscillation at $\omega_0 t = \pi/2$ (for example, at $t = \pi/2$ μs for $\omega_0 = 10^6$ rad/s) is

$$V_c = v_C \mid_{t=\frac{\pi}{2}\mu s} = \frac{10^{-3}}{10^6 \cdot 10^{-9}} e^{-5 \cdot 10^4 t} \approx 1\,V$$

After solving for $v_C(t)$, one can put v_C of Eq. 6.4 in Eq. 6.3 to obtain a second-order differential equation in $i_L(t)$:

$$\frac{d^2 i_L}{dt^2} + \frac{1}{RC} \frac{di_L}{dt} + \frac{1}{LC} i_L = \frac{1}{LC} I_S \tag{6.13}$$

This equation differs from Eq. 6.5 for $v_C(t)$ in that the right-hand side is nonzero. The solution consists of the damped sinusoidal term, which is the natural solution, and an

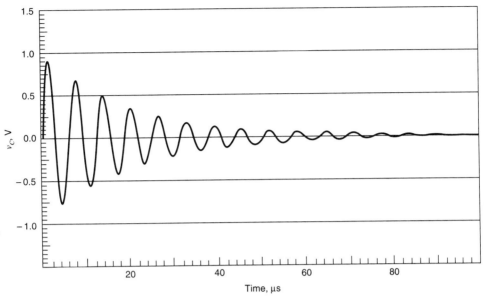

Figure 6.2   The waveform $v_c(t)$ computed by SPICE.

additional forced solution:

$$i_L(t) = Be^{-\alpha t} \sin \omega_d t + I_S \tag{6.14}$$

The SPICE waveform for $i_L(t)$ is shown in Figure 6.3. It can be seen that once the oscillations die out the inductor current assumes the forced solution, which is equal to the input current of 1 mA. The natural solution, therefore, represents the transient response, whereas the forced solution is the steady-state response.

A few comments can be made at this point. In order to observe the oscillations in a circuit, we use a step function at $t = 0+$ described as a PWL current source. An alternate way to achieve the same result is to use a DC current source at the input and to omit an INITIAL TRANSIENT SOLUTION by specifying the **UIC** option in the .**TRAN** statement. This second approach is equivalent to applying a step function at $t = 0+$ since all currents and voltages at $t = 0$ are zero.

Exercise:
Show that no oscillation can be observed in the SPICE solution if the **UIC** keyword is not used when I_S is a DC source. Explain the result.

The previous example demonstrated the use of the transient analysis in SPICE for computing the response of a linear RLC circuit. The example also outlined the

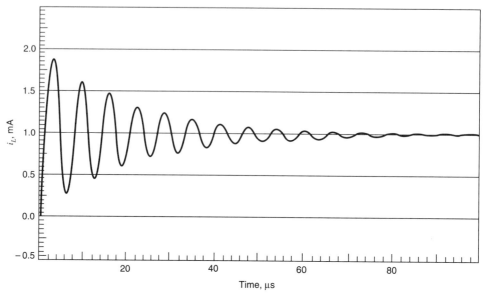

Figure 6.3   The waveform $i_L(t)$ computed by SPICE.

connection between the equations describing the electrical circuit and the analysis parameters and solution computed by SPICE. The importance of the different analysis parameters is exemplified best by oscillators. The following example demonstrates the use of SPICE for computing the response of a Colpitts oscillator.

EXAMPLE 6.3

Verify the oscillation condition and find the amplitude and frequency of the Colpitts oscillator shown in Figure 6.4 using SPICE.

Solution

The passive RLC circuit in Example 6.2 cannot sustain oscillations, which are damped by the factor $e^{-\alpha t}$. Oscillations can be sustained only if the real part of the natural frequencies computed from Eq. 6.7 is positive, or in other words, the circuit has right-hand-plane poles. This can be achieved with a gain block connected in a feedback configuration (Pederson and Mayaram 1990). In the frequency domain the overall transfer function, $A(s)$, of a gain block, $a(s)$ (where $s = \sigma + j\omega$ is the complex frequency), connected in a feedback loop, f, which is shown in Figure 6.5 is equal to the following (Gray and Meyer 1993):

$$A(s) = \frac{S_o}{S_i} = \frac{a(s)}{1 + a(s)f} \tag{6.15}$$

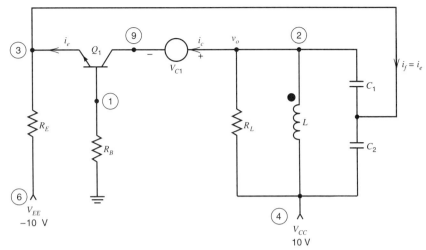

Figure 6.4 Colpitts oscillator.

The loop gain of the system, $T(s)$, where $T(s) = a(s)f$, must encircle the point $(-1, 0)$ in the complex frequency plane as shown in Figure 6.6 in order for the circuit to oscillate. The plot in Figure 6.6 is known as the *Nyquist diagram*.

The Colpitts oscillator shown in Figure 6.4 has the common-base (CB) transistor amplifier, Q_1 connected in a feedback loop. The feedback network consists of capacitors C_1 and C_2. The signal fed back to the input is

$$v_f = \frac{C_1}{C_1 + C_2} v_o = \frac{v_o}{n} \tag{6.16}$$

where n can be referred to as the capacitive turns ratio and is equal to the inverse of the feedback factor, f:

$$n = \frac{1}{f} = \frac{C_1 + C_2}{C_1} = 10 \tag{6.17}$$

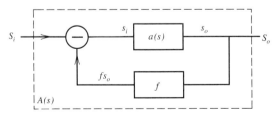

Figure 6.5 Feedback amplifier.

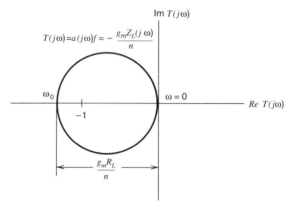

Figure 6.6   Nyquist diagram.

The small-signal gain $a(s)$ of the CB transistor is

$$a(s) = \frac{v_o}{v_i} = -g_m Z_L(s) = -g_m R_L \frac{Ls/R_L}{1 + Ls/R_L + LCs^2} \tag{6.18}$$

The circuit is unstable, and oscillations are initiated when the loop gain is

$$T(j\omega_0) = a(j\omega_0)f = -\frac{g_m R_L}{n} = -1 \tag{6.19}$$

where ω_0 is the resonant frequency of the tank circuit in the collector of transistor Q_1:

$$\omega_0 = \frac{1}{\sqrt{LC}} = \frac{1}{\sqrt{5 \cdot 10^{-6} \cdot 450 \cdot 10^{-12}}} \text{ rad/s} = 21.1 \cdot 10^6 \text{ rad/s} \tag{6.20}$$

with

$$C = \frac{C_1 C_2}{C_1 + C_2} = 450 \text{ pF} \tag{6.21}$$

This corresponds to an oscillation frequency $f_0 = 3.36$ MHz. When oscillations build up, the small-signal approximation is no longer valid and the equivalent large-signal G_m must be considered, which is less than g_m. A good approach to ensuring steady-state oscillations (Meyer 1979) is to dimension R_L so that the initial loop gain is

$$|T(j\omega_0)| = \frac{g_m R_L}{n} \approx 3 \tag{6.22}$$

The gain of the circuit including feedback is

$$A(s) = -g_m R_L \frac{Ls/R_L}{1 + (1 - g_m R_L/n)Ls/R_L + LCs^2} \tag{6.23}$$

The denominator can be compared with the characteristic equation of the parallel RLC circuit, Eq. 6.7. It has a pair of complex poles leading to a time-domain solution, Eq. 6.11, of the form

$$v_o(t) \propto e^{-\alpha t} \sin \omega_0 t \tag{6.24}$$

where, in this example,

$$\alpha = \frac{1}{2RC} \left(1 - \frac{g_m R_L}{n} \right) \tag{6.25}$$

and represents the damping factor. If $g_m R_L/n > 1$, corresponding to complex poles in the right half-plane, the predicted solution, Eq. 6.24, is a growing sinusoid that reaches a steady-state amplitude constrained by circuit biasing and loading.

The above equation is very important for understanding at what rate the oscillations build up. The analysis of oscillators with SPICE can be tricky for certain circuits because a large number of periods must be simulated before oscillations can be observed. The oscillation buildup can be related to the quality factor, Q, of a parallel tuned circuit, defined by

$$Q = \frac{R}{\omega_0 L} = \omega_0 CR \tag{6.26}$$

Putting Q in Eq. 6.24 leads to the following expression of the circuit response:

$$v_o(t) \propto e^{K\omega_0 t/Q} \sin \omega_0 t \tag{6.27}$$

where K is a constant dependent on the actual oscillator configuration. The above expression shows that the higher the Q of the tuned circuit, the longer it takes to reach steady-state oscillations. Eq. 6.27 is also valid for series resonant circuits with the appropriate change in the definition of Q.

In order to complete the circuit specification, we need to bias the circuit; this can be achieved by setting the emitter current, I_{EE}, using a negative supply, V_{EE}, and a resistor, R_E:

$$V_{EE} = -10 \text{ V}$$

$$R_E = 4.65 \text{ k}\Omega$$

$$g_m = 0.076 \text{ A}/\text{V}$$

$$I_C = \alpha_F I_{EE} = \alpha_F \frac{-V_{EE} - V_{BE}}{R_E} = 0.99 \frac{(10 - 0.8) \text{ V}}{4.65 \cdot 10^3 \text{ }\Omega} = 1.96 \text{ mA}$$

With the above values the minimum value of R_L for which, according to Eq. 6.22, the circuit oscillates is

$$R_L \approx \frac{3n}{g_m} = 395 \text{ }\Omega$$

The SPICE input COLPITTS OSCILLATOR is listed below; note that the negative supply is implemented as a step function using the **PULSE** source. It is always desirable to kick the circuit in order for the program to find the oscillatory solution. The step function used in simulation is similar to the real situation of connecting a circuit to a supply before proper operation can be observed. The value chosen for R_L is 750 Ω. The simulation is carried out for ten periods, to 3 μs, with results to be printed in the output file every 20 ns. Note that for the graphical output of Nutmeg or Probe, the time step is used only to set a default upper bound on the internal integration time interval.

```
COLPITTS OSCILLATOR
*
RB 1 0 1
Q1 9 1 3 MOD1
VC1 2 9 0
VCC 4 0 10
RL 4 2 750
C1 2 3 500P
C2 4 3 4.5N
L 4 2 5U
RE 3 6 4.65K
VEE 6 0 -10 PULSE -15 -10 0 0 0 1
*
.MODEL MOD1 NPN RC=10
*
.TRAN  20N 3U
.PLOT TRAN V(2) I(VC1)
.OPTIONS LIMPTS=5000 ITL5=0 ACCT
.END
```

The results of the simulation, the collector voltage, v_o, of Q_1, are shown in graphical form in Figure 6.7. The amplitude of oscillations at the collector, V_o, or V(2) in SPICE,

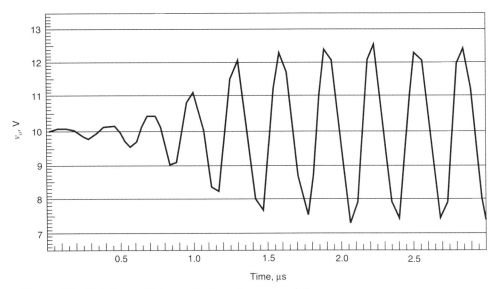

Figure 6.7   Colpitts oscillator: collector voltage, v_0, of Q_1.

can be verified at ω_0:

$$V_o = \left(\alpha - \frac{1}{n}\right)R_L I_e = \left(\alpha - \frac{1}{n}\right)R_L 2I_{EE}\frac{I_1(b)}{I_0(b)}$$

$$= 0.89 \cdot 750 \cdot 2 \cdot 1.9 \cdot 10^{-3} \text{ V} = 2.54 \text{ V} \qquad (6.28)$$

The incremental part of the collector current, I_c, is approximated by a Fourier series (Pederson and Mayaram 1990) in which the ratio of the modified Bessel functions, I_1/I_0, is a function of $b = V_t/V_{th}$ and is equal to 0.95 for $V_t/V_{th} > 6$. A power-series representation cannot be used because of the large value of V_i compared to V_{th}. The amplitude of V_o derived above is in good agreement with the waveform in Figure 6.7.

The waveform can be seen to be a piecewise linear approximation of a sinusoid. The points actually computed by SPICE are apparent on the graph. In order to obtain a smoother sinusoid, the maximum integration time step used by SPICE must be limited. This is achieved by specifying the *TMAX* parameter on the **.TRAN** statement:

```
.TRAN 20N 3U 0 10N
```

The new waveform resulting after replacement of the initial **.TRAN** line with the above line is shown in Figure 6.8 and can be observed to be much smoother.

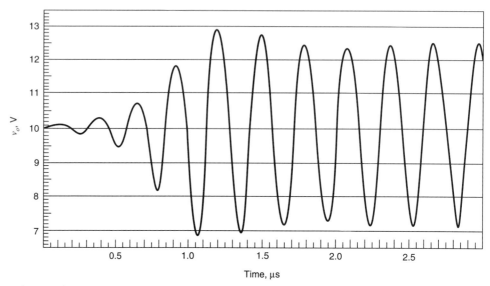

Figure 6.8   Colpitts oscillator: more time points used for $V(2)$.

Exercise
Verify that oscillations can be observed for $R_L = 395\,\Omega$ but not for $R_L = 100\,\Omega$. This exercise should prove the validity of the oscillation condition derived above.

6.3   INITIAL CONDITIONS

The solution of the time response of electric circuits starts with the time-zero values, or initial conditions, as shown in Example 6.2. A transient analysis in SPICE is preceded by an INITIAL TRANSIENT SOLUTION, which computes the initial conditions. The steady-state time-domain solution of more complex circuits is reached faster if the user initializes voltages across capacitors, currents through inductors, semiconductor-device junction voltages, and node voltages. As described in Sec. 4.6, the .NODESET statement helps the DC solution to be found faster. Similarly, user-defined initial conditions enhance the accuracy and offer quicker access to the desired solution. Most SPICE programs support two types of user-specified initial conditions.

First, initial values for node voltages can be set with the following statement:

.IC $V(node1)=value1$ <$V(node2)=value2 \ldots$>

This command sets the time-zero voltage at *node1* to *value1*, that at *node2* to *value2*, and so on. Initial values of charges on capacitors and semiconductor devices are also computed based on these initial voltages. Note that unlike voltages initialized by

.NODESET, which are used only as initial guesses for the iterative process and then released to converge to a final solution, voltages defined by the .IC statement do not change in the final INITIAL TRANSIENT SOLUTION.

The effect of the .IC statement differs depending on whether the UIC parameter is present on the .TRAN line. In the absence of UIC an INITIAL TRANSIENT SOLUTION for the entire circuit is computed with the initialized nodes kept at the specified voltages. If UIC is specified on the .TRAN line, all the values in the time-zero solution except the initialized node voltages are zero. As many node voltages should be initialized as possible when the UIC parameter is set.

Second, initial conditions for capacitors, inductors, controlled sources, transmission lines, and semiconductor devices can be set on a device-by-device basis using the IC keyword. These values are used only in conjunction with UIC and have no effect on the INITIAL TRANSIENT SOLUTION. When initial values are specified both on devices and in an .IC statement, the device-based values take precedence.

Table 6.1 summarizes the effect of the different combinations of initial conditions, namely, the .IC statement, the device-based IC, and the UIC keyword, on the circuit solution. In this table *ITS* stands for *initial transient solution*, which is different from the small-signal bias solution (SSBS), which is the result of an .OP request.

EXAMPLE 6.4

Use a device-based IC to set the initial current $i_L(0) = 1$ mA through the inductor of the parallel RLC circuit in Figure 6.1.

Solution
According to Table 6.1, a device-based IC effects the solution only when UIC is present in the .TRAN statement. Therefore the SPICE input file in Example 6.2 is modified as shown below. The third modification in the RLC description listed below is the replacement of the PWL source used for I_S with a DC current, which in the presence of UIC has the effect of a step function.

Table 6.1 Effects of IC Combinations

UIC	.IC	Device-based IC	SPICE2/3 Initialization
no	no	no	ITS is equivalent to SSBS
no	no	yes	ITS is equivalent to SSBS; device-based ICs have no influence
no	yes	no	ITS uses .IC voltages; is different from SSBS
no	yes	yes	ITS uses .IC voltages; is different from SSBS; device-based ICs have no influence
yes	no	no	No ITS; all initial values are zero
yes	no	yes	No ITS; uses device-based IC; rest of initial values are zero
yes	yes	no	No ITS; uses .IC; rest of initial values are zero
yes	yes	yes	No ITS; uses device-based IC first, .IC next; rest of initial values are zero

```
PARALLEL RLC CIRCUIT W/ INITIAL CONDITION
IS 0 1 1MA
L 1 0 1MH IC=1MA
C 1 0 1NF
R 1 0 10K
.TRAN 1US 100US UIC
.PLOT TRAN V(1)
.END
```

The SPICE analysis results in a constant current $i_L(t) = 1$ mA without the damped oscillations observed in Example 6.2. The explanation for this result is that the specified initial condition corresponds to the steady-state solution. This constitutes a very important observation: The fastest way to find the steady-state response of a circuit is to initialize as many elements as possible in the state they are expected to reach.

Note that omission of the keyword **UIC** from the **.TRAN** statement results in damped oscillations, because device-based **IC**s have no effect, as shown in Table 6.1.

The above example demonstrates the use of the device-based **IC** and its applicability for finding the steady-state response. In the analysis of oscillators initial conditions must be used in order to shorten the simulation time during the build-up phase (the higher the Q of the circuit, the longer this phase lasts, according to Eq. 6.27). The next example will demonstrate the use of **.IC** for the correct initialization of a ring oscillator.

EXAMPLE 6.5

Use SPICE to simulate the behavior of the three-stage enhancement-depletion (E-D) MOS ring oscillator shown in Figure 6.9. The enhancement and depletion transistors have the following device and model parameters:

Enhancement NMOS: $W = 40$ μm; $L = 10$ μm; $VTO = 1.8$ V;
$KP = 40$ μA/V^2; $LAMBDA = 0.001$ V^{-1}, $CGSO = 20$ pF/m.
Depletion NMOS: $W = 5$ μm; $L = 10$ μm; $VTO = -3$ V;
$KP = 40$ μA/V^2; $LAMBDA = 0.001$ V^{-1}.

Solution
Following is the SPICE input for this circuit:

```
RING OSCILLATOR MOS
*
VDD 11 0 5
M1 1 3 0 0 ENH L=10U W=40U
M2 2 1 0 0 ENH L=10U W=40U
M3 3 2 0 0 ENH L=10U W=40U
```

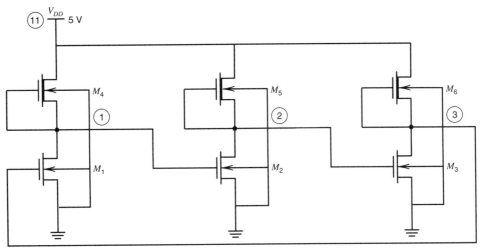

Figure 6.9 NMOS ring oscillator.

```
M4 11 1 1 0 DEP L=10U W=5U
M5 11 2 2 0 DEP L=10U W=5U
M6 11 3 3 0 DEP L=10U W=5U
*
.MODEL DEP NMOS LEVEL=1 VTO=-3 LAMBDA=.001 KP=.4E-4
.MODEL ENH NMOS LEVEL=1 VTO=1.8 CGSO=20N LAMBDA=.001 KP=.4E-4
*
.TRAN .01U .5U
.PLOT TRAN V(1) V(2) V(3) (0,5)
.END
```

If the circuit is analyzed as is, no oscillations are observed; the outputs of the three inverters, nodes 1, 2, and 3, settle at 2.5 V. An initial imbalance is necessary for oscillations to build up; this can be achieved by initializing the outputs of the inverters at high or low values, 5 V or 0 V, with an **.IC** line:

```
.IC V(1)=5 V(2)=0
```

Resimulation of the circuit including the above line produces the waveforms shown in the graph of Figure 6.10.

Note that the data in the **.IC** statement are used to compute the INITIAL TRANSIENT SOLUTION in the absence of the **UIC** parameter. The values in the initial solution are not always identical to the values in the **.IC** statement, because initial conditions are set up in SPICE by connecting a Thevenin equivalent with a voltage

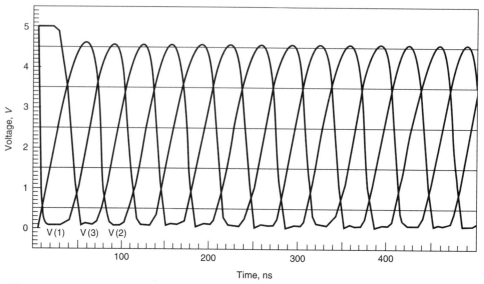

Figure 6.10   Waveforms at the outputs of the inverters in the ring oscillator.

equal to the initial value and a 1-Ω resistor to the initialized node. The Thevenin equivalent nets are removed only at the first time point in the transient analysis.

6.4   FOURIER ANALYSIS

A periodic signal can be decomposed into a number of sinusoidal components of frequencies that are multiples of the fundamental frequency. These components of the signal are also referred to as spectral or harmonic components. In other words, a periodic signal can be represented by a Fourier series (Nilsson 1990):

$$v(t) = \frac{1}{2}a_0 + \sum_{k=1}^{n}(a_k \cos k\omega t + b_k \sin k\omega t) \tag{6.29}$$

where $\frac{1}{2}a_0$ is the DC component and the coefficients a_k, b_k of the series are defined by

$$a_0 = \frac{2}{T}\int_{t}^{t+T} v(t)dt$$

$$a_k = \frac{2}{T}\int_{t}^{t+T} v(t)\cos(k\omega t)dt \tag{6.30}$$

$$b_k = \frac{2}{T}\int_{t}^{t+T} v(t)\sin(k\omega t)dt$$

The coefficients a_k and b_k give the magnitude of the signal of frequency $k\omega$, or the kth harmonic component. Only a single-tone sinusoid has a single spectral component, which is at the oscillation frequency.

In electrical engineering a different formulation, having only one periodic component, is used for the Fourier series:

$$v(t) = \frac{1}{2}a_0 + \sum_{k=1}^{n} A_k \cos(k\omega t - \phi_k) \tag{6.31}$$

where the amplitude, A_k, and the phase, ϕ_k, are given by

$$A_k = \sqrt{a_k^2 + b_k^2}$$

$$\phi_k = \arctan\frac{b_k}{a_k} \tag{6.32}$$

In the time-domain mode SPICE can compute the spectral components, magnitude and phase, of a given signal if the following line is present along with the **.TRAN** statement:

.FOUR   *freq OUT_var1* $<$*OUT_var2 . . .* $>$

In the above statement *freq* is the fundamental frequency and *OUT_var1, OUT_var2,* . . . are voltages and currents the spectral components of which are to be computed. SPICE2 and PSpice compute the first nine spectral components for each of the signals listed on the **.FOUR** line. SPICE3 allows the user to define the number of harmonics to be computed. Only one **.FOUR** line can be used during an analysis.

A few remarks are necessary about the accuracy of the Fourier analysis in SPICE. Because of the assumption of periodicity, the Fourier coefficients defined above are computed based on the values of *OUT_var* during the last period, that is for the interval (*TSTOP*-1/*freq, TSTOP*). For an accurate spectral analysis enough periods must be simulated that the circuit reaches the steady state. The Fourier coefficients defined in Eqs. 6.30 are evaluated based on the values for *OUT_var* computed at discrete time points; thus for good accuracy the maximum time step must be limited, using *TMAX* on the **.TRAN** line.

Example
```
.FOUR 1Meg V(3) I(VDD)
```

This line added to a SPICE deck causes the computation of the spectral components of the voltage at node 3 and of the current through the voltage source V_{DD}. The frequencies of the harmonics are multiples of the fundamental frequency of 1 MHz.

EXAMPLE 6.6

Verify the spectral values computed by SPICE2 for the output signal of the CMOS square-wave clock generator shown in Figure 6.11. The two transistors are described by the following model and device parameters:

NMOS: $VTO = 1$ V; $KP = 20$ $\mu A/V^2$; $CGSO = CGDO = 0.2$ nF/m; $CGBO = 2$ nF/m.

PMOS: $VTO = -1$ V; $KP = 10$ $\mu A/V^2$; $CGSO = CGDO = 0.2$ nF/m; $CGBO = 2$ nF/m.

M1: $W = 20$ μm; $L = 5$ μm.

M2: $W = 40$ μm; $L = 5$ μm.

A sinusoidal voltage source is applied at the input with a peak- to-peak amplitude of 5 V and a frequency of 20 MHz.

Solution
Because the CMOS inverter is nonlinear, the output signal contains harmonics of the 20-MHz input sinusoid. The following line requests the computation of the harmonics for the output signal V(2):

```
.FOUR 20MEG V(2)
```

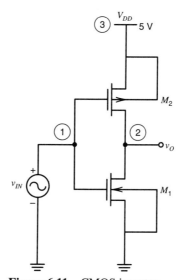

Figure 6.11   CMOS inverter.

The period of the output signal is 50 ns, and the analysis is requested for two periods, until 100 ns; simulation of two periods is sufficient for this circuit because no oscillations need to settle. The SPICE deck for this example is

```
CMOS INVERTER
*
M1 2 1 0 0 NMOS W=20U L=5U
M2 2 1 3 3 PMOS W=40U L=5U
VDD 3 0 5
VIN 1 0 SIN 2.5 2.5 20MEG
.MODEL NMOS NMOS LEVEL=1 VTO=1 KP=20U
+ CGDO=.2N CGSO=.2N CGBO=2N
.MODEL PMOS PMOS LEVEL=1 VTO=-1 KP=10U
+ CGDO=.2N CGSO=.2N CGBO=2N
.OP
.TRAN 1N 100N
.FOUR 20MEG V(2)
.PLOT TRAN V(2) V(1) (-1,5)
.PLOT TRAN I(VDD)
.END
```

The output waveform, V(2), is a square wave, as shown in Figure 6.12. Note that if the above deck were used the output voltage would display some ringing, which is due to numerical inaccuracy. This problem can be corrected by SPICE2 analysis option parameters, described in Secs 9.4.2 and 9.5. For this circuit it is necessary to add the line

```
.OPTION RELTOL = 1E-4
```

in order to obtain the waveform in Figure 6.12. The results of the Fourier analysis are listed in Figure 6.13. The DC component computed by SPICE2 is 2.5 V, the amplitude of the fundamental is 3.15 V, and all the even harmonics are negligible. The results of the Fourier analysis are listed according to the formulation in Eq. 6.31; the magnitudes of the spectral components, A_k, are listed under FOURIER COMPONENT, and the phases, ϕ_k, appear in the PHASE(DEG) column. In the Fourier analysis output two additional columns list the amplitudes of the spectral components normalized to the amplitude, A_1, of the fundamental, and the phases normalized to ϕ_1.

The Fourier series coefficients can be easily checked with Eqs. 6.29 and 6.30. The output signal, $v_o(t)$ (V(2)), shown in Figure 6.12 can be expressed as follows:

$$v_o(t) = 0 \quad \text{for } 0 \le t < \frac{T}{2}$$

$$v_o(t) = V_{DD} \quad \text{for } \frac{T}{2} \le t < T$$

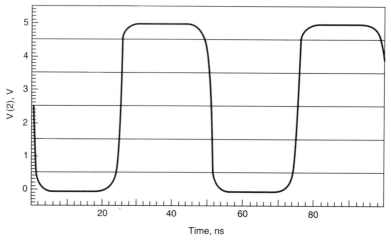

Figure 6.12 Square-wave signal $V(2)$ at the output of the CMOS inverter.

```
**** FOURIER ANALYSIS TEMPERATURE = 27.000 DEG C

FOURIER COMPONENTS OF TRANSIENT RESPONSE V(2)

DC COMPONENT =   2.503637E+00
```

HARMONIC NO	FREQUENCY (HZ)	FOURIER COMPONENT	NORMALIZED COMPONENT	PHASE (DEG)	NORMALIZED PHASE (DEG)
1	2.000E+07	3.147E+00	1.000E+00	1.794E+02	0.000E+00
2	4.000E+07	8.058E-03	2.561E-03	1.198E+02	- 5.962E+01
3	6.000E+07	9.612E-01	3.055E-01	1.791E+02	- 2.898E-01
4	8.000E+07	9.945E-03	3.160E-03	1.425E+02	- 3.694E+01
5	1.000E+08	4.951E-01	1.573E-01	1.791E+02	- 2.692E-01
6	1.200E+08	1.249E-02	3.970E-03	1.593E+02	- 2.009E+01
7	1.400E+08	2.946E-01	9.362E-02	1.799E+02	4.797E-01
8	1.600E+08	1.555E-02	4.941E-03	1.716E+02	- 7.748E+00
9	1.800E+08	1.907E-01	6.061E-02	-1.782E+02	- 3.576E+02

```
TOTAL HARMONIC DISTORTION =  3.613225E+01 PERCENT
```

Figure 6.13 Fourier analysis results for the square-wave voltage $V(2)$.

The DC component, A_0, is

$$A_0 = \frac{1}{2}a_0 = \frac{1}{T}\int_0^T v_o(t)dt = \frac{1}{2}5 \text{ V} = 2.5 \text{ V}$$

Before deriving the coefficients of the harmonics, note that the function is of odd symmetry:

$$f(t) = -f(-t)$$

Thus all coefficients a_k are zero. $v_o(t)$ also possesses half-wave symmetry; that is,

$$f(t) = -f(t - T/2)$$

making all coefficients b_k with even k equal to zero (Nilsson 1990). The first three harmonics, b_1, b_2, and b_3, are derived by solving the integral in Eqs. 6.30 for the two values of $v_o(t)$ corresponding to the half-periods of the waveform in Figure 6.12:

$$b_1 = \frac{2}{T}\int_0^T v_o(t)\sin(\omega t)dt = \frac{2}{T}V_{DD}\int_0^{T/2}\sin(\omega t)dt$$

$$= -\frac{2}{T}\cdot\frac{T}{2\pi}V_{DD}\cos\left(\frac{2\pi}{T}\right)t\ |_0^{T/2} = \frac{2\cdot 5}{\pi}\text{V} = 3.18 \text{ V}$$

$$b_2 = 0$$

$$b_3 = \frac{2V_{DD}}{3\pi} = 1.06 \text{ V}$$

The above coefficients scaled by the appropriate DC value are generally valid for any square wave. The small discrepancies with the SPICE2 Fourier coefficients can be attributed to the imperfection of the square wave V(2).

A useful application of the Fourier analysis is the evaluation of large-signal distortion. The TOTAL HARMONIC DISTORTION *(THD)* computed by SPICE is equal to

$$THD = \frac{\sqrt{A_2^2 + A_3^2 + \cdots + A_9^2}}{A_1} \cdot 100\% \qquad (6.33)$$

In the design of many circuits the *THD* must be kept below a specified limit. The second and third NORMALIZED COMPONENTS listed among the Fourier analysis results correspond to *HD2* and *HD3* in the AC small-signal distortion analysis presented in Chapter 5. If the results of the two analyses are compared, the Fourier components should be scaled by the reference power, P_{ref}, in the load resistor to match the values of *HD2* and *HD3*. More detail on the two types of distortion analysis can be found in Chapter 8.

Sinusoidal oscillators for various applications must have a small content of harmonics. It is instructive to compute the harmonic content in the output voltage of the Colpitts oscillator.

EXAMPLE 6.7

Use Fourier analysis to find the total harmonic distortion of the output signal of the Colpitts oscillator in Figure 6.4.

Solution

For an accurate estimate of the harmonics, the circuit needs to be simulated for more than the 10 periods used in Example 6.3. We will perform a transient analysis for 10 μs corresponding to 33 periods; the `.TRAN` line in the input file is replaced by the following line:

```
.TRAN 15N 10U 9.3U 15N
```

The waveform is saved for displaying only the last two periods, and limiting *TMAX* to 15 ns ensures that at least 20 time points are used in each period to evaluate the response.

The following statement defines the frequency of the fundamental and the output variable for which the spectral components are desired:

```
.FOUR 3.36MEG V(2)
```

The frequency of the fundamental must be specified as accurately as possible, because an error as small as 1% can make a difference in the values of the Fourier coefficients.

The output of the Fourier analysis from SPICE2 is listed in Figure 6.14. Note that the amplitude of the fundamental found by the Fourier analysis agrees with the value computed by hand in Example 6.3, Eq. 6.28. The *THD* of the sinusoidal signal produced is 8.25%.

A few comments are necessary regarding the implementation of Fourier analysis in SPICE3. Although the limit of only nine harmonics imposed by SPICE2 and most other SPICE versions is not a problem for most circuits, this limitation can become an impediment in finding the intermodulation (*IM*) terms for such circuits as mixers. In

```
****    FOURIER ANALYSIS    TEMPERATURE = 27.000 DEG C

FOURIER COMPONENTS OF TRANSIENT RESPONSE V(2)

DC COMPONENT =   1.000407E+001

HARMONIC  FREQUENCY    FOURIER   NORMALIZED     PHASE      NORMALIZED
   NO       (HZ)      COMPONENT  COMPONENT     (DEG)      PHASE (DEG)

    1     3.360E+006  2.523E+000  1.000E+000  -8.683E+001  0.000E+000
    2     6.720E+006  1.890E-001  7.490E-002  -1.545E+002 -6.769E+001
    3     1.008E+007  7.725E-002  3.062E-002  -1.471E+002 -6.027E+001
    4     1.344E+007  3.532E-002  1.400E-002  -1.374E+002 -5.054E+001
    5     1.680E+007  1.602E-002  6.420E-003  -1.257E+002 -3.886E+001
    6     2.016E+007  7.680E-003  3.044E-003  -1.111E+002 -2.426E+001
    7     2.352E+007  4.226E-003  1.675E-003  -9.404E+001 -7.206E+000
    8     2.688E+007  2.962E-003  1.174E-003  -7.896E+001  7.869E+000
    9     3.024E+007  2.516E-003  9.971E-004  -6.880E+001  1.803E+001

TOTAL HARMONIC DISTORTION =  8.245843E+000 PERCENT
```

Figure 6.14 Fourier analysis of the Colpitts oscillator.

SPICE3 the user can define the number of harmonics to be computed by issuing the following **set** command in the SPICE3 shell:

> spice3> **set nfreqs=**n

where **nfreqs** is the keyword and n is the desired number of harmonics. The default for n is 9.

Another variable that can be set by the user in SPICE3 is the degree of the polynomial used to interpolate the waveform. In order to request polynomial interpolation of higher degree, the following command must be issued at the SPICE3 shell prompt:

> spice3> **set polydegree=**n

where **polydegree** is the keyword and n is the degree.

6.5 SUMMARY

This chapter presented the analyses performed by SPICE in the time domain. The control statements for each analysis were introduced as well as the specifications of output variables and result-processing requests. Emphasis was placed on exemplifying the transient and steady-state responses of both a linear and a nonlinear circuit and comparing the manual derivation with SPICE simulations.

SPICE supports two analysis types in the time domain, transient and Fourier analysis, which are specified by the following control lines:

.TRAN *TSTEP TSTOP* <*TSTART* <*TMAX*>> <**UIC**>

.FOUR *freq OUT_var1* <*OUT_var2* ... >

The **.IC** (initial conditions) statement is a third control statement introduced in this chapter used for specifying the known node voltages at time $t = 0$:

.IC V(*node1*)=*value1* <**V**(*node2*)=*value2* ... >

Initial conditions can also be defined for individual elements; terminal voltages and initial currents can be used to initialize charge-storage and nonlinear elements. Element-based initial conditions are taken into account only in conjunction with the **UIC** (use initial conditions) option in the **.TRAN** statement. Table 6.1 summarizes the ways of setting initial conditions.

The waveforms of voltages and currents computed in a transient analysis must be saved by use of the **.PRINT** or **.PLOT** control statement, in tabular or line-printer-plot format, respectively. The general format of the output request that must accompany a **.TRAN** line is

.PRINT/PLOT TRAN *OUT_var1* <*OUT_var2* ... > <*plot_limits*>

The seven detailed examples in this chapter also highlighted the relation between large-signal time-domain analysis and small-signal AC analysis.

REFERENCES

Gray, P. R., and R. G. Meyer. 1993. *Analysis and Design of Analog Integrated Circuits,* 3d. ed. New York: John Wiley & Sons.

Meyer, R. G. 1979. Nonlinear integrated circuits. In *EE 240 Class Notes.* Berkeley: University of California.

Nilsson, J. W. 1990. *Electric Circuits,* 3d ed. Reading, MA: Addison-Wesley.

Pederson, D. O., and K. Mayaram. 1990. *Integrated Circuits for Communication.* Boston: Kluwer Academic Publishers.

Seven

FUNCTIONAL AND HIERARCHICAL SIMULATION

7.1 HIGH-LEVEL CIRCUIT DESCRIPTION

The example circuits presented in previous chapters use circuit elements, such as resistors, capacitors, and transistors, that have a one-to-one correspondence with components on electronic circuit boards or ICs. Such a description is generally referred to as a *structural* representation of the circuit. The simulation of a structural circuit produces very accurate results, but may take a long time. The analysis time grows proportionally to the number of components and is dominated by semiconductor elements, which are described by complex nonlinear equations. The analysis time sets a limit on the size of circuits that can be simulated at the structural level. Although circuits with several hundred to a few thousand components can be analyzed with SPICE on current PCs and engineering workstations, alternate ways of modeling circuits can increase design productivity.

The most common approach is to group several components in a block according to the function performed. According to this criterion, we can distinguish gain blocks, oscillators, integrators, differentiators, NAND and NOR blocks, adder blocks, and so on. Then, the SPICE description needs to be an equivalent circuit that achieves the same function as the component-level implementation. This functional model can be built with fewer components and with special SPICE elements, such as controlled sources. Simulation times for circuits with functional models are considerably shorter than those for detailed circuits.

SPICE provides a *subcircuit* capability, which allows a user to define a subnet or a block and then instantiate it repeatedly in the overall circuit. For example, the functional, or transistor-level, schematic of a NAND gate can be defined once and then instantiated repeatedly to form complex digital or mixed analog/digital circuits. This SPICE feature and its application for large circuits is described in Sec. 7.2.

When the SPICE input of large circuits is prepared, the netlist description can be very long and difficult to understand. A hierarchical approach to describing large circuits is recommended; with this approach a designer can quickly recognize the top-level block diagram of the circuit from the SPICE description. The subcircuit definition capability of the SPICE input language provides the means for hierarchical descriptions. An example of SPICE hierarchical definition is described in Sec. 7.2. In a hierarchical description various blocks can be described at different levels of accuracy.

The simplest representation of the function of a given block is an *ideal model*. Ideal functional blocks are introduced in Sec. 7.3 for both analog and digital circuits. Ideal blocks are very simple and result in short simulation times but may not provide sufficient accuracy or adequate SPICE convergence, as described in Chap. 10.

More complex models for SPICE simulation can be developed, which reproduce detailed characteristics of the circuit, such as limited output swing, finite bandwidth, and other range restrictions. These models combine SPICE primitives (Chap. 2) and arbitrary functions (Sec. 7.4.1) to form *functional models*. A few examples of functional models are presented in Sec. 7.4.

All details of the operation of circuit blocks or entire ICs can be built into SPICE primitives. The macro-model can incorporate all or a part of the first- and second-order effects of a circuit with a considerably smaller number of elements, resulting in significantly shorter simulation times. An operational amplifier macro-model commonly used by many suppliers of SPICE models for standard parts is described in Sec. 7.5.

7.2 SPICE SUBCIRCUIT AND CIRCUIT HIERARCHY

7.2.1 .SUBCKT Definition

A circuit block that appears more than once in the overall circuit and consists of SPICE primitives can be defined as a *subcircuit* (Vladimirescu, Zhang, Newton, Pederson, and Sangiovanni-Vincentelli 1981). The block can then be referenced as a single component, the *subcircuit instance,* and connected throughout the circuit. There is a similarity between the .SUBCKT definition and the .MODEL definition. Whereas a .MODEL statement defines a set of parameters to be collectively used by a number of devices, the .SUBCKT definition represents a circuit topology, which can be connected through its external pins or nodes anywhere in the circuit.

The elements that form the subcircuit block are preceded by the following control statement:

.SUBCKT *SUBname node1* <*node2* ...>

where *SUBname* uniquely identifies the subcircuit and *node1, node2, . . .* are the external nodes that can be connected to the external circuit. There is no limit to the number of external nodes. The rest of the nodes in the subcircuit definition are referred to as internal nodes. The internal nodes cannot be connected or referenced in the top-level circuit. The ground node, node 0, is global from the top circuit through all subcircuits.

The completion of the subcircuit definition is marked by the following line:

> **.ENDS** <*SUBname*>

Repetition of *SUBname* is not required, except for nested subcircuit definitions, but is recommended for the ease of checking the correctness of the circuit description.

In addition to SPICE elements, a number of control statements can be used within a subcircuit definition. Local **.MODEL** lines introduce models that can be referenced only by elements that belong to the subcircuit. Other **.SUBCKT** definitions can be nested inside a subcircuit; nested subcircuits can only be referenced from the subcircuit in which they are defined. No other control lines, that is, analysis, print/plot, or initialization requests, are allowed in a subcircuit definition. One difficulty created by this restriction is related to initializing node voltages. Although device initial conditions can be defined for elements inside a subcircuit, no **.NODESET** or **.IC** statement can be used to set the starting voltages on internal nodes. This limitation is overcome by declaring all nodes that require initialization as external nodes on the **.SUBCKT** line.

7.2.2 Subcircuit Instance

A subcircuit block is placed in the circuit by an **X**-*element* call, or subcircuit call, defined by the following line:

> **X***name xnode1* <*xnode2 . . .*> *SUBname*

The letter **X** must appear in the first column to identify a subcircuit instance; the number of nodes must be equal to those on the corresponding subcircuit definition, *SUBname*. *xnode1, xnode2, . . .*, are the numbers or names of the nodes that are to correspond with the nodes *node1, node2, . . .* of the **.SUBCKT** line, at the circuit level where *SUBname* is instantiated.

Subcircuit definition and calls can be exemplified by a hierarchical description of the three-stage ring oscillator in the previous chapter, Example 6.5. The new SPICE description, using inverters rather than the detailed schematic of the circuit as in Figure 6.9 is listed in Figure 7.1. The corresponding circuit diagram is shown in Figure 7.2.

7.2.3 Circuit Hierarchy

The SPICE subcircuit capability offers the designer the ability to describe a complex circuit in a hierarchical fashion. Any number of hierarchical levels can be defined. The

```
RING OSCILLATOR W/ MOS INVERTERS
*
X1 1 2 5 INVERTER
X2 2 3 5 INVERTER
X3 3 1 5 INVERTER
*
VDD 5 0 5
*
.SUBCKT INVERTER 1 2 3
* NODES: VIN, VOUT, VDD
M1 2 1 0 0 ENH L=10U W=40U
M2 3 2 2 0 DEP L=10U W=5U
*
.MODEL DEP NMOS LEVEL=1 VTO=-3 LAMBDA=.001 KP=.4E-4
.MODEL ENH NMOS LEVEL=1 VTO=1.8 CGSO=20N LAMBDA=.001 KP=.4E-4
*
.ENDS INVERTER
*
.IC V(1)=5 V(2)=0
.TRAN .01U .5U
.PLOT TRAN V(1) V(2) V(3) (0, 5)
.WIDTH OUT=80
.END
```

Figure 7.1 SPICE input for ring oscillator with MOS inverters using .SUBCKT.

hierarchical description of an adder built from NAND gates is presented in this section as an example of the proper application of the SPICE .SUBCKT statement.

EXAMPLE 7.1

Use subcircuits and hierarchy to create the SPICE input of the 4-bit adder built with TTL NAND gates that is shown in Figure 7.3 (Vladimirescu 1982). Partition the adder at the following levels: NAND gate, 1-bit adder, and 4-bit adder. Run SPICE to find the DC operating point of the 1-bit adder and interpret the results.

Solution
The first step is to write the SPICE netlist of the TTL NAND gate in Figure 7.3.*a*. This description is listed between the .SUBCKT NAND and .ENDS NAND lines. The external nodes, or terminals, of the NAND gate are the two inputs IN1 and IN2, the output, OUT, and the supply connection, VCC. These are the only pins needed for connecting a NAND gate in an external circuit and correspond to the pins available in a 7400-series TTL IC.

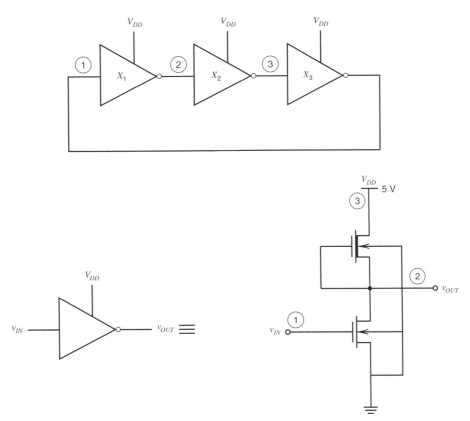

Figure 7.2   Ring oscillator with MOS inverters.

Next, a description of the 1-bit adder is created by specifying how the nine NAND gates in the schematic in Figure 7.3.*b* are connected. The NAND gates are instantiated in the 1-bit description using the **X** element. This new circuit is labeled as subcircuit ONEBIT and is used at the next level of the hierarchy to define the 4-bit adder. The external nodes of ONEBIT are the two inputs, A and B, the carry-in bit, CIN, the output, OUT, and carry-out bit, COUT, as well as the supply, VCC. When digital circuits are described in the following sections of this chapter, node names are uppercase, such as A and OUT, the voltages or analog signals at these nodes are indicated by an uppercase V, as in V_A and V_{OUT}, and the boolean (digital) variables associated with the terminals are lowercase, such as *a* and *out*.

Four ONEBIT subcircuits are connected according to Figure 7.3.*c* to form the 4-bit adder. Four instances (**X**) of ONEBIT are needed to define the FOURBIT subcircuit. The hierarchical SPICE definitions of the 4-bit and 1-bit adders and the NAND gate are listed in Figure 7.4. All the top-level input and output pins of the 4-bit adder are

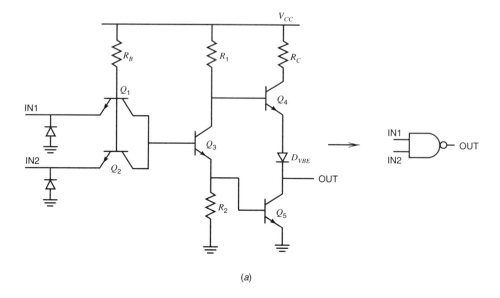

(a)

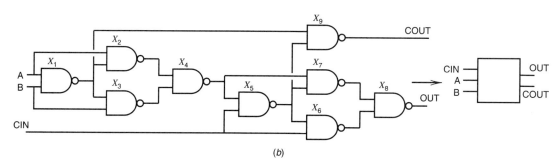

(b)

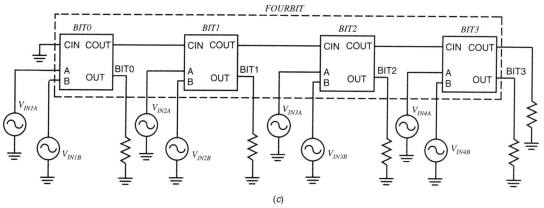

(c)

Figure 7.3   Hierarchy of 4-bit adder: (*a*) TTL NAND gate; (*b*) 1-bit adder with symbol; (*c*) 4-bit adder.

connected to signal sources or resistors, respectively. The 4-bit adder circuit defined in Figure 7.4 has three levels of hierarchy.

The DC operating point of the 1-bit adder can be found by running the SPICE deck shown in Figure 7.5 with the subcircuit definitions listed in Figure 7.4 for NAND and ONEBIT. The SPICE2 output in Figure 7.5 seems confusing at first. A long list of node

```
ADDER - 4 BIT ALL-NAND-GATE BINARY ADDER
*
.SUBCKT NAND 1 2 3 4
* NODES: IN1 IN2 OUT VCC
Q1 9 5 1 QMOD
D1CLAMP 0 1 DMOD
Q2 9 5 2 QMOD
D2CLAMP 0 2 DMOD
RB 4 5 4K
R1 4 6 1.6K
Q3 6 9 8 QMOD
R2 8 0 1K
RC 4 7 130
Q4 7 6 10 QMOD
DVBEDROP 10 3 DMOD
Q5 3 8 0 QMOD
.ENDS NAND
*
.SUBCKT ONEBIT 1 2 3 4 5 6
* NODES: A B CIN OUT COUT VCC
X1 1 2 7 6 NAND
X2 1 7 8 6 NAND
X3 2 7 9 6 NAND
X4 8 9 10 6 NAND
X5 3 10 11 6 NAND
X6 3 11 12 6 NAND
X7 10 11 13 6 NAND
X8 12 13 4 6 NAND
X9 11 7 5 6 NAND
.ENDS ONEBIT
*
.SUBCKT FOURBIT 1 2 3 4 5 6 7 8 9 10 11 12 13 14 15
* NODES: INPUT - BITO(2) / BIT1(2) / BIT2(2) / BIT3(2),
* OUTPUT - BIT0 / BIT1 / BIT2 / BIT3, CARRY-IN, CARRY-OUT, VCC
X1 1 2 13 9 16 15 ONEBIT
X2 3 4 16 10 17 15 ONEBIT
X3 5 6 17 11 18 15 ONEBIT
X4 7 8 18 12 14 15 ONEBIT
.ENDS FOURBIT
```

(continued on next page)

Figure 7.4 Hierarchical SPICE definition of a 4-bit adder.

```
***
*** DEFINE NOMINAL CIRCUIT
***
X1 1 2 3 4 5 6 7 8 9 10 11 12 0 13 99 FOURBIT
RBIT0 9 0 1K
RBIT1 10 0 1K
RBIT2 11 0 1K
RBIT3 12 0 1K
RCOUT 13 0 1K
VCC 99 0 DC 5V
VIN1A 1 O PULSE(0 3 0 10NS 10NS 10NS 50NS)
VIN1B 2 0 PULSE(0 3 0 10NS 10NS 20NS 100NS)
VIN2A 3 0 PULSE(0 3 0 10NS 10NS 40NS 200NS)
VIN2B 4 0 PULSE(0 3 0 10NS 10NS 80NS 400NS)
VIN3A 5 0 PULSE(0 3 0 10NS 10NS 160NS 800NS)
VIN3B 6 0 PULSE(0 3 0 10NS 10NS 320NS 1600NS)
VIN4A 7 0 PULSE(0 3 0 10NS 10NS 640NS 3200NS)
VIN4B 8 0 PULSE(0 3 0 10NS 10NS 1280NS 6400NS)
*
.MODEL DMOD D
.MODEL QMOD NPN(BF=75 RB=100 CJE=1PF CJC=3PF)
.OP
.OPT ACCT
.END
```

Figure 7.4   *(continued)*

```
ADDER - 1 BIT ALL-NAND-GATE BINARY ADDER

****     INPUT LISTING                 TEMPERATURE  =    27.000 DEG C

***********************************************************************

.SUBCKT NAND 1 2 3 4
*   NODES:  IN1, IN2, OUT, VCC
Q1 9 5 1 QMOD
D1CLAMP 0 1 DMOD
Q2 9 5 2 QMOD
D2CLAMP 0 2 DMOD
RB 4 5 4K
R1 4 6 1.6K
Q3 6 9 8 QMOD
R2 8 0 1K
RC 4 7 130
Q4 7 6 10 QMOD
DVBEDROP 10 3 DMOD
Q5 3 8 0 QMOD
.ENDS NAND
```

Figure 7.5   SPICE2 output for 1-bit adder.

```
.SUBCKT ONEBIT 1 2 3 4 5 6
*    NODES:  A, B, CIN, OUT, COUT, VCC
X1 1 2 7 6 NAND
X2 1 7 8 6 NAND
X3 2 7 9 6 NAND
X4 8 9 10 6 NAND
X5 3 10 11 6 NAND
X6 3 11 12 6 NAND
X7 10 11 13 6 NAND
X8 12 13 4 6 NAND
X9 11 7 5 6 NAND
.ENDS ONEBIT
***
***  DEFINE NOMINAL CIRCUIT
***
X1 1 2 9 0 13 99 ONEBIT
RBIT0 9 0 1K
RCOUT 13 0 1K
.MODEL DMOD D
.MODEL QMOD NPN(BF=75 RB=100 CJE=1PF CJC=3PF)
VCC 99 0 DC 5V
VIN1A 1 0 PULSE(0 3 0 10NS 10NS   10NS   50NS)
VIN1B 2 0 PULSE(0 3 0 10NS 10NS   20NS   100NS)
*
*
.OP
.OPT ACCT NODE
.WIDTH OUT=80
.END
******* 02/24/92 ********  SPICE  2G.6  3/15/83 ********  17:44:45  ****


ADDER - 1 BIT ALL-NAND-GATE BINARY ADDER

****      ELEMENT NODE TABLE            TEMPERATURE =    27.000 DEG C


***********************************************************************

     1    VIN1A    D1CLAMP* D1CLAMP* Q1.X1.X1 Q1.X2.X1
     2    VIN1B    D2CLAMP* D1CLAMP* Q2.X1.X1 Q1.X3.X1
     9    RBIT0    D1CLAMP* D1CLAMP* Q1.X5.X1 Q1.X6.X1
    13    RCOUT    DVBEDRO* Q5.X9.X1
    99    RB.X1.X1 R1.X1.X1 RC.X1.X1 RB.X2.X1 R1.X2.X1 RC.X2.X1 RB.X3.X1
          R1.X3.X1 RC.X3.X1 RB.X4.X1 R1.X4.X1 RC.X4.X1 RB.X5.X1 R1.X5.X1
          RC.X5.X1 RB.X6.X1 R1.X6.X1 RC.X6.X1 RB.X7.X1 R1.X7.X1 RC.X7.X1
          RB.X8.X1 R1.X8.X1 RC.X8.X1 RB.X9.X1 R1.X9.X1 RC.X9.X1 VCC
   100    DVBEDRO* D2CLAMP* D2CLAMP* D2CLAMP* Q5.X1.X1 Q2.X2.X1 Q2.X3.X1
          Q2.X9.X1
   101    DVBEDRO* D1CLAMP* Q5.X2.X1 Q1.X4.X1
   102    DVBEDRO* D2CLAMP* Q5.X3.X1 Q2.X4.X1
```

Figure 7.5 *(continued)*

```
103    DVBEDRO* D2CLAMP* D1CLAMP* Q5.X4.X1 Q2.X5.X1 Q1.X7.X1
104    DVBEDRO* D2CLAMP* D2CLAMP* D1CLAMP* Q5.X5.X1 Q2.X6.X1 Q2.X7.X1
       Q1.X9.X1
105    DVBEDRO* D1CLAMP* Q5.X6.X1 Q1.X8.X1
106    DVBEDRO* D2CLAMP* Q5.X7.X1 Q2.X8.X1
107    Q1.X1.X1 Q2.X1.X1 Q3.X1.X1
108    RB.X1.X1 Q1.X1.X1 Q2.X1.X1
109    R1.X1.X1 Q3.X1.X1 Q4.X1.X1
110    R2.X1.X1 Q3.X1.X1 Q5.X1.X1
111    RC.X1.X1 Q4.X1.X1
112    DVBEDRO* Q4.X1.X1
113    Q1.X2.X1 Q2.X2.X1 Q3.X2.X1
114    RB.X2.X1 Q1.X2.X1 Q2.X2.X1
115    R1.X2.X1 Q3.X2.X1 Q4.X2.X1
116    R2.X2.X1 Q3.X2.X1 Q5.X2.X1
117    RC.X2.X1 Q4.X2.X1
118    DVBEDRO* Q4.X2.X1
119    Q1.X3.X1 Q2.X3.X1 Q3.X3.X1
120    RB.X3.X1 Q1.X3.X1 Q2.X3.X1
121    R1.X3.X1 Q3.X3.X1 Q4.X3.X1
122    R2.X3.X1 Q3.X3.X1 Q5.X3.X1
123    RC.X3.X1 Q4.X3.X1
124    DVBEDRO* Q4.X3.X1
125    Q1.X4.X1 Q2.X4.X1 Q3.X4.X1
126    RB.X4.X1 Q1.X4.X1 Q2.X4.X1
127    R1.X4.X1 Q3.X4.X1 Q4.X4.X1
128    R2.X4.X1 Q3.X4.X1 Q5.X4.X1
129    RC.X4.X1 Q4.X4.X1
130    DVBEDRO* Q4.X4.X1
131    Q1.X5.X1 Q2.X5.X1 Q3.X5.X1
132    RB.X5.X1 Q1.X5.X1 Q2.X5.X1
133    R1.X5.X1 Q3.X5.X1 Q4.X5.X1
134    R2.X5.X1 Q3.X5.X1 Q5.X5.X1
135    RC.X5.X1 Q4.X5.X1
136    DVBEDRO* Q4.X5.X1
137    Q1.X6.X1 Q2.X6.X1 Q3.X6.X1
138    RB.X6.X1 Q1.X6.X1 Q2.X6.X1
139    R1.X6.X1 Q3.X6.X1 Q4.X6.X1
140    R2.X6.X1 Q3.X6.X1 Q5.X6.X1
141    RC.X6.X1 Q4.X6.X1
142    DVBEDRO* Q4.X6.X1
143    Q1.X7.X1 Q2.X7.X1 Q3.X7.X1
144    RB.X7.X1 Q1.X7.X1 Q2.X7.X1
145    R1.X7.X1 Q3.X7.X1 Q4.X7.X1
146    R2.X7.X1 Q3.X7.X1 Q5.X7.X1
147    RC.X7.X1 Q4.X7.X1
148    DVBEDRO* Q4.X7.X1
149    Q1.X8.X1 Q2.X8.X1 Q3.X8.X1
150    RB.X8.X1 Q1.X8.X1 Q2.X8.X1
```

Figure 7.5 *(continued)*

```
151     R1.X8.X1 Q3.X8.X1 Q4.X8.X1
152     R2.X8.X1 Q3.X8.X1 Q5.X8.X1
153     RC.X8.X1 Q4.X8.X1
154     DVBEDRO* Q4.X8.X1
155     Q1.X9.X1 Q2.X9.X1 Q3.X9.X1
156     RB.X9.X1 Q1.X9.X1 Q2.X9.X1
157     R1.X9.X1 Q3.X9.X1 Q4.X9.X1
158     R2.X9.X1 Q3.X9.X1 Q5.X9.X1
159     RC.X9.X1 Q4.X9.X1
160     DVBEDRO* Q4.X9.X1
```
******* 02/24/92 ******** SPICE 2G.6 3/15/83 ******** 17:44:45 ****

ADDER - 1 BIT ALL-NAND-GATE BINARY ADDER

****   SMALL SIGNAL BIAS SOLUTION   TEMPERATURE =   27.000 DEG C

****

NODE	VOLTAGE	NODE	VOLTAGE	NODE	VOLTAGE	NODE	VOLTAGE
(1)	0.0000	(2)	0.0000	(9)	0.7616	(13)	0.0178
(99)	5.0000	(100)	3.5278	(101)	3.6308	(102)	3.6308
(103)	0.0309	(104)	3.5339	(105)	3.4928	(106)	3.6303
(107)	0.0179	(108)	0.8268	(109)	4.9724	(110)	0.0000
(111)	4.8320	(112)	4.1899	(113)	0.0363	(114)	0.8448
(115)	4.9941	(116)	0.0000	(117)	4.9642	(118)	4.2529
(119)	0.0363	(120)	0.8448	(121)	4.9941	(122)	0.0000
(123)	4.9642	(124)	4.2529	(125)	1.9994	(126)	2.7685
(127)	1.1122	(128)	1.0745	(129)	4.9998	(130)	0.5122
(131)	0.0672	(132)	0.8752	(133)	4.9745	(134)	0.0000
(135)	4.8444	(136)	4.1940	(137)	0.7979	(138)	1.5919
(139)	4.8567	(140)	0.0870	(141)	4.9639	(142)	4.1152
(143)	0.0672	(144)	0.8752	(145)	4.9941	(146)	0.0000
(147)	4.9639	(148)	4.2526	(149)	1.9763	(150)	2.7459
(151)	1.0877	(152)	1.0500	(153)	4.9998	(154)	0.4845
(155)	1.9897	(156)	2.7590	(157)	1.1020	(158)	1.0642
(159)	4.9998	(160)	0.5005						

```
      VOLTAGE SOURCE CURRENTS

      NAME        CURRENT

      VCC        -1.876D-02
      VIN1A       2.075D-03
      VIN1B       2.075D-03

      TOTAL POWER DISSIPATION  9.38D-02  WATTS
```

Figure 7.5   *(continued)*

numbers appears in the SMALL-SIGNAL BIAS SOLUTION *(SSBS)* section. In order to understand the meaning of these newly created node numbers the user must request the **NODE** option on an **.OPTION** line:

```
.OPTION NODE
```

For more detail on SPICE options see Chap. 9, especially the summary in Sec. 9.5.

The ELEMENT NODE TABLE generated by SPICE2 is also part of the output listed in Figure 7.5. A practice that distinguishes the top-level circuit nodes from those generated by the program due to subcircuit expansion is to make the last node number in the top-level circuit easy to identify. In the adder example the last node number is 99; therefore all three-digit numbers are introduced due to subcircuit expansion and the meaning of each node can be derived from the ELEMENT NODE TABLE.

The name of an element connected at a node is formed by concatenating the names of the **X** calls at every level of hierarchy to the element name of a SPICE primitive appearing at the bottom of the hierarchy, in this case the NAND gate definition level. The composite names are limited to eight characters in SPICE2, making it difficult to trace the hierarchy path of elements having long names. The composite name starts with the component name followed by the subcircuit instance names that call it; the names at different levels of hierarchy are separated by periods. Component RB.X2.X1 is the resistor RB of the X2 NAND instance that is part of the 1-bit adder X1. PSpice uses the composite node names in the SSBS printout.

Once a hierarchy of a circuit is established, a designer can choose among several levels of detail and accuracy for each hierarchical block. Describing a block by the detailed structural schematic is always straightforward if the schematic is known, but it may not be economical. The operational amplifier, which is widely used in many designs and is considered a basic circuit element due to its availability in IC implementation can be described as an ideal, functional, macro-model or as a detailed model. Because the detailed design of an opamp is complex, including from 20 to 50 transistors, it is advisable to select the representation with only those characteristics that are relevant for a given design. This process is similar to using simpler or more complex transistor models by selectively specifying values of model parameters representing certain second-order effects.

The following sections present several approaches to defining SPICE models for more complex blocks, both analog and digital.

7.3   IDEAL MODELS

Ideal models are the simplest and computationally most efficient. As described in the following sections, for both analog and digital circuits an ideal model provides only the single most relevant function of a device; for example, an opamp is a gain block and a transistor is a switch. Although very efficient in simulation ideal models can cause problems in SPICE analyses due to the ideality.

7.3.1 Operational Amplifiers

The main characteristics of an opamp are very high gain, very high input resistance, and low output resistance. An ideal opamp can be reduced to a gain block with infinite input resistance and zero output resistance, as shown in Figure 7.6. The principle of the virtual short can be applied in the analysis of circuits with ideal opamps (Dorf 1989; Oldham and Schwartz 1987; Paul 1989; Sedra and Smith 1990). This assumption consists of

$$v_{id} = 0$$

$$i_{i+} = i_{i-} = 0 \tag{7.1}$$

A commonly used circuit configuration of the opamp is shown in Figure 7.7. The output voltage in the frequency domain is given by

$$\mathbf{V_o} = -\frac{\mathbf{Z_f}}{\mathbf{Z_i}}\mathbf{V_{id}} \tag{7.2}$$

This feedback connection is often used to implement integrators, differentiators, filters, and amplifiers. An ideal integrator can be built with the circuit in Figure 7.7 by replacing Z_f with a capacitor C_f and Z_i with a resistor R_i. The correct operation as an integrator can be checked with SPICE both in the time domain and the frequency domain.

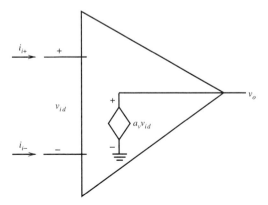

Figure 7.6 Ideal opamp.

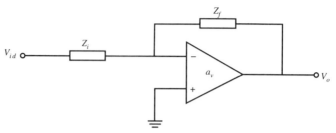

Figure 7.7   Opamp in feedback configuration.

EXAMPLE 7.2

Check the operation as a bandpass filter of the circuit shown in Figure 7.8. Use the ideal opamp description with $a_v = 10^5$ and the following values for the resistors and capacitors: $R_i = 100\ \Omega; C_i = 1\ \text{nF}; R_f = 10\ \text{k}\Omega; C_f = 1\ \text{nF}$.

Solution
The SPICE input is listed below. Note that the ideal opamp is defined as a subcircuit (**.SUBCKT**) that contains just a voltage-controlled voltage source, as shown in Figure 7.6.

```
BAND-PASS FILTER W/ IDEAL OPAMP
*
XOP1 0 1 2 OPAMP
RI 3 4 100
CI 1 3 1N
RF 1 2 10K
CF 1 2 1N
VID 4 0 AC 1
*
```

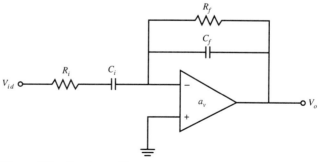

Figure 7.8   Bandpass filter.

```
.SUBCKT OPAMP 1 2 3
EGAIN 3 0 1 2 1E5
.ENDS
.AC DEC 10 10 1G
.PRINT AC VDB(2) VP(2)
.WIDTH OUT=80
.END
```

The transfer function of this active filter is obtained through substitution of the expressions for Z_i and Z_f in Eq. 7.2:

$$\mathbf{H}(j\omega) = \frac{\mathbf{V_o}}{\mathbf{V_{id}}} = -\frac{R_f}{R_i} \cdot \frac{j\omega}{R_f C_f \left(j\omega + \dfrac{1}{R_f C_f}\right)\left(j\omega + \dfrac{1}{R_i C_i}\right)} \tag{7.3}$$

In the pass-band, where $1/(R_f C_f) << \omega << 1/(R_i C_i)$, the transfer function becomes

$$|\mathbf{H}(j\omega)| = \frac{C_i}{C_f} \tag{7.4}$$

The limits of the pass-band for this active filter are defined by the two poles:

$$p_1 = -\frac{1}{R_f C_f} = -10^5 \text{ rad/s}$$

$$p_2 = -\frac{1}{R_i C_i} = -10^7 \text{ rad/s} \tag{7.5}$$

The magnitude and phase resulting from the SPICE frequency analysis are plotted in Figure 7.9. The Bode plot produced by SPICE is in agreement with the behavior predicted by Eqs. 7.3–7.5; that is, the pass-band extends from 15.9 kHz to 1.59 MHz.

The ideal opamp model presented above is a useful concept for instructional purposes and quick hand calculations. Its use in SPICE simulations should be limited due to the potential numerical problems that can be caused by the approximations involved.

A very important nonideality factor to consider in large-signal analyses is the supply voltage, which limits the excursion of the output signal, as shown by the transfer characteristic, v_o versus v_{id}, in Figure 7.10.a. The output voltage of the open-loop ideal opamp model, however, can rise to thousands of volts, possibly causing simulation problems depending on the circuit. The desired large-signal transfer characteristic can be achieved by adding a voltage limiter to the output stage.

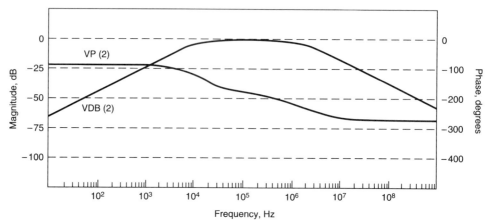

Figure 7.9   Magnitude and phase of V_o of bandpass filter.

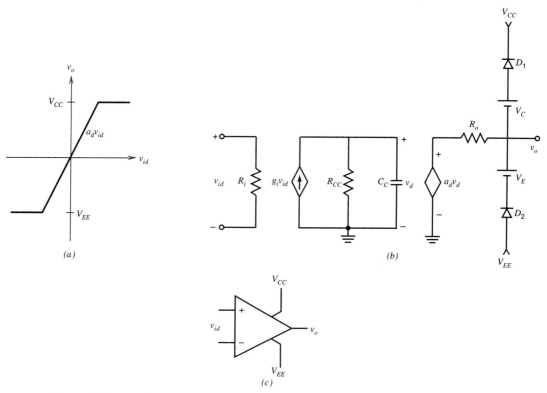

Figure 7.10   Nonideal operational amplifier: (a) v_o versus v_{id} transfer characteristic; (b) opamp model; (c) opamp symbol.

Two additional nonideality factors that should be added to the opamp model in order to avoid very high currents are input and output resistances, R_{in} and R_o.

The ideal opamp also assumes that the frequency bandwidth is infinite; in reality the bandwidth of the opamp is finite and is defined by an internal compensation capacitor.

A simple but nonideal opamp model that includes the above properties is presented in Figure 7.10.*b*, and the revised symbol is in Figure 7.10.*c*. An intermediate stage has been added to this model to include the single-pole roll-off of the frequency characteristic. The pole is defined by the intermediate stage at 10 Hz, which corresponds to a value $C_C = 1.59 \cdot 10^{-2}$ F and $R_C = 1\ \Omega$. Note that the simple nonideal opamp model has five terminals, which include the two power supplies, V_{CC} and V_{EE}. The corresponding SPICE subcircuit description using typical values is listed below.

```
.SUBCKT OPAMP 1 2 3 4 5
* TERMINALS: IN+ IN- OUT VCC VEE
RIN 1 2 1MEG
* GAIN STAGE
GI 0 6 1 2 1
RCC 6 0 1
CC 6 0 1.59E-2
* OUTPUT STAGE
EGAIN 7 0 6 0 1E5
RO 7 3 100
VC 8 3 0.7
VE 3 9 0.7
D1 8 4 DMOD
D2 5 9 DMOD
.MODEL DMOD D RS=1
.ENDS
```

7.3.2 Logic Gates and Digital Circuits

Many circuits simulated with SPICE have both analog and digital functions. If the analog blocks are required to be characterized with very high accuracy, the digital functions might not be critical to the performance of the circuit but might need to be included for verifying the correct operation of the overall circuit. In this case simplified versions of the logic blocks can be used. The simplest and computationally most efficient are the ideal models.

Ideal models of logic gates can be implemented with switches. Ideal logic gate models can be easily derived from the NMOS implementation (Hodges and Jackson 1983). The MOS transistor in this implementation acts as a voltage-controlled switch. The ideal models of the NAND, AND, NOR, OR, and XOR gates using voltage-controlled switches and positive logic are shown in Figure 7.11. These ideal models of logic gates can be used to simulate circuits in SPICE3 and PSpice but not in SPICE2.

Computationally efficient models for logic gates accepted by SPICE2 can be implemented as functional models using controlled sources. These models are presented in Sec. 7.4.

EXAMPLE 7.3

Derive the most efficient implementation of the 1-bit adder (Figure 7.3.*b*) using the ideal gates shown in Figure 7.11. Compute the response to signals *a*, *b*, and c_i in Figure 7.12. using SPICE; compare the results and the run time with those obtained with the detailed model of Figure 7.3.*b*, the SPICE input of which is listed in Fig. 7.5.

Solution
The Boolean function, *s*, implemented by the 1-bit adder with carry-in c_i and carry-out c_o, is

$$s = a + b + c_i$$
$$c_o = (a + b)c_i + a \cdot b \tag{7.6}$$

The logic diagram of the adder is drawn in Figure 7.13 and the corresponding SPICE3 input is given in Figure 7.14. The computed waveforms *s* and c_o for both the ideal and the detailed implementations (see listing in Figure 7.5) are displayed with the input signals on the same graph in Figure 7.12, for convenient verification of correct operation.

The analysis time for the ideal circuit in SPICE3 on a SUN 4/110 workstation is 1.2 s (no hysteresis), whereas that of the all-NAND implementation is 146 s. In PSpice on an IBM PS/2-80 with a 386 processor at 16 MHz the run times 37.9 s and

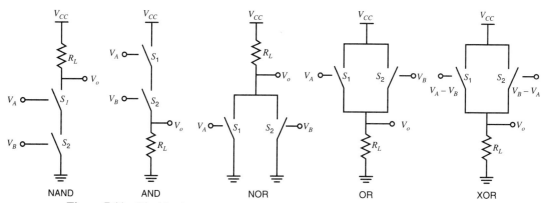

Figure 7.11　Ideal logic gates.

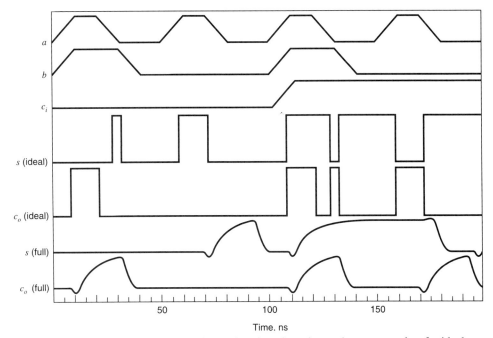

Figure 7.12   1-bit adder simulation: input signals *a, b,* and c_i and output *s* and c_o for ideal and full models.

617 s, respectively. Note that the waveforms obtained with the two representations are different; the ideal model predicts more changes in the output signals, *s* and c_o, than the detailed circuit. The difference between the operations of the two circuits can be traced to the instantaneous switching of the ideal version after each change in input, compared to the delayed response of the transistor-level version: the ideal circuit lacks of charge

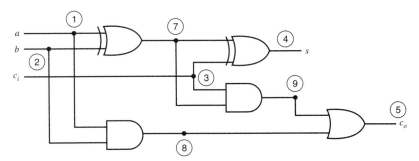

Figure 7.13   Logic diagram of the 1-bit adder.

```
1-BIT ADDER WITH SWITCHES
*
.SUBCKT OR 1 2 3 4
* TERMINALS A B OUT VCC
RL 3 0 1K
S1 3 4 1 0 SW
S2 3 4 2 0 SW
.ENDS OR
*
.SUBCKT AND 1 2 3 4
* TERMINALS A B OUT VCC
RL 3 0 1K
S1 4 5 1 0 SW
S2 5 3 2 0 SW
.ENDS AND
*
.SUBCKT XOR 1 2 3 4
* TERMINALS A B OUT VCC
RL 3 0 1K
S1 4 3 1 2 SW
S2 4 3 2 1 SW
.ENDS XOR
*
.SUBCKT ONEBIT 1 2 3 4 5 6
* TERMINALS: A B CIN OUT COUT VCC
X1 1 2 7 6 XOR
X2 1 2 8 6 AND
X3 7 3 4 6 XOR
X4 3 7 9 6 AND
X5 8 9 5 6 OR
.ENDS ONEBIT
*
* MAIN CIRCUIT
*
X1 1 2 3 9 13 99 ONEBIT
RINA 1 0 1K
RINB 2 0 1K
RCIN 3 0 1K
RBIT0 9 0 1K
RCOUT 13 0 1K
VCC 99 0 5
VINA 1 0 PULSE 0 3 0 10N 10N 10N 50N
VINB 2 0 PULSE 0 3 0 10N 10N 20N 100N
VCIN 3 0 PULSE 0 3 100N 10N 10N 100N 200N
*
.MODEL SW VSWITCH RON=1 ROFF=1MEG VON=2.5 VOFF=2.5
*
.TRAN 2N 200N
.END
```

Figure 7.14   SPICE3 description of 1-Bit adder.

storage. Besides predicting incorrect behavior, ideal circuit models can also lead to an analysis failure in SPICE.

This section has introduced the concept of ideal elements. With the hierarchical specification features of SPICE input language, it is possible to have one description of the top-level circuit and use different levels of detail in the component building blocks or subcircuits. This approach was exemplified above for a full adder implemented with ideal gates rather than the all-NAND TTL transistor-level implementation used in Example 7.1. The savings in simulation time are very important. On the other hand, ideal models can be inappropriate for some analyses, as pointed out by the shortcomings of the ideal opamp.

7.4 FUNCTIONAL MODELS

Several ideal subcircuit definitions were introduced in the previous section that perform the main function of the circuit blocks they model in a SPICE analysis. These subcircuits are built with just a few SPICE primitives, such as controlled sources, the equivalent of a gain block, switches, and other components. The circuit structure, or topology, of these subcircuits is similar to that of the transistor-level implementation.

This section describes how the function of a circuit block can be modeled using controlled sources that are nonlinear or of arbitrary functional form. In a departure from the above circuit models, which are structural, functional models achieve the same operation as the blocks they represent with just a few controlled sources connected in a network having no resemblance to the original one.

In the following subsections a few examples of functional models are provided that are developed based on the different capabilities of the controlled sources available in the versions of SPICE under consideration.

7.4.1 Nonlinear (Arbitrary-Function) Controlled Sources in SPICE3

A single type of nonlinear controlled source, which covers all possible cases, is supported in SPICE3. Linear sources are treated in the same way as in SPICE2. The general format for nonlinear controlled sources is

> B*name node1 node2* **V/I**=*expr*

B in the first column identifies the controlled source as nonlinear, and *name* can be an arbitrary long string, as can any element name in SPICE3. The controlled source is connected between nodes *node1* and *node2* and is either a voltage or a current source depending on whether the character at the left of the equal sign is **V** or **I**, respectively.

The variable *expr* is an arbitrary function of the node voltages and currents in the circuit. All common elementary functions can be used in *expr*: transcendental, exp, `log`, `ln`, and `sqrt`; trigonometric, `sin`, `cos`, `tan`, `asin`, `acos`, and `atan`; and hyperbolic, `sinh`, `cosh`, `tanh`, `asinh`, `acosh`, and `atanh`. The nonlinear function applies only to the time domain; in the frequency domain the source assumes a constant voltage or current value equal to the small-signal value in the DC operating point (see Sec. 7.4.2, Eqs. 7.13 to 7.15). PSpice supports, in addition to polynomial sources, nonlinear controlled sources equivalent to those in SPICE3 and identified as **G** or **E** sources depending on whether the output is current or voltage, respectively. These sources are described in the "Analog Behavioral Modeling" section in the PSpice manual (Microsim Corporation 1991).

Controlled sources described by arbitrary functions enable the user to define functional or behavioral models. It is important to understand that the SPICE algorithms require nonlinear functions to be continuous and bound. Rapidly growing functions, such as $\exp(x)$ and $\tan(x)$, can cause convergence problems, and discontinuous functions, such as $1/x$, can cause an arithmetic exception. A sound approach is to check that the variables of the functions are in a safe range.

The following example describes how a voltage-controlled oscillator can be modeled in SPICE3 using **B**-type controlled sources.

EXAMPLE 7.4

Define a functional model of a voltage-controlled oscillator (VCO) in SPICE3 using **B** sources.

Solution
A VCO generates a signal whose frequency, ω_{osc}, is controlled by an input voltage, v_C:

$$\omega_{osc}(t) = \omega_0 + K v_C(t) = \omega_0 \left(1 + K' v_C(t)\right) \tag{7.7}$$

where K is the VCO gain in rad $\cdot$ V^{-1}s^{-1} and ω_0 is the signal frequency of the free-running oscillator when the controlling voltage is zero.

The function for the **B** source must express the instantaneous value at any time point; therefore, for a VCO this function is

$$v_{osc} = V_{osc} \sin \theta_{osc} \tag{7.8}$$

where

$$\theta_{osc} = \int_0^t \omega_{osc}\, dt = \omega_0 t + K \int_0^t v_C\, dt = \omega_0 \int_0^t \left(1 + K' v_C\right) dt \tag{7.9}$$

Assume that the controlling voltage, v_C, ramps up from 0 V to 1 V in the first half of the time interval and then decreases back to 0 V in the second half of the interval. v_C is implemented as an independent PWL source whose value must be integrated to obtain the second term of Eq. 7.9. The SPICE3 **B**-source expression does not allow explicit *time* dependence; a PWL voltage source, which varies linearly with time, can be used.

Under the assumption that $f_0 = 10 \, \text{kHz}$ and $K' = 1 \, \text{V}^{-1}$, the SPICE3 input circuit is listed below:

```
VCO FUNCTIONAL MODEL FOR SPICE3
*
* CONTROL VOLTAGE
*
VC 1 0 PWL 0 0 0.5 1 1 0
RIN 1 0 1
*
* V(2) IS INTEGRAL IN EQ. 7.9
*
BINT 0 2 I=1+V(1)
CINT 2 0 1
*
BVCO 3 0 V=5*SIN(2*PI*10*V(2))
ROUT 3 0 1
*
.PLOT TRAN V(3) V(2)
.TRAN 0.01 1 0 2M UIC
.END
```

Note that $V(2)$ in the expression of BVCO equals the integral of the current through capacitor CINT and represents the time integral of $(1 + K'v_C)$ (see Eq. 7.9). The model introduced in this example for a VCO is known as a *functional*, or *behavioral*, model. The waveforms of the controlling voltage, $v_C = V(1)$, and the VCO output, $v_{osc} = V(3)$, are shown in Figure 7.15.

7.4.2   Analog Function Blocks

The circuits presented in these sections implement the desired function with the polynomial controlled source (introduced in Chap. 2) in SPICE2 and the arbitrary-function controlled source in SPICE3 and PSpice.

The nonlinear controlled source can be used to implement a number of arithmetic functions, such as addition, subtraction, multiplication, and division (Epler 1987). Note that the results of these operations are analog, rather than digital, or binary, as was the case in Examples 7.1 and 7.3. Addition and multiplication can be achieved with the following polynomial, VCVS:

```
EADD 3 0 POLY(2) 1 0 2 0 0 1 1
EMULT 3 0 POLY(2) 1 0 2 0 0 0 0 1
```

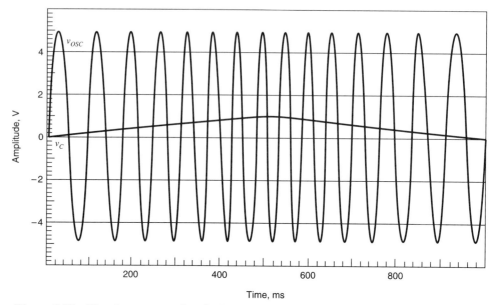

Figure 7.15   Waveforms v_{osc} and v_C for VCO functional model.

The functions implemented by the two elements are

$$V_3 = V_1 + V_2 \tag{7.10}$$
$$V_3 = V_1 V_2 \tag{7.11}$$

where V_1, V_2, and V_3 are the voltages at nodes 1, 2 and 3, referenced to ground, which is node 0. For details on the **POLY** coefficient specification see Sec. 2.3.2.

A divider is slightly more complex and requires two poly sources, as shown in Figure 7.16. The SPICE specification of the two VCCSs needed is

```
GV1 3 0 1 0 1
GV2V3 0 3 POLY(2) 2 0 3 0 0 0 0 0 1
```

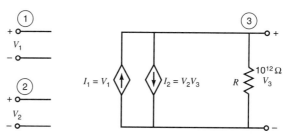

Figure 7.16   Divider circuit.

The resulting output voltage, V_3, is the quotient V_1/V_2 because of the equality of the currents of the two VCCSs imposed by the KCL:

$$I_1 = I_2$$
$$V_1 = V_2 V_3 \tag{7.12}$$

Note that the large output resistor is needed to satisfy the SPICE topology checker, which would otherwise flag the two controlled current sources in series as an error.

An important observation regarding the use of nonlinear controlled sources is that the desired function, such as multiplication or division, is performed only in a DC or time-domain large-signal analysis but not in a small-signal AC analysis. The cause of this is the small-signal nature of AC analysis, which uses the differential of a nonlinear function in the DC operating point and not the function itself. The value of the function expressed by a nonlinear controlled source $f(x + \Delta x)$ when a small signal Δx is added to the quiescent value of the controlling signal x is

$$f(x + \Delta x) = f(x) + \frac{df}{dx}\Delta x = f + \Delta f \tag{7.13}$$

Therefore, the controlled source function Δf in small-signal analysis is a linear function given by

$$\Delta f = \frac{df}{dx}\Delta x$$

The derivative of f is computed in DC bias-point analysis in a way similar to the evaluation of small-signal conductances of semiconductor devices (see Secs. 3.2, 3.3.2, 3.4, and 3.5.2). A controlled source with two arguments, x_1 and x_2, is evaluated in AC analysis according to

$$\Delta f = \frac{\partial f}{\partial x_1}\Delta x_1 + \frac{\partial f}{\partial x_2}\Delta x_2 \tag{7.14}$$

The addition and multiplication functions implemented with polynomial controlled sources are evaluated according to the following equalities in AC analysis:

$$f_S = x_1 + x_2, \qquad \Delta f_S = \Delta x_1 + \Delta x_2 \tag{7.15}$$
$$f_M = x_1 x_2, \qquad \Delta f_M = x_2 \Delta x_1 + x_1 \Delta x_2$$

Another useful function that can be implemented with polynomial sources is a meter for the instantaneous and cumulative power over the time interval of transient analysis. The setup is shown in Figure 7.17 and uses a CCCS to transform the current of the bias source, V_{DD}, into a voltage with the correct sign and a VCVS to obtain the product $-i_{DD}V_{DD}$. The voltage v_3 represents the instantaneous power, and v_4 is the cumulative

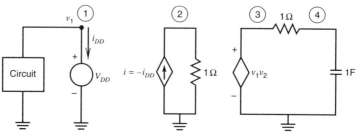

Figure 7.17   Power-measuring circuit.

power from 0 to the current time:

$$v_4 = \int_0^t V_{DD} i_{DD} d\tau \tag{7.16}$$

The average power at any time point can be obtained by dividing v_4 as given in Eq. 7.16, by the current time, which can be measured by a PWL voltage source whose value increases proportionally with the analysis time.

The ease of defining functional models is greatly enhanced by the arbitrary controlled sources available in SPICE3 and PSpice. A good example is the limiter circuit that was used in the previous section at the output of a nonideal opamp block. The output characteristic shown in Figure 7.18 is similar to that of a differential amplifier (Gray and Meyer 1993) and is expressed by

$$v_o = V_{CC} \tanh \frac{a_v v_{id}}{V_{CC}} \tag{7.17}$$

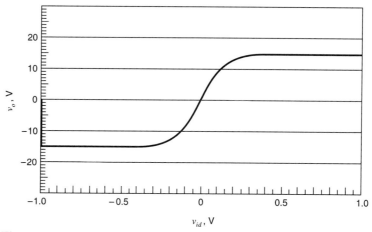

Figure 7.18   Output of tanh x limiter functional block.

In general the function $\tanh x$ is ideally suited to implementing a limiter; for small values of the argument it is approximately equal to the argument, according to the series expansion:

$$\tanh x = x - \frac{1}{3}x^3 + \frac{2}{15}x^5 - \cdots \tag{7.18}$$

Its absolute value is limited at 1 for large values of the argument. The limiter built with the arbitrary controlled source defined by Eq. 7.17 allows the output voltage v_o to follow the input voltage v_{id} amplified by a_v as long as $a_v v_{id}$ is smaller than the supply voltage and then limits v_o to the supply when $a_v v_{id}$ surpasses the supply voltage. The functional description of a differential amplifier built with an emitter-coupled pair (Gray and Meyer 1993) is as follows:

$$v_o = \alpha R_C I_{EE} \tanh\left(-\frac{v_{id}}{2V_{th}}\right) \tag{7.19}$$

where I_{EE} is the sum of the emitter currents, α is the CB gain factor, and R_C is the collector resistance, which is equal for the two input transistors of a differential pair.

A functional limiter can be built also for SPICE2, but it takes more components (Mateescu 1991). The behavior of $\arctan x$ is similar to that of $\tanh x$, assuming values between 0 and π when x varies from $-\infty$ to ∞, and it can be represented by conventional polynomial sources, because the derivative of the arctangent is

$$\frac{d}{dt}\arctan u = \frac{u'}{1+u^2} \tag{7.20}$$

Therefore the limiter can be built by implementing the right-hand side of Eq. 7.20 and then integrating it to obtain

$$v_o = \frac{2V_{CC}}{\pi}\left(\int_0^t \frac{u'}{1+u^2}d\tau - \frac{\pi}{2}\right) \tag{7.21}$$

where

$$u = \frac{\pi}{2}\cdot\frac{a_v v_{id}}{V_{CC}}$$

The functional limiter for SPICE2 is shown in Figure 7.19. The behavior of $\arctan x$ is identical to that of $\tanh x$, that is, for small values of the argument it can be approximated by the value of the argument, and for large values of the argument it is limited to a constant value.

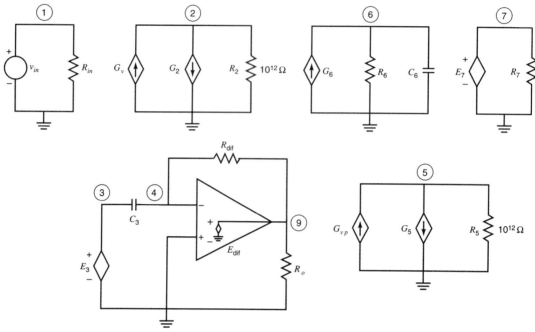

Figure 7.19 Circuit diagram of arctan x limiter.

The voltages at nodes 2, 5, 6, 7, and 9 are given in the following equations.

$$v_2 = u = \frac{\pi}{2} \cdot \frac{a_v v_{id}}{V_{CC}}$$

$$v_9 = u' = \frac{\pi}{2} \cdot \frac{a_v}{V_{CC}} \cdot \frac{dv_{id}}{dt}$$

$$v_5 = \frac{u'}{1 + u^2}$$

$$v_6 = \int_0^t \frac{u'}{1 + u^2} d\tau$$

$$v_7 = \frac{2V_{CC}}{\pi} \left(\int_0^t \frac{u'}{1 + u^2} d\tau - \frac{\pi}{2} \right)$$

Each section in Figure 7.19 performs a specific function, such as division, differentiation, or integration, toward the realization of Eq. 7.21. The differentiator is built with an ideal opamp. Each section is also identified by comments in the SPICE2 input listed in Figure 7.20. The voltages representing the values of the integral, its arguments, and the limiting output function are plotted in Figure 7.21.

Functional models must be used with care to avoid arithmetic exceptions such as division by zero. Another important point is to initialize differentiators and integra-

```
ARCTAN(X) LIMITER
*
VIN 1 0 PWL 0 -1M 2 1M
RIN 1 0 1MEG
*
* COMPUTE THE ARGUMENT U, EQ. 7.21
*
GV 0 2 1 0 1E5

G2 2 0 2 0 9.54
R2 2 0 1E12
*
* DIFFERENTIATION CKT TO COMPUTE U'
*
E3 3 0 2 0 1
C3 3 4 1 IC=-10.48
R3 4 0 1E12
EDIF 9 0 0 4 1E3
RDIF 4 9 1
R0 9 0 1E12
*
* COMPUTE U'/(1+U^2)
*
GVP 0 5 9 0 1
G5 5 0 POLY(2) 2 0 5 0 0 0 1 0 0 0 0 1
R5 5 0 1E12
*
* INTEGRATION CKT, COMPUTES INTEGRAL OF V(5)
*
G6 0 6 5 0 1
R6 6 0 1E12
C6 6 0 1
*
* LEVEL-SHIFT AND SCALE
*
E7 7 0 POLY(1) 6 0 -15 9.54
R7 7 0 1
*
.TRAN .01 2 0 UIC
.PRINT TRAN V(5) V(6) V(7)
.END
```

Figure 7.20 SPICE deck for arctan x limiter.

tors in the expected initial states in order to avoid convergence failures. The capacitor of the differentiator, C_3, used in the arctan x function is initialized with the appropriate voltage at $t = 0$ in order to prevent a convergence failure due to a large voltage at the input of the ideal opamp.

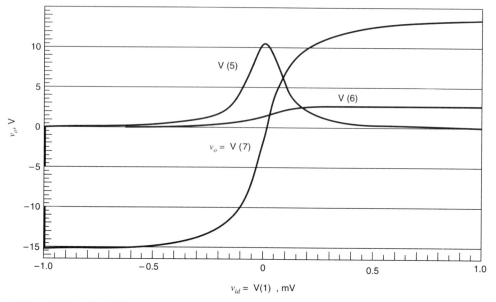

Figure 7.21   Simulation results of SPICE2 limiter.

So far we have defined a number of functional blocks, such as the adder, the multiplier, the integrator, and the differentiator; these can be saved in a library and then used in any circuit where such blocks are needed.

A functional block that proves very useful in many simulations is a frequency-domain transfer function. Some proprietary SPICE simulators, such as PSpice and SpicePLUS (Valid Logic Systems 1991), support a frequency-domain transfer function. With this capability a variety of filters can be described by the locations of poles and zeros.

With an arbitrary frequency-domain transfer function block one could define the parallel RLC circuit of Example 6.2 by the following transfer function:

$$\frac{V_o}{I_i} = \frac{Ls}{LCs^2 + Ls/R + 1} \tag{7.22}$$

The PSpice input specification for this block is

```
.PARAM L=1MH C=1NF R=10K
EFILTR 1 0 LAPLACE {I(VIN)}={L*s/(L*C*s*s+L/R*s+1)}
```

This block can replace the RLC lumped-element representation in Example 6.2. Both a frequency- and a time-domain analysis can be performed with this block.

Exercise

Show that the time-domain response of the above block to a current step function cal-culated by PSpice is identical to the response of the lumped RLC circuit in Figure 6.2.

7.4.3 Digital Function Blocks

In Sec. 7.3, on ideal building blocks, digital gates were implemented with simple cir-cuits that are structurally similar to actual implementations but use ideal elements, such as switches. All digital gates can be described at a functional level with polynomial sources (Sitkowski 1990). The elements presented below implement the analog opera-tion of digital circuits, that is, logic levels 0 and 1 are expressed in voltages correspond-ing to positive logic.

The structure of a universal gate is shown in Figure 7.22. The supply and input resistors are chosen to model the behavior of a TTL gate. The operation of this model is based on the correspondence of the logical AND operation to multiplication. The AND function is implemented by the VCVS EAND, which contains only the term $V_A V_B$:

```
EAND OUT 0 POLY(2) A 0 B 0 0 0 0 0 0 0.2
```

Assuming that V_A and V_B take values between 0 and 5 V, the product must be scaled by 0.2 so that the output voltage V_{OUT} is 5 V when both inputs are at a logical 1. In order to model the OR function, DeMorgan's theorem (Mano 1976) is applied to transform the OR function into an AND function, which can be implemented as above:

$$a + b = \overline{\overline{a} \cdot \overline{b}} \tag{7.23}$$

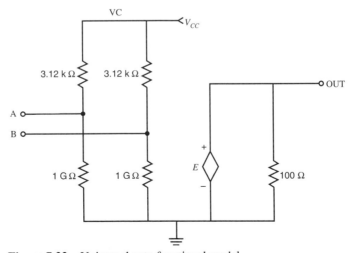

Figure 7.22   Universal gate functional model.

The inversion of the input signals is achieved by referencing the two input signals, V_A and V_B, to V_{CC} rather than to ground; the inversion of the output function, the NAND function, is achieved by inverting the polarity of the controlled source and subtracting 5 V from the result. According to Eq. 7.23, the OR function becomes

```
EOR  0 OUT POLY(2) A VC B VC -5 0 0 0 0.2
```

The NAND function needs only to have the output inverted, and the NOR function needs only the inputs inverted, with respect to the AND function, according to Eq. 7.23:

```
ENAND 0 OUT POLY(2) A 0 B 0 -5 0 0 0 0.2
ENNOR OUT 0 POLY(2) A VC B VC 0 0 0 0 0.2
```

The XOR and XNOR functions are implemented according to the definition

$$a \oplus b = \bar{a}b + a\bar{b} \tag{7.24}$$

These functions need four variables, $a, \bar{a}, b,$ and $\bar{b}$, and therefore a four-dimensional polynomial is used. Each variable and its inverse introduce two controlling voltages; that is, input a is V_A and input $\bar{a}$ is $V_{CC} - V_A$. The SPICE definitions of the VCVS for the XOR and XNOR functions are, respectively,

```
EXOR  OUT 0 POLY(4) VC A B 0 A 0 VC B 0 0 0 0 0 0 0.2 0 0 0 0 0 0 0.2
EXNOR 0 OUT POLY(4) VC A B 0 A 0 VC B -5 0 0 0 0 0 0.2 0 0 0 0 0 0 0.2
```

The order of the coefficients follows the rule introduced in Sec. 2.3.2. The only terms of the polynomial used in both functions are the product of variables 1 and 2, $\bar{a}b$, and the product of variables 3 and 4, $a\bar{b}$, leading to the following expression of the XOR function

$$V_{OUT} = 0.2(V_{CC} - V_A)V_B + 0.2(V_{CC} - V_B)V_A$$

The two terms require the definition of coefficients p_6 and p_{13} for both XOR and XNOR. The XNOR function is implemented by switching the polarity of the VCVS. Note that for SPICE2 the node names used above must be replaced by numbers.

7.4.4 Equation Solution

The operation of electrical systems as well as nonelectric systems, such as mechanical, hydraulic, and electro-mechanical systems, are described by complex nonlinear differential equations. In order to simulate these devices with SPICE it is necessary to create a model or, equivalently, to cast the component equation in a form that the program can understand and solve. The functional models defined above can be combined in creative ways to solve most complex equations describing not only electrical but also other physical systems.

EXAMPLE 7.5

With the functional blocks defined above develop a circuit that can be used in SPICE to model the turn-on I-V characteristic of a fluorescent discharge lamp. The relationship between the current, I, and the voltage, V, of the lamp is defined by the following equation:

$$\frac{dG}{dt} = aI^2 + bIV + cV^2 + d\left(\frac{I}{V}\right)^3 + e\left(\frac{I}{V}\right)^2 + f\frac{I}{V} \tag{7.25}$$

where G is the conductance of the lamp. Create a model for both an arbitrary controlled source as well as a polynomial controlled source implementation. After designing the complete model defined by Eq. 7.25, simulate the turn-on characteristic for a lamp with the following parameters:

$$a = 2.422, \qquad e = 5.19 \cdot 10^4, \qquad f = 271.7$$

The rest of parameters, b, c, and d, equal zero.

Solution

The main part of the model is the defining equation of the conductance G, Eq. 7.25; it can be implemented as a nonlinear controlled voltage source, `BRHS`, which sums the terms on the right-hand side of Eq. 7.25. The lamp is represented by a nonlinear controlled current source, `BLAMP`, the current of which is

$$I = G(V, I) \cdot V$$

where $G(V, I)$ is the voltage at node G1, equal to the integral of the right-hand side of Eq. 7.25. The circuit is shown in Figure 7.23. The corresponding SPICE3 description is shown in Figure 7.24.

  A SPICE2 model can be developed using functional blocks, such as multipliers and dividers, which implement the terms of the right-hand side of Eq. 7.25, which are then added. The result is integrated to obtain $V(G1)$. Note that for PSpice the definitions of G and of the arbitrary controlled sources `BRHS` and `BLAMP` need to be changed to the following syntax:

```
ERHS 6 0 VALUE={2.422*PWR(I(VA),2)-5.186E4*PWR((I(VA)/V(2)),2)
+  +271.7*(I(VA)/V(2))}
GLAMP 4 5 VALUE={ V(G1)*V(IN) }
```

  The I-V characteristic of the lamp obtained from SPICE3 is plotted in Figure 7.25 and corresponds to the operation known from text books.

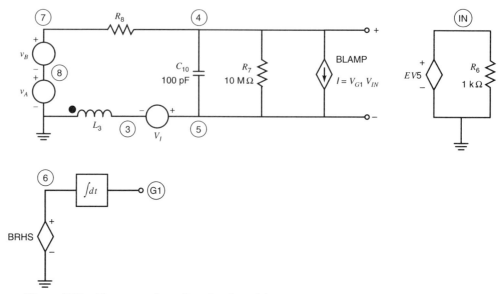

Figure 7.23   Fluorescent lamp functional model.

```
FLUORESCENT LAMP
VA 8 0 DC 201 SIN ( 2.01E+02 2E+02 6E+01 -1.2E-02 0 )
VB 7 8 PWL ( 0 0 0.1 100 )
C10 4 5 100P
BRHS  6  0 V=2.422*(I(VA)^2)-5.186E04*((I(VA)/V(2))^2)+271.7*(I(VA)/V(2))
R8 7 4 978M
R7 5 4 10MEG
R6 0 IN 1K
BLAMP 4 5 I=V(G1)*V(IN)
EV5 IN 0 4 5 1
L3 0 3 55.7M
XF2 6 G1 INT1M
.SUBCKT INTEG 1 2
RIN1 1 4 2
RIN2 1 4 2
G1 2 0 4 0 -1
R1 2 0 1E12
C1 2 0 1
.ENDS INTEG
VI 5 3 0.0
.IC V(G1)=1 V(IN)=201
.TRAN 0.005 .1 0 .5M UIC
.END
```

Figure 7.24   SPICE3 input for the fluorescent lamp functional model.

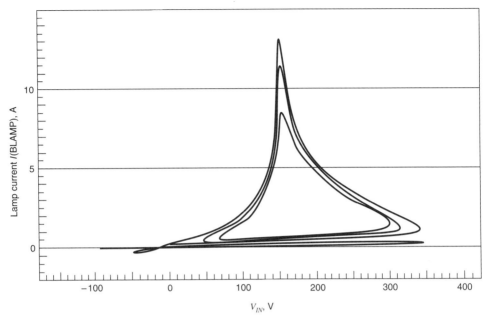

Figure 7.25 I-V characteristic of fluorescent lamp.

7.5 MACRO-MODELS

A number of concepts have been introduced so far in this chapter on higher-level modeling. The approaches have varied from structural simplifications of the circuit using ideal elements to pure functional blocks, which synthesize the input/output relationship of the circuit.

This section introduces the most general and powerful modeling concept, called *macro-modeling*. Macro-models combine functional elements with accurate nonlinear models for some elements, such as transistors and diodes, in order to obtain the behavior of the original circuit from a simplified circuit that can be simulated several times faster.

This concept was first applied to opamps (Boyle, Cohn, Pederson, and Solomon 1974), which contain ten to a hundred transistors in a detailed representation. Since the introduction in IC realization in the late 1960s, opamps have been widely used in such circuits as active filters, phase-locked loops, and control systems. A SPICE simulation of a circuit with more than a couple of opamps can be very expensive, and the development of macro-models that implement all or part of the data-book characteristics of a given opamp is crucial. The next section describes a common opamp macro-model. The components used to model specific characteristics are identified, and their evaluation from the data sheet is highlighted.

7.5.1 The Opamp Macro-Model

The macro-model developed by Boyle et al. (1974) is described in this section. Unlike the ideal opamp described in Sec. 7.3.1, the macro-model attempts to provide an accurate representation of differential and common-mode gain versus frequency characteristics, input and output characteristics, and accurate large-signal characteristics, such as offset, voltage swing, short-circuit current limiting, and slew rate. The approach described below can be applied to model a broad class of opamps.

The macro-model derivation for the μA741 is outlined below according to Boyle et al. (1974). The detailed circuit, containing 26 BJTs, is shown in Figure 7.26. The schematic of the macro-model is shown in Figure 7.27. The macro-model of the μA741 opamp is divided into three distinct stages, the input, gain, and output stages; the three stages are modeled separately to implement all the data-sheet characteristics. These characteristics are presented in Table 7.1.

The number of transistors in the macro-model has been reduced to only two, which are necessary to model the correct input characteristics. The gain stage is built with linear elements, and four diodes are used in the output stage for limiting the voltage excursion and the short-circuit current. At the conclusion of this section the switching and frequency-domain characteristics of the macro-model compared with those of the detailed circuit are shown in Figures 7.31 and 7.33.

The input stage must be designed to reproduce characteristics such as DC offset, slew rate, and frequency roll-off. The element values and model parameters of Q_1 and Q_2 must be derived from the opamp data sheet, Table 7.1. The gain factor, β_F, of the two transistors, model parameter BF, is derived from the DC offset and the slew rate. I_{C1}, equal to I_{C2}, is established by the positive-going slew rate, S_R^+ and compensation capacitor C_2:

$$I_{C1} = \tfrac{1}{2}C_2 S_R^+ \qquad (7.26)$$

I_B of Q_1 and Q_2 is taken from the data sheet, and an imbalance is introduced to account for the offset input current, I_{Bos}:

$$I_{B1} = I_B + \frac{I_{Bos}}{2}, \qquad I_{B2} = I_B - \frac{I_{Bos}}{2} \qquad (7.27)$$

β_{F1} and β_{F2} are derived from Eqs. 7.26 and 7.27; the input offset voltage, V_{os}, is due to an inequality in the two saturation currents, I_{S1} and I_{S2}, according to the following equation:

$$I_{S2} = I_{S1}e^{V_{os}/V_{th}} \qquad (7.28)$$

which leads to the values of the model parameter IS of Q_1 and Q_2.

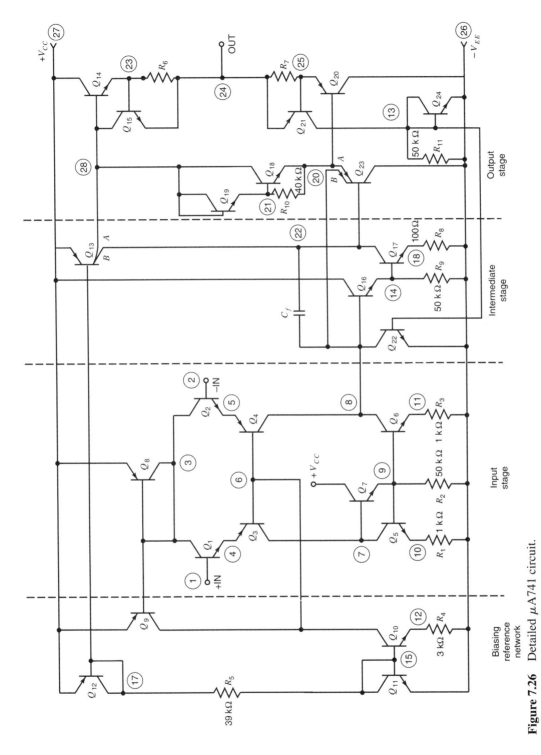

Figure 7.26 Detailed μA741 circuit.

229

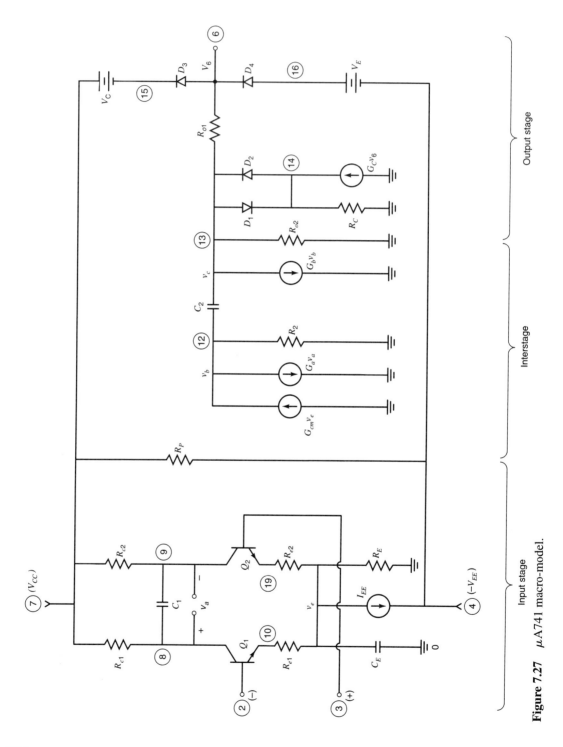

Figure 7.27 μA741 macro-model.

Table 7.1 μA741 Performance Characteristics

Parameter	Data Sheet	Device-Level Model	Macro-Model
C_2 (pF)	30	30	30
S_R^+ (V/μs)	0.67	0.9	0.899
S_R^- (V/μs)	0.62	0.72	0.718
I_B (nA)	80	256	255
I_{Bos} (nA)	20	0.7	<1
V_{os} (mV)	1	0.283	0.265
a_{vd}	$2 \cdot 10^5$	$4.17 \cdot 10^5$	$4.16 \cdot 10^5$
a_{vd} (1 kHz)	10^3	$1.219 \cdot 10^3$	$1.217 \cdot 10^3$
$\Delta\phi$ (degrees)	20	16.8	16.3
CMRR (dB)	90	106	106
R_{out} (Ω)	75	566	566
$R_{o\text{-}sc}$ (Ω)	—	76.8	76.8
I_{sc}^+ (mA)	25	25.9	26.2
I_{sc}^- (mA)	25	25.9	26.2
V^+ (V)	14	14.2	14.2
V^- (V)	-13.5	-12.7	-12.7

The emitter resistor, R_E is chosen so that it matches the output resistance of a transistor with an Early voltage, VAF, of 200 V:

$$R_E = \frac{VAF}{I_{EE}} \tag{7.29}$$

where I_{EE} is the sum of the collector and base currents of Q_1 and Q_2.

The frequency response is established by resistors R_{c1} and R_{c2} in the input stage and components of the gain stage. The frequency of the dominant pole, f_{3dB}, set by the compensation capacitor C_2 and by $G_a = 1/R_{c1}$, is given by

$$f_{3dB} = \frac{1}{2\pi a_{vd} R_{c1} C_2} \tag{7.30}$$

R_{c1} and R_{c2} can be derived from the equality

$$R_{c1} = R_{c2} = \frac{1}{2\pi f_{0dB} C_2} \tag{7.31}$$

where f_{0dB} is the 0-dB frequency from the data sheet; f_{0dB} can be approximated for a roll-off of 6dB/octave by

$$f_{0dB} = a_{vd} f_{3dB} \tag{7.32}$$

There are two more capacitors in the input stage, C_E and C_1. The first is included to model a smaller negative-going slew rate S_R^- according to the following equality:

$$C_E = \frac{2I_{C1}}{S_R^-} - C_2 \tag{7.33}$$

C_1 models the excess phase at f_{0dB} by introducing a second pole at

$$p_2 = -\tfrac{1}{2} R_{c1} C_1 \tag{7.34}$$

With the phase shift $\Delta\phi$ given in the data sheet, C_1 results as

$$C_1 = \frac{C_2}{2} \tan \Delta\phi \tag{7.35}$$

The gain stage is designed to provide the correct differential mode, DM, and common mode, CM, gains. This is achieved by a correct sizing of the transconductances G_{cm} and G_b. The two gains, a_{vd} and a_{vc}, are defined by the following equalities:

$$a_{vd} = \frac{R_2}{R_{c1}} \tag{7.36}$$

$$a_{vc} = G_{cm}R_2 \tag{7.37}$$

where $R_2 = 100\,\text{k}\Omega$ and $G_a = 1/R_{c1}$. The value of G_{cm} is based on the common-mode rejection ratio, $CMRR = a_{vd}/a_{vc}$, and is equal to

$$G_{cm} = \frac{1}{(CMRR)R_{c1}} \tag{7.38}$$

Transconductance G_b defines the correct open-loop DM gain, a_{vd}, and is given by

$$G_b = \frac{a_{vd}R_{c1}}{R_2 R_{o2}} \tag{7.39}$$

where R_{o2} is approximately equal to the DC output resistance given in the data sheet.

The design of the output stage of the macro-model requires the dimensioning of diodes D_1 and D_2 and resistor R_{o1} for current limiting and diodes D_3 and D_4 and bias sources V_C and V_E for proper output voltage limiting.

The complete list of the parameters of the macro-model is provided in Table 7.2, reproduced from Boyle et al. (1974); for the detailed derivation of this model see that paper.

EXAMPLE 7.6

Compare the voltage-follower slew-rate performance and the open-loop frequency response to f_{0dB} between the complete $\mu A741$ circuit shown in Figure 7.26 and the macro-model in Figure 7.27 with the component values from Table 7.2.

Solution
First, write the SPICE netlist for the detailed $\mu A741$, called subcircuit UA741, and for the macro-model, called UA741MAC. The two subcircuit definitions are listed in Figure 7.28. Both subcircuits have five external terminals, the two inputs, the output, and the two supplies.

The circuit for the slew-rate performance is shown in Figure 7.29, and the SPICE input is shown in Figure 7.30. Note that two unconnected voltage followers, X1 and X2, are used in order to compare the output voltages, V_3 and V_6, for the two opamp models; X1 is represented by the macro-model, UA741MAC, and X2 is described by the complete circuit, UA741, listed in Figure 7.28. A 20-μs pulse from -5 V to 5 V is applied to the positive input ports of both opamps. The offset voltages V_{os} corresponding to each model are applied as DC biases at the inputs of the two opamps. The two waveforms of v_{OUT} are plotted in Figure 7.31 and can be seen to match very well.

Table 7.2 $\mu A741$ Macro-Model Parameters

Parameter	Value	Parameter	Value
I_{S1} (A)	$8 \cdot 10^{-16}$	$R_P(k\Omega)$	15.363
I_{S2} (A)	$8.0925 \cdot 10^{-16}$	$G_a(\mu mho)$	229.774
β_1	52.6726	$G_{cm}(nmho)$	1.1516
β_2	52.7962	$G_b(mho)$	37.0978
I_{EE} (μA)	27.512	$R_{o1}(\Omega)$	76.8
R_{c1} (Ω)	4352	$R_{o2}(\Omega)$	489.2
R_{e1} (Ω)	2391.9	I_{SD1} (A)	$3.8218 \cdot 10^{-32}$
R_E ($M\Omega$)	7.2696	I_{SD3} (A)	$8 \cdot 10^{-16}$
C_E (pF)	7.5	$R_C(m\Omega)$	0.1986
C_1 (pF)	4.5288	G_C (mho)	5034.3
R_2 ($K\Omega$)	100	V_C (V)	1.6042
C_2 (pF)	30	V_E (V)	3.1042

```
**************************************
* DETAILED CIRCUIT FROM BOYLE PAPER *
**************************************
.SUBCKT UA741 1 2 24 27 26
* NODES: IN+ IN- OUT VCC VEE
R1 10 26 1K
R2 9 26 50K
R3 11 26 1K
R4 12 26 3K
R5 15 17 39K
R6 23 24 50
R7 24 25 25
R8 18 26 100
R9 14 26 50K
R10 21 20 40K
R11 13 26 50K
COMP 22 8 30PF
Q1 3 1 4 BNP1
Q2 3 2 5 BNP1
Q3 7 6 4 BPN1
Q4 8 6 5 BPN1
Q5 7 9 10 BNP1
Q6 8 9 11 BNP1
Q7 27 7 9 BNP1
Q8 3 3 27 BPN1
Q9 6 3 27 BPN1
Q10 6 15 12 BNP1
Q11 15 15 26 BNP1
Q12 17 17 27 BPN1
Q13A 28 17 27 BPN3
Q13B 22 17 27 BPN4
Q14 27 28 23 BNP2
Q15 28 23 24 BNP1
Q16 27 8 14 BNP1
Q17 22 14 18 BNP1
Q18 28 21 20 BNP1
Q19 28 28 21 BNP1
Q20 26 20 25 BPN2
Q21 13 25 24 BPN1
Q22 8 13 26 BNP1
Q23A 26 22 20 BPN5
Q23B 26 22 8 BPN6
Q24 13 13 26 BNP1
*
.MODEL BNP1 NPN BF=209 BR=2.5 RB=670 RC=300 CCS=1.417P TF=1.15N TR=405N
+ CJE=0.65P CJC=0.36P IS=1.26E-15 VA=178.6 C2=1653 IK=1.611M NE=2 PE=0.6
+ ME=0.33 PC=0.45 MC=0.33
```

Figure 7.28   Detailed description and macro-model of the μA741 opamp.

```
.MODEL BNP2 NPN BF=400 BR=6.1 RB=185 RC=15 CCS=3.455P TF=0.76N TR=243N
+ CJE=2.8P CJC=1.55P IS=0.395E-15 VA=267 C2=1543 IK=10M NE=2 PE=0.6
+ ME=0.33 PC=0.45 MC=0.33
.MODEL BPN1 PNP BF=75 BR=3.8 RB=500 RC=150 CCS=2.259P TF=27.4N TR=2540N
+ CJE=0.10P CJC=1.05P IS=3.15E-15 VA=55.11 C2=1764 IK=270U NE=2 PE=0.45
+ ME=0.33 PC=0.45 MC=0.33
.MODEL BPN2 PNP BF=117 BR=4.8 RB=80 RC=156 TF=27.4N TR=2540N
+ CJE=4.05P CJC=2.80P IS=17.6E-15 VA=57.94 C2=478.4 IK=590.7U NE=2 PE=0.6
+ ME=0.25 PC=0.6 MC=0.25
.MODEL BPN3 PNP BF=13.8 BR=1.4 RB=100 RC=80 CCS=2.126P TF=27.4N TR=55N
+ CJE=0.10P CJC=0.3P IS=2.25E-15 VA=83.55 C2=84.37K IK=5M NE=2 PE=0.45
+ ME=0.33 PC=0.45 MC=0.33
.MODEL BPN4 PNP BF=14.8 BR=1.5 RB=160 RC=120 CCS=2.126P TF=27.4N TR=220N
+ CJE=0.10P CJC=0.90P IS=2.25E-15 VA=83.55 C2=84.37K IK=171.8U NE=2 PE=0.45
+ ME=0.33 PC=0.45 MC=0.33
.MODEL BPN5 PNP BF=80 BR=1.5 RB=1100 RC=170 TF=26.5N TR=9550N
+ CJE=0.10P CJC=2.40P IS=0.79E-15 VA=79.45 C2=1219 IK=80.55U NE=2 PE=0.6
+ ME=0.25 PC=0.6 MC=0.25
.MODEL BPN6 PNP BF=19 BR=1.0 RB=650 RC=100 TF=26.5N TR=2120N
+ CJE=1.90P CJC=2.40P IS=0.63E-17 VA=167.1 C2=57.49K IK=80.55U NE=2 PE=0.6
+ ME=0.25 PC=0.6 MC=0.25
*
.ENDS UA741
***********************
* UA741 MACRO-MODEL *
***********************
.SUBCKT UA741MAC 3 2 6 7 4
* NODES: IN+ IN- OUT VCC VEE
Q1 8 2 10 QMOD1
Q2 9 3 11 QMOD2
RC1 7 8 4352
RC2 7 9 4352
RE1 1 10 2391.9
RE2 1 11 2391.9
RE 1 0 7.27MEG
CE 1 0 7.5P
IEE 1 4 27.512U
C1 8 9 4.5288P
RP 7 4 15.363K
GCM 0 12 1 0 1.1516N
GA 12 0 8 9 229.774U
R2 12 0 100K
C2 12 13 30P
GB 13 0 12 0 37.0978
R02 13 0 489.2
R01 13 6 76.8
D1 13 14 DMOD1
```

Figure 7.28   *(continued)*

```
D2 14 13 DMOD1
*EC 0 14 6 0 1
GC 0 14 6 0 5034.3
RC 14 0 0.1986M
D3 6 15 DMOD3
D4 16 6 DMOD3
VC 7 15 1.6042
VE 16 4 3.1042
*
.MODEL QMOD1 NPN IS=8E-16 BF=52.6726
.MODEL QMOD2 NPN IS=8.0925E-16 BF=52.7962
.MODEL DMOD1 D IS=3.8218E-32
.MODEL DMOD3 D IS=8E-16
.ENDS
```

Figure 7.28   *(continued)*

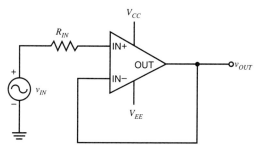

Figure 7.29   Voltage follower configuration for slew-rate comparison.

```
UA741 SLEW RATE
*
* MACRO-MODEL
X1 1 3 3 4 5 UA741MAC
* FULL MODEL
X2 2 6 6 4 5 UA741
VIN1 10 0 DC -.2765M PULSE -5 5 0 10N 10N 20U 100U AC 1
VIN2 11 0 DC -.2834M PULSE -5 5 0 10N 10N 20U 100U AC 1
RIN1 10 1 100
RIN2 11 2 100
VCC 4 0 15
VEE 5 0 -15
*
.TRAN .25U 50U
.OPTIONS ACCT
.END
```

Figure 7.30   SPICE input for voltage-follower configuration for slew-rate comparison.

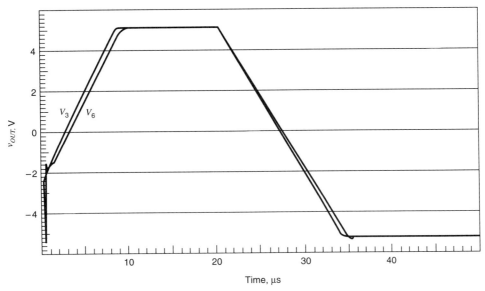

Figure 7.31   Slew-rate results for opamp models.

The SPICE input for the frequency response is modified, as shown in Figure 7.32, so that the two opamps, X1 and X2, are in an open-loop configuration, the pulse is replaced by an AC source, and the frequency range is set from below the dominant pole defined by the compensation capacitor to six orders of magnitude above that value. The magnitudes of the two output signals are plotted in Figure 7.33. The two models track

```
UA741 FREQUENCY RESPONSE
*
* MACRO-MODEL
X1 1 0 3 4 5 UA741MAC
* FULL MODEL
X2 2 0 6 4 5 UA741
*
VIN1 1 0 DC -.2765M AC 1
VIN2 2 0 DC -.2834M AC 1
VCC 4 0 15
VEE 5 0 -15
*
*
.TF V(3) VIN1
.TF V(6) VIN2
.AC DEC 10 .1 1G
.END
```

Figure 7.32   SPICE input for open-loop frequency response comparison.

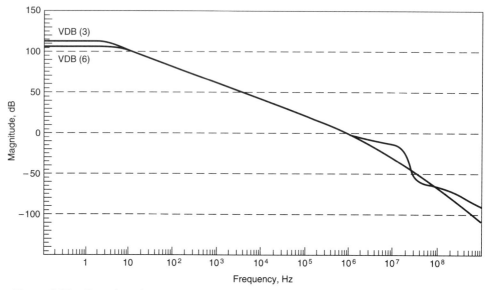

Figure 7.33 Open-loop frequency response comparison.

each other very well up to 1 MHz, after which secondary poles and zeros come into play.

The above characterization is useful for identifying the range of the macro-model's validity for different applications. The reductions in circuit complexity and simulation time are important advantages for macro-models. Separate runs for individual opamps show that the complete circuit has 79 nodes and 38 components, versus 16 nodes and 26 elements for the macro-model. In addition, the complete circuit has 26 BJTs, versus only 2 for the macro-model. This ratio is very important, because in SPICE the transistor is the most time-consuming component to evaluate.

The accounting information option of SPICE, **ACCT** (see Sec. 9.5.1), offers detailed information on the two circuits. This option also shows that the run time for the open-loop DC operating point is 1.17 s and 21 iterations for the complete μA741 circuit and 0.25 s and 32 iterations for the macro-model. The gains in analysis time become significant for the transient analysis, where the complete circuit requires 56 s versus only 5 s for the macro-model, a performance gain of 11. These results were obtained from SPICE2 running on a SUN 4/110 workstation. The same analysis performed by PSpice on an IBM PS/2 Model 80 equipped with a 16 MHz 386 processor requires 209 s for the complete circuit versus 14 s for the macro-model. The performance gain is achieved in the analysis of the macro-model circuit partly because fewer time points and iterations are required by this solution; see Chap. 9 for more details. The simpler circuit takes only half the number of iterations used by the full opamp circuit.

An important conclusion to the modeling practices described in this chapter is that SPICE can accommodate different levels of accuracy in the representations of circuits. It is up to the user to decide which behavior of a circuit block is important and needs to be modeled for the correct operation of an entire circuit. For the opamp macro-model varying degrees of accuracy can be implemented, from the ideal model in Figure 7.6 to the complex macro-model of Figure 7.27.

With the reduction in complexity, the savings in run time are of major significance. Whereas the detailed macro-model can save up to an order of magnitude in run time, a nearly ideal model of an opamp can run up to two orders of magnitude faster.

7.6 SUMMARY

This chapter has introduced several concepts pertaining to modeling and simulating electric circuits and systems, which are applicable as well to nonelectrical physical systems. This approach to modeling and simulation is known as *high-level*, or *behavioral*, in contrast to the *structural* representation common to circuit schematics.

The first step in defining circuit representational levels is to identify a hierarchy when describing a more complex circuit. The SPICE language provides constructs for identifying subcircuits and instantiating them in the circuit description.

The *subcircuit definition*, a global statement, and the *subcircuit instance* are defined by the following syntax:

.SUBCKT *SUBname node1* <*node2 . . .* >
    circuit elements

.ENDS

X*name xnode1* <xnode2 . . .> *SUBname*

These statements can be nested to create several levels in a hierarchical circuit description, as exemplified by several digital building blocks.

High-level circuit representations have been presented in three categories depending on the level of accuracy. Ideal models describe the single most relevant function of a device and can be used for computational efficiency in circuits where the ideal blocks are not critical for the circuit performance. Ideal opamps can be used in active filters, and ideal logic gates save analysis time in mixed analog/digital circuits.

Functional models implement complex analytical current-voltage expressions with one or more nonlinear controlled sources. The arbitrary controlled source introduced in SPICE3 and the equivalent or more powerful capabilities in PSpice, Spice-PLUS/Profile, HSPICE (Meta-Software Inc. 1991) and Saber (Analogy Inc.,1991) make it possible to model circuit blocks such as limiters, voltage-controlled oscillators, and phase-locked loops with just a few components. Devices described by complex equations, which need not be electrical, can also be simulated using arbitrary controlled sources. A number of useful functional blocks can be defined using the polynomial controlled sources available in all SPICE simulators.

Macro-models can achieve the accuracy of a detailed circuit with important savings in analysis time by combining structural elements, such as transistors and capacitors, with functional blocks implemented with controlled sources. Well-designed macro-models can be simulated up to an order of magnitude faster than the detailed model with little if any loss in accuracy.

Both analog and digital circuit examples have been presented in this chapter to highlight the diverse approaches to high-level modeling.

A cautionary note at the conclusion of this chapter regarding the accuracy of the results using high-level representations: the analysis is only as accurate as the models used! Erroneous operation as well as analysis difficulties can result from using idealized models, as shown in one of the examples, as well as in Secs. 10.2.2, 10.3, and 10.4.

REFERENCES

Boyle, G. R., B. M. Cohn, D. O. Pederson, and J. E. Solomon. 1974. Macromodeling of integrated circuit operational amplifiers. *IEEE Journal of Solid-State Circuits* (December).

Dorf, R. C. 1989. *Introduction to Electric Circuits.* New York: John Wiley & Sons.

Epler, B. 1987. SPICE2 application notes for dependent sources. *IEEE Circuits and Devices Magazine* 3 (September).

Gray, P. R., and R. G. Meyer. 1993. *Analysis and Design of Analog Integrated Circuits,* 3d ed. New York: John Wiley & Sons.

Hodges, D. A., and H. G. Jackson. 1983. *Analysis and Design of Digital Integrated Circuits.* New York: McGraw-Hill.

Mano, M. M. 1976. *Computer System Architecture.* Englewood Cliffs, NJ: Prentice Hall.

Mateescu, A. 1991. Personal communication.

Meta-Software Inc. 1991. *HSPICE User's Guide.* Campbell, CA: Author.

Microsim Corporation. 1991. *PSpice, Circuit Analysis User's Guide, Version 5.0.* Irvine, CA: Author.

Oldham, W. G., and Schwartz. 1987. *Introduction to Electronics.* New York: McGraw-Hill.

Paul, C. R. 1989. *Analysis of Linear Circuits.* New York: McGraw–Hill.

Saber User's Guide. Release 3.0. Analogy Inc. 1990. Beaverton, OR: Author. Sedra, A. S., and K. C. Smith. 1990. *Microelectronic Circuits.* Philadelphia: W. B. Saunders.

Sitkowski, M. 1990. Simulation and modeling—The macro-modeling of logic functions for the SPICE simulator. *IEEE Circuits and Devices Magazine* 6 (September).

Valid Logic Systems. 1991. *Analog Workbench Reference.* Vol. 2. San Jose, CA: Author.

Vladimirescu, A. 1982. LSI circuit simulation on Vector computers. Univ. of California, Berkeley, ERL Memo UCB/ERL M82/75 (October).

Vladimirescu, A., K. Zhang, A. R. Newton, D. O. Pederson, and A. L. Sangiovanni-Vincentelli. 1981. *SPICE version 2G User's Guide.* Dept. of Electrical Engineering and Computer Science, Univ. of California, Berkeley (August).

Eight

DISTORTION ANALYSIS

8.1 DISTORTION IN SEMICONDUCTOR CIRCUITS

This chapter describes in detail the evaluation of distortion in electronic circuits with SPICE. Emphasis is placed on the small-signal distortion analysis in SPICE2, which is a powerful tool for designing electronic circuits and ICs. The main features of the small-signal distortion analysis were presented in Sec. 5.4.

The distortion analysis presented in Chap. 5 was limited to frequency-independent circuits. A more rigorous analysis of distortion components for a single-transistor amplifier and a mixer circuit is carried out in this chapter, in Sec. 8.2. SPICE2 provides more distortion information about a circuit than shown in Chap. 5. The total distortion measures are broken down in the SPICE2 output into contributions for each nonlinearity of each diode or BJT in the circuit as exemplified in Sec. 8.2.2.

The signal amplitude is often larger than the limit for small signals. Therefore, it is important to understand how to use large-signal time-domain analysis to derive the same distortion measures computed by the small-signal analysis. Sec. 8.3 describes the application of Fourier analysis for verifying the results of the one-transistor amplifier and for deriving the intermodulation component for a mixer.

8.2 SMALL-SIGNAL DISTORTION ANALYSIS

In Chap. 5 the distortion measures were derived for a *nearly linear frequency-independent circuit*. In this section the effect of frequency-dependent circuit elements is considered in the derivation of the distortion measures.

The SPICE2 capability of reporting individual distortion sources is explained and the additional feature of the AC analysis that provides a frequency sweep of the distortion coefficients is exemplified.

8.2.1 High-Frequency Distortion

In the absence of frequency-dependent elements, a power series is used to express the high-order terms of the signal at the output of a nonlinear circuit. A more general series expansion of a signal is derived below for computing the distortion terms. The approach consists of expressing the transfer function of a circuit containing both small nonlinearities and frequency-dependent elements (Meyer 1979; Pederson and Mayaram 1990).

In Figure 8.1 two frequency-dependent linear circuit blocks, both filters, are added to the nonlinear gain block A. The two linear filters have transfer functions $F(j\omega)$ and $J(j\omega)$, respectively.

The higher-order terms in the output signal, S_o, are obtained by taking into account all three transfer functions. The output, S_{oa}, of the nonlinear gain block, A, is a power series of the input to A, S_{ia}:

$$S_{oa} = a_1 S_{ia} + a_2 S_{ia}^2 + \cdots \tag{8.1}$$

S_{ia} is the result of passing the input signal, S_i,

$$S_i = S_1 \cos \omega_1 t + S_2 \cos \omega_2 t \tag{8.2}$$

through $F(j\omega)$ to get

$$S_{ia} = |F(j\omega_1)|S_1 \cos(\omega_1 t + \phi_{\omega_1}) + |F(j\omega_2)|S_2 \cos(\omega_2 t + \phi_{\omega_2}) \tag{8.3}$$

where $|F(j\omega)|$ is the magnitude and ϕ_ω is the phase of the transfer function. A new operator is introduced, $\circ$, to simplify the above expression of S_{ia} to

$$S_{ia} = F(j\omega) \circ S_i \tag{8.4}$$

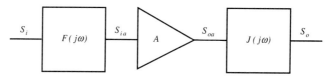

Figure 8.1   Nonlinear gain block with input and output filters.

The meaning of applying F through the operator $\circ$ on the input signal S_i is to multiply the amplitude S_n of each frequency component in S_i by $|F(j\omega_n)|$ and shift its phase by $\phi_{\omega n}$.

The signal at the output of the nonlinear gain block is a power series of the input signal; limited to three terms and up to three signal frequencies, ω_a, ω_b, and ω_c, the output becomes

$$\begin{aligned} S_{oa} &= S_{oa1} + S_{oa2} + S_{oa3} \\ &= a_1 F(j\omega_a) \circ S_i + a_2 F(j\omega_a)F(j\omega_b) \circ S_i^2 + a_3 F(j\omega_a)F(j\omega_b)F(j\omega_c) \circ S_i^3 \quad (8.5) \end{aligned}$$

The detailed expressions of S_{oa1}, S_{oa2}, and S_{oa3}, which are listed below for an input signal with two frequencies, ω_1 and ω_2, are useful for deriving the distortion components in the following section:

$$\begin{aligned} S_{oa1} =\ & a_1 \big[|F(j\omega_1)|S_1 \cos\alpha_1 + |F(j\omega_2)|S_2 \cos\alpha_2 \big] \\ S_{oa2} =\ & \frac{a_2}{2} \Big\{ \big[|F(j\omega_1)|^2 S_1^2 (\cos 2\alpha_1 + 1) + |F(j\omega_2)|^2 S_2^2 (\cos 2\alpha_2 + 1) \big] \\ & + 2|F(j\omega_1)||F(j\omega_2)|S_1 S_2 \big[\cos(\alpha_1 + \alpha_2) + \cos(\alpha_1 - \alpha_2) \big] \Big\} \\ S_{oa3} =\ & \frac{a_3}{4} \Big\{ \big[|F(j\omega_1)|^3 S_1^3 (\cos 3\alpha_1 + 3\cos\alpha_1) + |F(j\omega_2)|^3 S_2^3 (\cos 3\alpha_2 + 3\cos\alpha_2) \big] \\ & + 3|F(j\omega_1)|S_1|F(j\omega_2)|^2 S_2^2 \big[2\cos\alpha_1 + \cos(2\alpha_2 - \alpha_1) + \cos(2\alpha_2 + \alpha_1) \big] \\ & + 3|F(j\omega_1)|^2 S_1^2 |F(j\omega_2)|S_2 \big[2\cos\alpha_2 + \cos(2\alpha_1 - \alpha_2) + \cos(2\alpha_1 + \alpha_2) \big] \Big\} \end{aligned}$$

$$(8.6)$$

In the above equations α_1 and α_2 represent

$$\begin{aligned} \alpha_1 &= \omega_1 t + \phi_{\omega 1} \\ \alpha_2 &= \omega_2 t + \phi_{\omega 2} \end{aligned} \qquad (8.7)$$

The signal at the output, S_o, is obtained by including the contribution of the second frequency-dependent block, $J(j\omega)$:

$$\begin{aligned} S_o =\ & a_1 F(j\omega_a)J(j\omega_a) \circ S_i + a_2 F(j\omega_a)F(j\omega_b)J(j\omega_a, j\omega_b) \circ S_i^2 \\ & + a_3 F(j\omega_a)F(j\omega_b)F(j\omega_c)J(j\omega_a, j\omega_b, j\omega_c) \circ S_i^3 \quad (8.8) \end{aligned}$$

The argument of J in Eq. 8.8 is the frequency of a given spectral component which in turn is a linear combination of the input frequencies ω_a, ω_b, ω_c. The above result is generally valid for circuits with memoryless nonlinearities and linear frequency-dependent elements. The above power series expansion of the output signal is also known as a Volterra series (Narayanan 1967).

The different distortion measures introduced in Chap. 5 can be derived as ratios of the second- or third-order terms in Eqs. 8.8 and by an appropriate assignment of $\pm\omega_1$ and $\pm\omega_2$ to ω_a, ω_b, and ω_c.

In conclusion, the distortion terms of a signal at the output of a frequency-dependent circuit with small nonlinearities can be found by writing the overall transfer function and replacing the nonlinear signals by series expansions. Transfer functions of first, second, and third order can be isolated to compute the distortions of corresponding order. The following section exemplifies the use of this approach.

8.2.2   Distortion in a One-Transistor Amplifier

Consider the one-transistor amplifier of Figure 5.8 with a series base resistance, **RB**, of 100 Ω and junction capacitances **CJE** and **CJC** added to the .**MODEL** statement, as listed in Figure 8.2. The DC bias solution and the small-signal parameters of Q_1 are recomputed and the results are in Figure 8.2. The linear equivalent of the one-transistor

```
******* 04/07/89 ******** SPICE 2G.6     3/15/83******** 16:14:42 *****

ONE-TRANSISTOR CIRCUIT (Figure 5.8)

*****    INPUT LISTING              TEMPERATURE =   27.000 DEG C

************************************************************************
*
Q1 2 1 0 QMOD
RL 2 3 1K
VCC 3 0 5
*
VBE 4 0 793.4M
VI 1 4 AC 1
*
.MODEL QMOD NPN RB=100
+ CJE=1P
+ CJC=2P
.OP
.AC LIN 1 1MEG 1MEG
.DISTO RL 1
*
.PRINT DISTO HD2 HD3 SIM2 DIM2 DIM3
.WIDTH OUT=80
.OPTION NOPAGE
.END
```

Figure 8.2   One-transistor amplifier input and DC bias point.

```
****      BJT MODEL PARAMETERS                    TEMPERATURE = 27.000 DEG C
              QMOD
TYPE          NPN
IS          1.00D-16
BF          100.000
NF            1.000
BR            1.000
NR            1.000
RB          100.000
CJE         1.00D-12
CJC         2.00D-12


****      SMALL SIGNAL BIAS SOLUTION         TEMPERATURE = 27.000 DEG C

NODE    VOLTAGE      NODE    VOLTAGE      NODE      VOLTAGE

( 1)     0.7934     ( 2)     3.0522     ( 3)      5.0000

    VOLTAGE SOURCE CURRENTS

    NAME              CURRENT
    VCC            -1.948D-03
    VBE            -1.948D-05
    VI             -1.948D-05


    TOTAL POWER DISSIPATION 9.75D-03 WATTS

****      OPERATING POINT INFORMATION    TEMPERATURE = 27.000 DEG C

**** BIPOLAR JUNCTION TRANSISTORS
              Q1
MODEL       QMOD
IB          1.95E-05
IC          1.95E-03
VBE            0.793
VBC           -2.259
VCE            3.052
BETADC      100.000
GM          7.53E-02
RPI         1.33E+03
RX          1.00E+02
RO          1.00E+12
CPI         1.72E-12
CMU         1.26E-12
CBX         0.00E+00
CCS         0.00E+00
BETAAC      100.000
FT          4.02E+09
```

Figure 8.2 *(continued)*

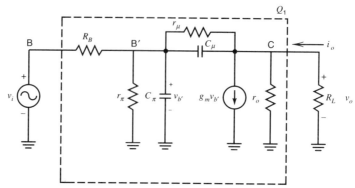

Figure 8.3   Small-signal equivalent of the one-transistor amplifier.

amplifier is shown in Figure 8.3. More detail than just the total distortion figures is obtained from SPICE2 when a summary interval, equal to 1 in the deck shown in Figure 8.2, is specified in the `.DISTO` statement; see Sec. 5.4. The distortion contribution in the load resistor is computed for every element in Figure 8.3 associated with a nonlinearity of the transistor.

The distortion summary computed by SPICE2 at 1 MHz is listed in Figure 8.4. Note that the addition of the base resistance and the two charge nonlinearities do not significantly affect the absolute values of the second- and third-order distortion terms, `HD2` and `HD3`, and the second-order intermodulation terms, `SIM2` and `DIM2`, which are listed under the `AC ANALYSIS` header, as compared to the results obtained at 1 kHz in Sec. 5.4 (Figure 5.9).

The `DISTORTION ANALYSIS` summary report lists five groups of `BJT DISTORTION COMPONENTS`. Each group starts with a header identifying the type of distortion, such as `2ND HARMONIC DISTORTION`, and the fundamental frequency, `FREQ1`, for one input signal or `FREQ1` and `FREQ2` for two input signals; the `DISTORTION FREQUENCY` and `MAG` and `PHS` of the transfer functions for the input signals are listed on the following line.

The `BJT DISTORTION COMPONENTS` are due to the nonlinear voltage dependencies of I_C, I_B, Q_{BE}, and Q_{BC}, as defined in Eqs. 3.10 through 3.14. The total currents, i_C and i_B, can be expressed as Taylor series around the DC operating values, I_C and I_B, in the same manner as derived in Eqs. 5.19 through 5.21 for an approximation of i_C. The elements g_π, g_μ, g_m, g_o, C_π, and C_μ in the linear equivalent in Figure 8.3 represent the partial derivatives of I_B, I_C, Q_{BE}, and Q_{BC} in the first-order terms of the corresponding Taylor expansions.

The higher-order terms of this series constitute distortion sources, as described in Chap. 5. The contribution of each second or third derivative in the Taylor series to the output distortion is listed in the corresponding column of the SPICE2 printout.

Note that the current equations for I_C and I_B in distortion analysis use not the Gummel-Poon formulation given in Appendix A, but the more simple Ebers-Moll

```
****    DISTORTION ANALYSIS                  TEMPERATURE =   27.000 DEG C

   2ND HARMONIC DISTORTION                          FREQ1 =  1.00D+06  HZ

      DISTORTION FREQUENCY   2.00D+06 HZ            MAG 6.989D+01   PHS  176.26

BJT DISTORTION COMPONENTS
NAME          GM         GPI        GO         GMU        GMO2
Q1     MAG 1.800D-01 1.261D-02 1.356D-08 1.000D-20 3.600D-10
       PHS  166.19    -14.02    -14.73      0.00    -14.27       0.00          0.00
           CB         CBR        CJE        CJC       TOTAL
       MAG  1.000D-20 1.000D-20 4.473D-06 5.016D-03 1.674D-01
       PHS   75.98     75.10    164.49

   HD2    MAGNITUDE   1.674D-01     PHASE    164.49    =   -15.53 DB

   3RD HARMONIC DISTORTION                          FREQ1 =  1.00D+06  HZ

      DISTORTION FREQUENCY   3.00D+06  HZ           MAG 6.989D+01   PHS  176.26

BJT DISTORTION COMPONENTS
NAME          GM         GPI        GO         GMU        GMO2
Q1     MAG 1.789D-02 1.253D-03 1.267D-07 1.000D-20 5.598D-11
       PHS  132.40    -47.92    -22.30      0.00    -38.68
           CB         CBR        CJE        CJC       GM203     GMO23     TOTAL
       MAG 1.000D-20 1.000D-20 3.678D-07 4.019D-03 6.468D-11 4.871D-09 1.842D-02
       PHS   0.00      0.00    -55.64     62.75    -21.10    -21.56    120.61

   HD3    MAGNITUDE   1.842D-02     PHASE    120.61    =   -34.69  DB

   2ND ORDER INTERMODULATION DIFFERENCE COMPONENT       FREQ1 =  1.00D+06  HZ
     FREQ2 =  9.00D+05  HZ
      DISTORTION FREQUENCY   1.00D+05  HZ           MAG 6.989D+01   PHS  176.26
 MAG 6.992D+01   PHS   176.63

BJT DISTORTION COMPONENTS
NAME          GM         GPI        GO         GMU        GMO2
Q1     MAG 3.632D-01 2.544D-02 2.735D-08 1.000D-20 7.264D-10
       PHS  179.31     -0.70     -0.74      0.00     -0.71
           CB         CBR        CJE        CJC       TOTAL
       MAG 1.000D-20 1.000D-20 4.512D-07 5.061D-04 3.377D-01
       PHS   0.00      0.00     89.30     89.25    179.22

   IM2D   MAGNITUDE   3.377D-01     PHASE    179.22    =    -9.43 DB

   2ND ORDER INTERMODULATION SUM COMPONENT              FREQ1 =  1.00D+06  HZ
     FREQ2 =  9.00D+05  HZ

      DISTORTION FREQUENCY   1.90D+06  HZ           MAG 6.989D+01   PHS  176.26
 MAG 6.992D+01   PHS   176.63

BJT DISTORTION COMPONENTS

NAME          GM         GPI        GO         GMU        GMO2
Q1     MAG 3.604D-01 2.524D-02 2.714D-08 1.000D-20 7.209D-10
       PHS  166.88    -13.32    -14.00      0.00    -13.56
           CB         CBR        CJE        CJC       TOTAL
       MAG 1.000D-20 1.000D-20 8.509D-06 9.543D-03 3.352D-01
       PHS   0.00      0.00     76.68     75.84    165.26

   IM2S   MAGNITUDE   3.352D-01     PHASE    165.26    =    -9.50  DB
```

(continued on next page)

Figure 8.4 SPICE2 `.DISTO` summary for the one-transistor amplifier.

```
3RD ORDER INTERMODULATION DIFFERENCE COMPONENT           FREQ1 =  1.00D+06  HZ
       FREQ2 =  9.00D+05  HZ

    DISTORTION FREQUENCY   1.10D+06  HZ                    MAG 6.989D+01   PHS  176.26
  MAG 6.992D+01   PHS  176.63

BJT DISTORTION COMPONENTS

NAME        GM         GPI        GO         GMU        GMO2
Q1    MAG 5.164D-02 3.616D-03 3.867D-07 1.000D-20 1.685D-10
      PHS  162.25    -17.86     -8.22      0.00     -14.24
            CB         CBR        CJE        CJC       GM203      GMO23      TOTAL
      MAG 1.000D-20 1.000D-20 2.388D-07 4.515D-03 1.973D-10 1.486D-08 4.883D-02
      PHS   0.00      0.00     -66.39     79.97     -7.78      -7.95     157.01

  IM3    MAGNITUDE   4.883D-02      PHASE   157.01     =   -26.23  DB

    APPROXIMATE CROSS MODULATION COMPONENTS

  CMA    MAGNITUDE   1.844D-01                         =   -14.68  DB
  CMP    MAGNITUDE   6.440D-02                         =   -23.82  DB

****     AC ANALYSIS                       TEMPERATURE =   27.000 DEG C

   FREQ        HD2        HD3        SIM2       DIM2        DIM3

 1.000E+06    1.674E-01  1.842E-02  3.352E-01  3.377E-01   4.883E-02
```

Figure 8.4 *(continued)*

formulation including the Early effect from Eqs. 3.10 through 3.12. For completeness
the equations of I_C and I_B used in SPICE2 for computing distortion are included here:

$$I_C = (I_{CC} - I_{CE})\left(1 - \frac{V_{BC}}{VAF}\right) - I_{BC}$$

$$= I_S\left(e^{V_{BE}/V_{th}} - e^{V_{BC}/V_{th}}\right)\left(1 - \frac{V_{BC}}{VAF}\right) - \frac{I_S}{\beta_R}\left(e^{V_{BC}/V_{th}} - 1\right) \qquad (8.9)$$

$$I_B = I_{BE} + I_{BC}$$

where I_{CC}, I_{CE}, I_{BC}, and I_{BE} have the same meanings as in Eqs. 3.10 and 3.11.

The current components for low- and high-level injection have been neglected in
order to simplify the derivation of higher-order terms and focus the results on the main
nonlinear distortion sources. Also, because of model simplicity at the time of the implementation, the emission coefficients *NF* and *NR* are set equal to 1.

The base and collector currents are functions of two variables, v_{BE} and v_{BC}. The
total currents and voltages, such as v_{BE}, are made up of a DC component, such as V_{BE},
and an incremental component, such as v_{be}:

$$v_{BE} = V_{BE} + v_{be} \qquad (8.10)$$

Similarly,

$$i_B = I_B + i_b \tag{8.11}$$

where I_B is defined by Eqs. 8.9. The incremental component i_b has the following Taylor series expansion:

$$i_b = i_{g\pi} + i_{g\mu} = g_\pi v_{be} + g_{\pi 2} v_{be}^2 + g_{\pi 3} v_{be}^3 + \cdots + g_\mu v_{bc}$$
$$+ g_{\mu 2} v_{bc}^2 + g_{\mu 3} v_{bc}^3 + \cdots \tag{8.12}$$

where

$$
\begin{aligned}
g_\pi &= \frac{\partial I_B}{\partial V_{BE}} = \frac{1}{\beta_F} \frac{dI_{CC}}{dV_{BE}} \\[2mm]
g_\mu &= \frac{\partial I_B}{\partial V_{BC}} = \frac{1}{\beta_R} \frac{dI_{CE}}{dV_{BC}} \\[2mm]
g_{\pi 2} &= \frac{1}{2} \frac{\partial^2 I_B}{\partial V_{BE}^2} = \frac{1}{2} \frac{d^2 I_{BC}}{dV_{BE}^2} \\[2mm]
g_{\mu 2} &= \frac{1}{2} \frac{\partial^2 I_B}{\partial V_{BC}^2} = \frac{1}{2} \frac{d^2 I_{BE}}{dV_{BC}^2}
\end{aligned}
\tag{8.13}
$$

and so on. The values listed under `GPI` and `GMU` in the `HD2` group represent the contributions to the total distortion due to the second-order terms $g_{\pi 2}$ and $g_{\mu 2}$ in i_b. Note that a value of 10^{-20} represents zero distortion contribution.

In the same way distortion contributions can be associated with the Taylor coefficients of the expansion of i_C:

$$
\begin{aligned}
i_c &= \frac{\partial I_C}{\partial V_{BE}} v_{be} + \frac{\partial I_C}{\partial V_{BC}} v_{bc} + \frac{1}{2} \left(\frac{\partial^2 I_C}{\partial V_{BE}^2} v_{be}^2 + 2 \frac{\partial^2 I_C}{\partial V_{BE} \partial V_{BC}} v_{be} v_{bc} + \frac{\partial^2 I_C}{\partial V_{BC}^2} v_{bc}^2 \right) \\
&= i_c^{(1)} + i_c^{(2)}
\end{aligned}
\tag{8.14}
$$

where $i_c^{(1)}$ and $i_c^{(2)}$ represent the first- and second-order terms, respectively, of the i_C series. Note that v_{be} and v_{bc} are the independent variables in the above equation; the input and output voltages of the transistor equivalent network shown in Figure 8.3, however, are v_{be} and v_{ce}, respectively. A change of variable

$$v_{bc} = v_{be} - v_{ce} \tag{8.15}$$

leads to the following expression for the first- and second-order components of i_c:

$$i_c^{(1)} = \left(\frac{\partial I_C}{\partial V_{BE}} + \frac{\partial I_C}{\partial V_{BC}} \right) v_{be} - \frac{\partial I_C}{\partial V_{BC}} v_{ce} = g_m v_{be} + g_o v_{ce} - g_\mu v_{bc}$$

$$i_c^{(2)} = \frac{1}{2} \left(\frac{\partial^2 I_C}{\partial V_{BE}^2} + 2 \frac{\partial^2 I_C}{\partial V_{BE} \partial V_{BC}} + \frac{\partial^2 I_C}{\partial V_{BC}^2} \right) v_{be}^2 \qquad (8.16)$$

$$- \left(\frac{\partial^2 I_C}{\partial V_{BE} \partial V_{BC}} + \frac{\partial^2 I_C}{\partial V_{BC}^2} \right) v_{be} v_{ce} + \frac{1}{2} \frac{\partial^2 I_C}{\partial V_{BC}^2} v_{ce}^2$$

where

$$g_m = \frac{\partial I_{CT}}{\partial V_{BE}} + \frac{\partial I_{CT}}{\partial V_{BC}} = \frac{\partial I_C}{\partial V_{BE}} - g_o$$

$$g_o = -\frac{\partial I_{CT}}{\partial V_{BC}} = -\frac{\partial I_C}{\partial V_{BC}} - g_\mu$$

GM, GO, and GMO2 in the SPICE2 output represent the second-order distortions

$$g_{m2} = \frac{1}{2} \frac{\partial^2 I_C}{\partial V_{BE}^2} - g_{o2} - g_{mo2}$$

$$g_{o2} = \frac{1}{2} \frac{\partial^2 I_C}{\partial V_{BC}^2} + g_{\mu2} \qquad (8.17)$$

$$g_{mo2} = -\frac{\partial^2 I_C}{\partial V_{BE} \partial V_{BC}} - 2 g_{o2}$$

respectively.

Next consider the distortion due to the charge components. The nonlinear charge formulations are described by Eqs. 3.2 and 3.4. The current through a nonlinear capacitance can be expressed in a power series:

$$i_Q = \frac{dQ}{dt} = \frac{dQ}{dV} \frac{dv}{dt} = C(V) \frac{dv}{dt} = C_0 \frac{dv}{dt} + C_1 v \frac{dv}{dt} + C_2 v^2 \frac{dv}{dt}$$

$$= C_0 \frac{dv}{dt} + \frac{1}{2} C_1 \frac{dv^2}{dt} + \frac{1}{3} C_2 \frac{dv^3}{dt} \qquad (8.18)$$

Here $C(V)$ has two possible formulations,

$$C_d = \tau \frac{dI_D}{dV_J} \qquad (8.19)$$

for the diffusion capacitances defined by Eq. 3.3 and

$$C_j = \frac{C_{JO}}{[1 - (V_J + v_j)/\phi_J]^{MJ}}$$ (8.20)

for the junction capacitances, respectively. I_D represents the diffusion current, I_{CC} or I_{CE}, and V_J is the junction voltage. In the small-signal equivalent C_π and C_μ include both types of capacitances; the printout of the distortion components separates them, however. CB and CBR represent the contributions due to C_{DE} and C_{DC}, defined by Eqs. 3.18:

$$C_{de2} = \frac{\partial^2 Q_{DE}}{\partial V_{BE}^2} = TF \cdot g_{m2}$$

$$C_{dc2} = \frac{\partial^2 Q_{DC}}{\partial V_{BC}^2} = TR \cdot g_{o2}$$ (8.21)

CJE and CJC represent the distortion contributions due to the junction capacitances, defined by Eqs. 3.14:

$$C_{je2} = \frac{dC_{JE}}{dV_{BE}}$$

$$C_{jc2} = \frac{dC_{JC}}{dV_{BC}}$$ (8.22)

respectively.

It is instructive to derive by hand the different distortion components for the one-transistor amplifier in order to understand the meaning of the SPICE2 distortion summary report. The approach used in Sec. 8.2.1 to find the transfer function of the frequency-dependent and nonlinear blocks of Figure 8.1 is used in this derivation. A source of small distortion can be associated with each nonlinear term (Chisholm and Nagel 1973) in Eqs. 8.13 and 8.17, namely, $I_{BE}, I_{BC}, I_{CC}, I_{CE}$ and the different charge terms in Eqs. 8.21 and 8.22. The distortion in the load resistor, R_L, due to each non-linearity is evaluated by considering only one distortion source at a time. The total distortion at the output of the linear circuit is obtained by superposition of all individual contributions. Note that the circuit distortion is additive in a vector sense; that is, two distortion generators of equal magnitude and opposing phase cancel each other.

Start out by writing the transfer function for the small-signal circuit in Figure 8.3. Kirchhoff's current law (KCL) is applied at nodes B′ and C to yield

$$\frac{v_i - v_{b'}}{R_B} + i_\mu - i_\pi = 0$$

$$i_o - i_{gm} - i_\mu = 0$$ (8.23)

These are general equations, which must be satisfied for the terms of different orders in the Taylor series expansion of the currents and voltages. Similarly to S_i in Eq. 8.2 the input signal, v_i, consists of two sinusoids of frequencies ω_1 and ω_2 and amplitudes V_{i1} and V_{i2}:

$$v_i = V_{i1} \cos \omega_1 t + V_{i2} \cos \omega_2 t \tag{8.24}$$

The derivation is simplified by splitting the circuit in Figure 8.3 into two disconnected subnetworks, shown in Figure 8.5. The input equivalent of g_μ, C_μ, is multiplied by the voltage gain, a_{v1}, from the output to the internal base; g_μ is very small and can be neglected.

In the following derivation second-order distortion figures are computed first. Each distortion source contribution derived below has a one-to-one correspondence with a distortion component in the SPICE2 summary.

Consider the nonlinearity due to I_{BC}, represented by $i_{g\pi}$ in Eq. 8.12, the distortion contribution of which appears in the GPI column in the printout. The current i_π, in Eqs. 8.23, has a nonlinear component, $i_{g\pi}$ (Eqs. 8.12), and a linear component, $i_{C\pi}$:

$$i_{C\pi} = j\omega C_\pi v_{b'} \tag{8.25}$$

Because $i_{g\pi}$ is nonlinear $v_{b'}$ can be expressed as a power series of the input signal, v_i:

$$v_{b'} = a_1 v_i + a_2 v_i^2 + a_3 v_i^3 + \cdots \tag{8.26}$$

where a_1 represents the transfer function from the input to the internal base, B'. The coefficients of the above series are derived by satisfying the KCL (Eqs. 8.23) for first-, second-, and third-order terms.

Substitution of the first-order terms in Eqs. 8.23 yields

$$\frac{1 - a_1}{R_B} - g_\pi a_1 - j\omega(C_\pi + a_{v1} C_\mu) a_1 = 0$$

from which the following expression for a_1 can be obtained:

$$a_1(j\omega) = \frac{1}{1 + g_\pi R_B + j\omega R_B(C_\pi + a_{v1} C_\mu)} \tag{8.27}$$

It is calculated by substituting the following circuit component values:

$R_B = 100 \ \Omega$

$g_\pi = 7.5 \times 10^{-4}$ mho

$C_\pi = 1.72$ pF

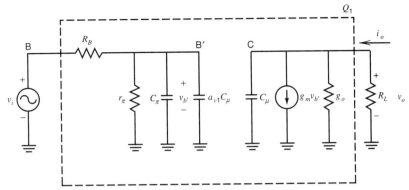

Figure 8.5   Small-signal equivalent using Miller's theorem.

$$C_\mu = 1.26 \text{ pF}$$
$$a_{v1} = g_m R_L = 0.075 \times 10^3 = 75$$

Next, the first of Eqs. 8.23 is written for the second-order terms of the currents and voltages:

$$\frac{-a_2}{R_B} - g_{\pi 2}a_1^2 - a_2 g_\pi - j(\omega_1 + \omega_2)(C_\pi + a_{v1}C_\mu)a_2 = 0$$

Note that v_i has no terms of second or higher order; a_2 can be expressed from the above equation:

$$a_2(j\omega_1, j\omega_2) = -\frac{g_{\pi 2}R_B a_1(j\omega_1)a_1(j\omega_2)}{1 + g_\pi R_B + j(\omega_1 + \omega_2)R_B(C_\pi + a_{v1}C_\mu)} \tag{8.28}$$

where a_2 is a function of ω_1 and ω_2, and different combinations of the two must be used depending on which distortion measure is evaluated.

The values of a_2 represent the second-order distortion at the internal base due to g_π, that is, the I_{BC} nonlinearity. The contribution at the output, in resistor R_L, can be obtained by multiplying the second-order component in the $v_{b'}$ series by the gain:

$$v_{og\pi}^{(2)} \approx -a_{v1}v_{b'}^{(2)} = -a_{v1}a_2(j\omega_1, j\omega_1)v_i^2 = -a_2(j\omega_1, j\omega_2)g_m R_L v_i^2 \tag{8.29}$$

The expressions of *HD2*, *SIM2*, and *DIM2*, due to g_π, result from the second-order terms in Eqs. 8.6 and 8.8 evaluated for $a_2(j\omega_1, j\omega_1)$, $a_2(j\omega_1, j\omega_2)$, and $a_2(j\omega_1, -j\omega_2)$, respectively.

$$HD2_{g\pi} = \frac{1}{2}\frac{v_{og\pi}^{(2)}}{v_o^2}v_{on} \tag{8.30}$$

where v_o is the gain referenced to the input for an input signal of amplitude 1 V, and v_{on} is the normalized output voltage corresponding to 1 mW power in R_L. $a_2(j\omega_1, j\omega_1)$ is

$$a_2(j\omega_1, j\omega_1) = -\frac{0.0145 \cdot 10^2(0.93 - j0.053)^2}{1.075 + j6.28 \cdot (2 \cdot 10^6) \cdot 10^2(1.72 + 75.3 \cdot 1.26)10^{-12}}$$

$$= -1.24(0.92 - j0.212)$$

where

$$g_{\pi 2} = \frac{1}{2}\frac{g_\pi}{V_{th}} = 0.0145 \text{ A}/\text{V}^2$$

Substitution in Eq. 8.30 yields the amplitude and phase of $HD2_{g\pi}$:

$$|HD2_{g\pi}| = 0.0126$$

$$\angle HD2_{g\pi} = -13°$$

These values are in excellent agreement with the values computed by SPICE2 in the GPI column of the 2ND HARMONIC DISTORTION category.

The second-order intermodulation components are calculated similarly; for $SIM2$ evaluate

$$a_2(j\omega_1, j\omega_2) = -\frac{1.45(0.93 - j0.053)^2}{1.075 + j6.28 \cdot (1.9 \cdot 10^6) \cdot 10^2 \cdot (1.72 + 75.3 \cdot 1.26)10^{-12}}$$

$$\approx a_2(j\omega_1, j\omega_1)$$

where $\omega_1 + \omega_2 = 1.9\omega_1$, corresponding to 1.9 MHz. The distortion signal at the output, $v_{og\pi}$, is calculated for $a_2(j\omega_1, j\omega_2)$, and its value is substituted into the definition of $SIM2$ of Eq. 5.16:

$$SIM2_{g\pi} = \frac{1}{2}\frac{v_{og\pi}^{(2)}(j\omega_1, j\omega_2)}{v_o^2}v_{on} \tag{8.31}$$

The magnitude and phase of $SIM2$ are

$$|SIM2_{g\pi}| = 0.0256$$

$$\angle SIM2_{g\pi} = -13°$$

The difference intermodulation component is derived similarly by substituting

$$a_2(j\omega_1, -j\omega_2) = -\frac{1.45(0.93 - j0.053)(0.93 + j0.047)}{1.075 + j6.28 \cdot (0.1 \cdot 10^6) \cdot 10^2(1.72 + 75.3 \cdot 1.26)10^{-12}}$$

$$= -1.25(0.93 - j0.012)$$

in Eq. 8.29, where $\omega_1 - \omega_2$ is $2\pi \times 0.1$ MHz. The magnitude and phase of *DIM2* are

$$|DIM2_{g\pi}| = 0.0255$$

$$\angle DIM2_{g\pi} = -0.7°$$

Note that the *IM2* components agree with the SPICE2 results and the magnitudes are equal to $2 \cdot HD2$ as predicted in the frequency-independent case in Example 5.7.

The harmonic distortion due to the nonlinearity of I_C is summarized in the SPICE2 output in the columns GM, GO, and GMO2. The equivalent distortion sources generating these three contributions to the second harmonic are defined in Eqs. 8.17, based on the Taylor series expansion of i_C around the operating point I_C, Eq. 8.14.

The following equations are used to derive $HD2_{gm}$, the second harmonic component due to the g_m distortion source. i_c is expressed as a power series of $v_{b'}$:

$$i_c = i_{gm} = g_m v_{b'} + g_{m2} v_{b'}^2 + g_{m3} v_{b'}^3 \tag{8.32}$$

and v_o is a power series of $v_{b'}$:

$$v_o = -i_o R_L = b_1 v_{b'} + b_2 v_{b'}^2 + b_3 v_{b'}^3 \tag{8.33}$$

with

$$v_{b'} = a_1 v_i$$

The first- and second-order terms of i_c and v_o are used in the KCL, the second of Eqs. 8.23, to find b_1 and b_2:

$$b_1(j\omega_1) = -\frac{g_m R_L}{1 + j\omega_1 C_\mu R_L}$$

$$b_2(j\omega_1, j\omega_2) = -\frac{g_{m2} R_L}{1 + j(\omega_1 + \omega_2) C_\mu R_L} \tag{8.34}$$

where g_{m2} is computed according to the Taylor series coefficients, Eqs. 8.17:

$$g_{m2} = \frac{1}{2}\frac{g_m}{V_{th}} = 1.45 \text{ A}/\text{V}^2$$

b_2 represents the contribution of g_m to the second harmonic in the output voltage:

$$v_{ogm}^{(2)} = b_2 v_{b'}^2$$

The distortion at the output is computed according to Eq. 8.29:

$$HD2_{gm} = \frac{1}{2}\frac{v_{ogm}^{(2)}}{v_o^2} v_{on} \tag{8.35}$$

resulting in the following values:

$$|HD2_{gm}| = 0.181$$
$$\angle HD2_{gm} = 166°$$

At first glance the distortion due to g_o should be zero since no value has been specified for the Early voltage, VAF. The second harmonic contribution is found in the GO column of the SPICE2 output and is equal to 1.36×10^{-8}. This result can be explained by the conductance $GMIN$ added in parallel to each junction (see Sec. 10.2.1), equal by default to 10^{-12} mho. The resulting value can be verified replacing the second-order terms in the KCL (Eq. 8.23) at the output node and using only the g_o terms of the i_c series:

$$i_c = g_m v_{b'} + g_o v_o + g_{o2} v_o^2 + g_{o3} v_o^3 \tag{8.36}$$

A new value is obtained for b_2, which now represents the distortion at the output due to g_o:

$$b_2(j\omega_1, j\omega_2) = \frac{g_{o2}R_L b_1(j\omega_1)b_1(j\omega_2)}{1 + g_o + j(\omega_1 + \omega_2)C_\mu R_L} \tag{8.37}$$

where the g_o term has been neglected and

$$g_{o2} = \frac{1}{2}\frac{g_o}{V_{th}} = 2 \times 10^{-11} \text{ A}/\text{V}^2$$

The second harmonic distortion coefficient, $HD2_{go}$, follows:

$$|HD2_{go}| = 1.4 \times 10^{-8}$$

The distortion contribution due to the second-order cross-term g_{mo2} in the i_c series, Eq. 8.16, is nonzero because of the *GMIN* conductance. The value listed under GMO2 in Figure 8.4 can be verified through the above approach.

The distortion components due to Q_{BE} and Q_{BC}, represented by C_π and C_μ in Figure 8.3, are evaluated next. Only the junction capacitances **CJE** and **CJC** are defined for transistor Q_1 in this example; that is, no **TF** or **TR** parameters are specified, and therefore the nonlinearity in C_π or C_μ can be expressed in the single power series

$$C_j = \frac{C_{JO}}{(\phi_J + V_{JR})^{MJ}} \left(1 + \frac{v_j}{\phi_J + V_{JR}}\right)^{-MJ} = C_{j0} + C_{j1}v_j + C_{j2}v_j^2 \qquad (8.38)$$

based on Eqs. 8.18 and 8.20 and the series

$$\frac{1}{(1+x)^\alpha} = 1 + \alpha x + \frac{1}{2!}\alpha(\alpha - 1)x^2 + \cdots$$

For the evaluation of the distortion contribution of C_{je}, Eqs. 8.23 and 8.26, which were used in calculating the distortion contributions of g_π, remain valid, but Eqs. 8.12 and 8.25, expressing currents $i_{g\pi}$ and $i_{C\pi}$, must be changed to

$$i_{g\pi} = g_\pi v_{b'} \qquad (8.39)$$

$$i_{C\pi} = C_{\pi 0}\frac{dv_{b'}}{dt} + C_{\pi 1}\frac{dv_{b'}^2}{dt} + C_{\pi 2}\frac{dv_{b'}^3}{dt} \qquad (8.40)$$

The values of $C_{\pi 0}$, $C_{\pi 1}$, and $C_{\pi 2}$, which correspond to C_{j0}, C_{j1}, and C_{j2} in Eq. 8.38, are not computed according to Eq. 8.38 because the BE junction is forward biased, with $V_{BE} > FC \cdot VJE$, and SPICE uses the following approximation:

$$C_{\pi 0} = C_{je} = \frac{CJE}{(1 - FC)^{MJE}}\left[1 + \frac{MJE}{VJE(1 - FC)}(V_{BE} - FC \cdot VJE)\right] \qquad (8.41)$$

described in Sec. 10.2.2; the graphical interpretation of the above approximation is shown in Figure 10.7. The coefficients C_{j1} and C_{j2} in the power series of C_j, Eq. 8.38, are the first and second derivatives of C_j with respect to V_J, respectively. The derivatives of Eq. 8.41 lead to the coefficients for Eq. 8.40:

$$C_{\pi 1} = C_{je1} = CJE\frac{MJE}{VJE(1 - FC \cdot VJE)^{(MJE+1)}} \qquad (8.42)$$

$$C_{\pi 2} = C_{je2} = 0$$

Eqs. 8.23 and 8.40 are used to compute the values of the coefficients a_1, a_2, and a_3 of the $v_{b'}$ series, Eq. 8.26. The first-order equation and, therefore, a_1 are the same as for g_π. The second-order terms inserted into the first of Eqs. 8.23 yield the following expression for a_2:

$$a_2(j\omega_1, j\omega_2) = -\frac{j(\omega_1 + \omega_2)C_{\pi 1}R_B}{(1 + j\omega_1 C_{\pi 0}R_B)(1 + j\omega_2 C_{\pi 0}R_B)[1 + j(\omega_1 + \omega_2)C_{\pi 0}R_B]} \quad (8.43)$$

The amplitude and phase of the second-order distortion, $HD2_{C\pi}$, due to C_π are

$$|HD2_{C\pi}| = 4.5 \cdot 10^{-6}$$

$$\angle HD2_{C\pi} = 77°$$

which are in excellent agreement with the SPICE2 values.

The distortion contribution due to C_μ will be derived next. First, calculate the distortion due to $C_{\mu i}$, the reflection of C_μ to the input circuit, as shown in Figure 8.5. The first of Eqs. 8.23 is used with i_μ expressed by the following series due to the nonlinearity in the junction capacitance C_μ:

$$i_\mu = C_{\mu 0}a_{v1}\frac{dv_{b'}}{dt} + C_{\mu 1}a_{v1}^2\frac{dv_{b'}^2}{dt} + C_{\mu 2}a_{v1}^3\frac{dv_{b'}^3}{dt}$$

$$= j\omega_1 C_{\mu 0}a_{v1}v_{b'} + j(\omega_1 + \omega_2)C_{\mu 1}a_{v1}^2 v_{b'}^2 + j(\omega_1 + \omega_2 + \omega_3)C_{\mu 2}a_{v1}^3 v_{b'}^3 \quad (8.44)$$

where v_o has been replaced by $a_{v1}v_{b'}$.

$C_{\mu 0}$ is equal to C_{jc}, the capacitance of the reverse-biased BC junction, and has the following derivatives C_{jc1} and C_{jc2}:

$$C_{\mu 1} = C_{jc1} = \frac{1}{2}C_{\mu 0}\frac{MJC}{VJC - V_{BC}}$$

$$C_{\mu 2} = C_{jc2} = \frac{2}{3}C_{\mu 1}\frac{1 + MJC}{VJC - V_{BC}} \quad (8.45)$$

Substitution of second-order terms in Eqs. 8.23 yields the following value for a_2 in Eq. 8.26:

$$a_2(j\omega_1, j\omega_2) = -\frac{j(\omega_1 + \omega_2)C_{\mu 1}R_B a_{v1}^2 a_1(j\omega_1)a_1(j\omega_2)}{1 + g_\pi R_B + j(\omega_1 + \omega_2)R_B(C_\pi + a_{v1}C_\mu)} \quad (8.46)$$

The second-order distortion component $HD2_{C\mu i}$ has the following magnitude:

$$|HD2_{C\mu i}| = 4.35 \times 10^{-3}$$

The second part of the distortion due to C_μ is evaluated from the output circuit in Figure 8.5. The nonlinearity of i_μ is assumed to be due only to v_o, which is a power series of $v_{b'}$:

$$v_o = b_1 v_{b'} + b_2 v_{b'}^2 + b_3 v_{b'}^3,$$

$$i_\mu = j\omega_1 C_{\mu 0} v_o + j(\omega_1 + \omega_2) C_{\mu 1} v_o^2 + j(\omega_1 + \omega_2 + \omega_3) C_{\mu 2} v_o^3$$

(8.47)

The assumption that the input circuit is linear represents an approximation made in order to avoid the more complex solution, which involves the power series of both $v_{b'}$ and v_o simultaneously. Substitution of the first- and second-order terms in Eqs. 8.23 yields the following values of the coefficients b_1 and b_2:

$$b_1(j\omega_1) = -\frac{g_m R_L}{1 + j\omega_1 C_\mu R_L}$$

(8.48)

$$b_2(j\omega_1, j\omega_2) = -\frac{1}{2} \cdot \frac{j(\omega_1 + \omega_2) C_{\mu 2} R_L b_1(j\omega_1) b_1(j\omega_2)}{1 + j(\omega_1 + \omega_2) C_\mu R_L}$$

(8.49)

The second-order distortion due to the C_μ component connected at the output node is obtained from Eqs. 8.47 after inserting the value of b_2 from Eq. 8.49:

$$|HD2_{C\mu o}| = 7 \times 10^{-4}$$

The contribution due to C_μ at the output is the sum of the two coefficients

$$|HD2_{C\mu}| = |HD2_{C\mu i}| + |HD2_{C\mu o}| = 5.05 \times 10^{-3}$$

This value is in agreement with the value found in the SPICE listing of Figure 8.4 in the CJC column.

The section 2ND HARMONIC DISTORTION is concluded by the line HD2 MAGNITUDE 1.674D-01 PHASE 164.49 = -15.53 DB, which gives the total second-harmonic distortion in the load resistor obtained by adding all the complex numbers representing the individual distortion components. The magnitude 0.167 is very close to the value of *HD2* obtained in Sec. 5.4 for the transistor with no frequency-dependent elements. Because 1 MHz is a relatively low frequency, the distortion due to the nonlinear capacitances C_μ and C_π is negligible compared to the nonlinear resistive contributions, g_m and g_π. Similarly, the total second-order intermodulation sum and difference components track the values of Chap. 5 and equal $2 \times HD2$.

The third-order distortion can be computed following the same steps as above. Because the number of terms involved is higher, the derivation is more difficult and prone to errors due to some approximations made in hand calculations.

The third-order distortion due to the I_{BC} component of the base current can be computed by replacing the third-order terms of Eqs. 8.25 and 8.26 into KCL, Eqs. 8.23.

Coefficient a_3 in the $v_{b'}$ series, Eq. 8.26, results:

$$a_3(j\omega_1, j\omega_2, j\omega_3)$$
$$= \frac{-g_{\pi3}R_B a_1(j\omega_1)a_1(j\omega_2)a_1(j\omega_3) - 2g_{\pi2}R_B a_1(j\omega_1)a_2(j\omega_1, j\omega_2)}{1 + g_\pi R_B + j(\omega_1 + \omega_2 + \omega_3)R_B(C_\pi + a_{v1}C_\mu)} \quad (8.50)$$

where the possible values of ω_3 are $\pm\omega_1$ or $\pm\omega_2$. The *HD3* contribution at the output is obtained by multiplying a_3 by the gain a_{v1}:

$$v_{og\pi}^{(3)} \approx -a_{v1}v_{b'}^{(3)} = -a_{v1}a_3(j\omega_1, j\omega_1, j\omega_1)v_i^3 = -a_3 g_m R_L v_i^3 \quad (8.51)$$

and the third-order distortion due to g_π is given by

$$HD3_{g\pi} = \frac{1}{4}\frac{v_{og\pi}^{(3)}}{v_o^3}v_{on}^2 \quad (8.52)$$

The magnitude of this distortion yields

$$|HD3_{g\pi}| = 1.21 \times 10^{-3}$$

which is in agreement with the SPICE2 value for `GPI` in the `3RD HARMONIC DISTORTION` section.

The third-order distortion component due to g_m is computed similarly, by the insertion of the third-order terms in Eqs. 8.32 and 8.33 into KCL, Eq. 8.23, for the output section of the transistor model. The third-order coefficient in the power series for v_o results:

$$b_3(j\omega_1, j\omega_2, j\omega_3) = -\frac{g_{m3}R_L}{1 + j(\omega_1 + \omega_2 + \omega_3)C_\mu R_L} \quad (8.53)$$

where ω_3 can be $\pm\omega_1$ or $\pm\omega_2$ depending on the distortion component to be derived, and

$$g_{m3} = \frac{1}{3}\frac{g_{m2}}{V_{th}} = 18.73 \text{ A}/\text{V}^3 \quad (8.54)$$

The third-order distortion contribution due to g_m in the output voltage is

$$v_{ogm}^{(3)} = b_3 v_{b'}^3 \quad (8.55)$$

which translates into the following magnitude for *HD3:*

$$|HD3_{gm}| = 2.22 \times 10^{-2}$$

$HD3_{gm}$ is evaluated according to Eq. 8.52 where the g_m contribution replaces that of g_π in v_o. This value is larger than the 1.79×10^{-2} predicted by SPICE; for the third-order distortion it is harder to separate in hand calculations the different distortion components contributed by I_C.

Two additional distortion components, GM2O3 and GMO23, can be noticed in the summary report for third-order distortion. They represent the distortion due to the following partial derivatives in the Taylor series expansion of i_C:

$$g_{m2o3} = \frac{\partial^3 I_C}{\partial V_{BE}^2 \partial V_{BC}} \tag{8.56}$$

$$g_{mo23} = \frac{\partial^3 I_C}{\partial V_{BE} \partial V_{BC}^2} \tag{8.57}$$

The total third-order distortion $HD3$ is the magnitude of the vector sum of all components, equal to 1.84×10^{-2}, and is very close to the value obtained in Chap. 5 for the circuit with pure resistive nonlinearities and without parasitic base resistance.

An important distortion measure is the third-order intermodulation, denoted $IM3$, which represents the spectral component at $2\omega_1 - \omega_2 = 2\pi \times 1.1$ MHz. The term with this frequency in Eqs. 8.6 represents the $IM3$ component. Assuming that $F(j\omega_1) = F(j\omega_2)$, the $IM3$ distortion can be related to $HD3$ as follows:

$$IM3 = 3\frac{S_2}{S_1} HD3 \tag{8.58}$$

as long as frequency-dependent effects are not important. In our example this assumption is generally valid, and the value $IM3 = 4.88 \times 10^{-2}$ computed by SPICE relates to $HD3$ approximately according to Eq. 8.58 with S_2 equal to 1 if not otherwise specified on the .DISTO line.

The last distortion category computed by SPICE as part of the summary is that of the APPROXIMATE CROSS MODULATION COMPONENTS. Cross modulation (Meyer, Shensha, and Eschenbach 1972) occurs when one of the two input signals in Eq. 8.2 is modulated in amplitude and the other is not:

$$v_i = V_{i1} \cos \omega_1 t + V_{i2}(1 + m \cos \omega_m t) \cos \omega_2 t \tag{8.59}$$

where m is the modulation index. Because of the nonlinearities in the circuit, amplitude modulation is transferred to the signal ω_1, the carrier signal. In the third-order term in Eqs. 8.6 the following cross-modulation term is generated:

$$3a_3 V_{i1} V_{i2}^2 m \cos \omega_m t \cos \omega_1 t \tag{8.60}$$

The *cross modulation index, CM,* is defined as the ratio of the transferred amplitude modulation to the original fractional modulation,

$$CM = 3\frac{a_3}{a_1}V_i^2 \tag{8.61}$$

for equal amplitudes of the two input signals, $V_{i1} = V_{i2}$. This definition is valid for memoryless, or resistive, nonlinearities only. For frequency-dependent nonlinear circuits the phase shifts of the different transfer functions must be considered in defining a frequency-domain cross-modulation factor, *CMF*:

$$CMF = 3\frac{|H_3(j\omega_1, j\omega_2, -j\omega_2)|}{|H_1(j\omega_1)|}V_i^2 \tag{8.62}$$

where $H_3(j\omega_1, j\omega_2, -j\omega_2)$ is the third-order transfer function for the modulated ω_1 component (see Eqs. 8.6 and 8.8). At high frequencies the amplitude crossmodulation, CMA, corresponds to *CM* at low frequencies (defined by Eq. 8.61) and is equal to

$$CMA = CMF \cdot \cos\phi \tag{8.63}$$

where

$$\phi = \angle H_3 - \angle H_1$$

A *phase cross-modulation factor,* CMP, can be defined as the ratio of the transferred phase modulation to the original fractional amplitude modulation:

$$CMP = CMF \sin\phi \tag{8.64}$$

The values computed by SPICE2 are based on the relation between *CMF* and the third-order intermodulation distortion, *IM3*. From comparison of Eq. 8.62 and the definition of *IM3*, Eq. 8.58, the following equality results:

$$CMF = 4 \cdot IM3 \tag{8.65}$$

The above equality is used in SPICE2 to compute the amplitude and phase cross-modulation terms:

$$CMA = 4|IM3|\cos(\phi_{IM3} - \phi_o) \tag{8.66}$$

$$CMP = 4|IM3|\sin(\phi_{IM3} - \phi_o) \tag{8.67}$$

where ϕ_{IM3} and ϕ_o are the phase of *IM3* and the phase of the signal at the output, respectively.

8.3 LARGE-SIGNAL DISTORTION ANALYSIS

A good approach for estimating the total harmonic distortion is to run a large-signal time-domain analysis and then use the `.FOUR` analysis introduced in Sec. 6.4. The Fourier analysis computes the first 10 spectral components in SPICE2 and a user-specified number of harmonics in SPICE3. Large-signal analysis provides more accurate results than AC small-signal analysis because of the removal of the linearity, or small-signal, assumption. The computation of the spectral components, however, can introduce errors due to the approximation of the waveform based on the values stored for the discrete timepoints used in the transient analysis. Also, the Fourier analysis does not offer the evaluation of intermodulation distortion in the presence of two input signals and the distortion contribution by type of nonlinearity available with the small-signal `.DISTO` analysis.

8.3.1 One-Transistor Amplifier Distortion

It is a useful exercise to check the distortion measures computed by SPICE2 using `.DISTO` for the one-transistor amplifier in the above section.

The most important issue is to scale the amplitude of the sinusoidal input signal properly so that the circuit dissipates the same power in the load resistor as specified in the `.DISTO` statement. The distortion components for the above circuit were computed for 1 mW power in R_L, or 0 dBm. The amplitude of the output voltage, V_o, corresponding to this specification is

$$V_o = \sqrt{2P_{ref}R_L} = 1.41 \text{ V} \qquad (8.68)$$

The gain predicted by the AC analysis for this circuit (see the results in Figure 8.4) is

$$a_v = \frac{V_o}{V_i} = 69.9$$

which leads to an input amplitude of

$$V_i = \frac{V_o}{a_v} = 20.2 \text{ mV} \qquad (8.69)$$

```
ONE-TRANSISTOR CIRCUIT (Figure 5.8)
*
Q1 2 1 0 QMOD
RL 2 3 1K
VCC 3 0 5
```

```
*
VBE 4 0 793.4M
VI 1 4 SIN 0 20.2M 1MEG AC 1
*
.MODEL QMOD NPN
+ RB=100
+ CJE=1P
+ CJC=2P
*
.OP
*.AC LIN 1 1MEG 1MEG
*.DISTO RL 1
*
.TRAN 1N 2U 0 1N
.PLOT TRAN V(2)
.FOURIER 1MEG V(2)
.WIDTH OUT=80
.OPT NOPAGE RELTOL=1E-4 ITL5=0 LIMPTS=5000
.END
```

The SPICE input is shown above; the circuit has the sinusoidal input signal V_i with a DC offset, V_{BE}, of 793.4 mV necessary to bias the circuit, an amplitude of 20.2 mV, and a frequency of 1MHz. Note the large number of time steps specified on the **.TRAN** line, 2000, as well as the value of *TMAX*, the maximum time step the program is allowed to use in order to estimate the spectral components over the last period. Before checking the results of the Fourier analysis, one needs to double-check that the value of the gain computed in the AC analysis is accurate, that is, the amplitude of V_o should be 1.41 V. The measurements show that this value is 1.46 V. This value points to a gain of 72.7, which inserted into Eq. 8.69 leads to the following value of V_i:

$$V_i = 19.4 \, \text{mV}$$

The discrepancy in the value of the gain can be explained by the fact that the input voltage is very close to the value of V_{th}, the thermal voltage, where the accuracy of the linear approximation declines.

The results of the analysis with $V_i = 19.4 \, \text{mV}$ are shown in Figure 8.6. First check the value of the first spectral component at 1 MHz, which must be 1.41 V for proper calibration of the spectral components. The second- and third-order harmonic distortions are

$$HD2 = 0.154$$

$$HD3 = 0.0162$$

which are close to but slightly less than the values obtained from the **.DISTO** analysis.

```
****    FOURIER ANALYSIS                    TEMPERATURE =    27.000 DEG C


FOURIER COMPONENTS OF TRANSIENT RESPONSE V(2)
DC COMPONENT =    2.831D+00
HARMONIC    FREQUENCY    FOURIER      NORMALIZED     PHASE     NORMALIZED
   NO         (HZ)      COMPONENT    COMPONENT      (DEG)    PHASE (DEG)
    1       1.000D+06   1.412D+00    1.000000     175.453        0.000
    2       2.000D+06   2.179D-01    0.154288      71.477     -103.976
    3       3.000D+06   2.288D-02    0.016205     -69.935     -245.388
    4       4.000D+06   5.976D-03    0.004232     127.044      -48.409
    5       5.000D+06   1.831D-03    0.001296      -4.229     -179.682
    6       6.000D+06   5.241D-04    0.000371    -131.263     -306.715
    7       7.000D+06   1.484D-04    0.000105     102.947      -72.505
    8       8.000D+06   4.055D-05    0.000029     -21.846     -197.299
    9       9.000D+06   1.062D-05    0.000008    -146.013     -321.466


    TOTAL HARMONIC DISTORTION =      15.520066  PERCENT
```

Figure 8.6 Fourier analysis results.

8.3.2 Single-Device Mixer Analysis

Mixer circuits are commonly used in radio receivers. The radio signal, ω_s, and a local signal generated in the receiver, ω_{lo}, are fed to a mixer circuit. The local oscillator frequency is tunable so that the difference between the two signal frequencies is approximately constant.

The mixer operates as an analog multiplier by generating spectral components of frequencies $\omega_{lo} \pm \omega_s$. The mixer is followed by a bandpass filter tuned to the difference frequency, which rejects the sum and other frequency components. The difference frequency is called the intermediate frequency, ω_{if}.

A typical single-transistor mixer circuit is shown in Figure 8.7. C_1, L_1, and the load resistor, R_1, in parallel form a bandpass filter at the output of the mixer. Two voltage sources are connected at the input: v_{LO} sets the bias at the base of transistor, Q_1, and generates the local oscillator signal (LO), and v_s is defined as the input signal (S):

$$v_s = V_s \cos \omega_s t$$

$$v_{lo} = V_{lo} \cos \omega_{lo} t$$

The two signals add up to an input voltage v_i consisting of two frequencies:

$$v_i = v_{lo} + v_s = V_{lo} \cos \omega_{lo} t + V_s \cos \omega_s t \tag{8.70}$$

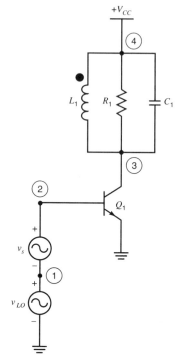

Figure 8.7   Single-transistor
mixer.

The transfer characteristic of the BJT is approximately equal to

$$i_C = I_s e^{(V_{BE}+v_s+v_{lo})/V_{th}} = I_C e^{v_i/V_{th}} = I_C e^{v_s/V_{th}} e^{v_{lo}/V_{th}} \tag{8.71}$$

where I_C is the quiescent DC value of the collector current. There are two possible series expansions for an exponential. In Eq. 5.21 and in the above sections a power series was used under the assumption of a small signal, namely $V_i/V_{th} \ll 1$.

A difference-frequency component is present in the output current, i_C, according to the small-signal derivation of Sec. 8.2.1:

$$i_C = I_C\left(1 + a_1 v_i + a_2 v_i^2 + a_3 v_i^3 + \cdots\right)$$

$$= I_C\left[1 + \cdots + a_2 V_{lo} V_s \cos(\omega_{lo} - \omega_s)t + \cdots\right] \tag{8.72}$$

where v_i is the total input signal for the mixer circuit. This *IM2* component represents the IF signal, which is the output of interest in a mixer:

$$i_{Cif} = a_2 V_{lo} V_s \cos(\omega_{lo} - \omega_s)t \qquad (8.73)$$

In the case of a mixer or a multiplier the small-signal assumption is not always true. If the amplitude of an input sinusoid is not small with respect to V_{th}, the exponential can be expanded in a Fourier series with Bessel function coefficients (Pederson and Mayaram 1990):

$$\exp(V_s \cos \omega_s t / V_{th}) = I_0(s) + 2I_1(s) \cos \omega_s t + \cdots$$
$$\exp(V_{lo} \cos \omega_{lo} t / V_{th}) = I_0(l) + 2I_1(l) \cos \omega_{lo} t + \cdots \qquad (8.74)$$

where

$$s = \frac{V_s}{V_{th}}$$

$$l = \frac{V_{lo}}{V_{th}}$$

and $I_n(x)$ are modified Bessel functions of order n, which can be found in mathematical tables. After substitution of the series of Eq. 8.74 into Eq. 8.71 and use of trigonometric identities, the output collector current is

$$
\begin{aligned}
i_C &= I_C I_0(s) I_0(l) + \cdots + 2I_C I_1(s) I_1(l) \cos(\omega_{lo} - \omega_s)t + \cdots \\
&= I_{DC} \left[1 + \cdots + 2 \frac{I_1(s) I_1(l)}{I_0(s) I_0(l)} \cos(\omega_{lo} - \omega_s)t + \cdots \right] \qquad (8.75)
\end{aligned}
$$

where

$$I_{DC} = I_C I_0(s) I_0(l)$$

is the dynamic average value of the collector current. A difference-frequency spectral component $(\omega_{lo} - \omega_s)$, the IF component, is generated as well as other components not included above; all spectral components except the IF component must be rejected by the output filter. The IF component in Eq. 8.75 derived through a Fourier series expansion is more general than the expression in Eq. 8.72 or that in 8.73, obtained from small-signal analysis. For small signals Eqs. 8.72 and 8.75 are identical.

Assume the following values for the amplitudes and frequencies of the two signals: $V_{lo} = 100\,\text{mV}$, $f_{lo} = 1.1\,\text{MHz}$, $V_s = 1\,\text{mV}$, and $f_s = 1\,\text{MHz}$. V_{lo} is a large signal and therefore Eq. 8.75 must be used to calculate the IF component. The ratios of Bessel coefficients are found in math tables:

$$\frac{I_1(l)}{I_0(l)} = 0.95$$

$$\frac{I_1(s)}{I_0(s)} = \frac{s}{2} = \frac{1}{2}\frac{V_s}{V_{th}}$$

The difference component of the collector current becomes

$$i_{Cif} = 0.95 I_{DC}\frac{V_s}{V_{th}}\cos\omega_{if}t \tag{8.76}$$

The correct design of a mixer requires that the amplitude of the IF output voltage V_{if} be much larger than any other voltage. The selection of this value is made difficult by the fact that the amplitude of the local oscillator, V_{lo}, is much larger than the radio signal, V_s. The Q of the parallel RLC filter must be dimensioned so that the rejection, Rej, at ω_{lo} is sufficient to boost the IF component.

From the Fourier series,

$$i_{Clo} = 2I_{DC}\frac{I_1(l)}{I_0(l)}\cos\omega_{lo}t \tag{8.77}$$

and using i_{Cif} as given in Eq. 8.76 the ratio of the LO component to the IF component is given by

$$\frac{i_{Clo}}{i_{Cif}} = 2\frac{V_{th}}{V_s} \tag{8.78}$$

which in this example is over 50.

The rejection of a given frequency component, $Rej(\omega)$, for a single-tuned parallel RLC filter is given by

$$Rej(\omega) = \left[\frac{|Z(j\omega)|}{R}\right]^{-1} \approx Q\left|\frac{\omega_o}{\omega} - \frac{\omega}{\omega_o}\right| \tag{8.79}$$

where

$$\omega_o = \frac{1}{\sqrt{LC}}$$

is the resonant frequency and Q is the quality factor defined in Eq. 6.26.

The ratio of interest between the voltage amplitudes of the IF and LO components is

$$\frac{V_{oif}}{V_{olo}} = 2\frac{V_{th}}{V_s}Rej(\omega_{lo}) \qquad (8.80)$$

If $Q = 40$ for the tuned circuit of Figure 8.7, the rejection of the LO component is $Rej = 440$, leading to an IF voltage component 8.49 times larger than the amplitude of the local oscillator at the output V_{olo}.

The SPICE circuit description and the resulting DC operating point are shown in Figure 8.8. The IF and LO frequency components can be checked by requesting a .**FOUR** analysis. There is a limit of only nine harmonics printed by SPICE2 including the fundamental. In order to have SPICE2 compute the amplitudes of both the IF and LO components, two separate runs must be performed. The first set of input commands is

```
.TRAN .2U 620U 600U 45N
.FOUR 100K V(3)
.PLOT TRAN V(3)



*******12/13/90 ********  SPICE 2G.6    3/15/83 ********18:44:38*****


ONE TRANSISTOR MIXER CIRCUIT
****      INPUT LISTING              TEMPERATURE =   27.000 DEG C


********************************************************************


Q1 3 2 0 MOD1
R1 3 4 15K
C1 3 4 4.244NF
L1 3 4 596.83UH
.MODEL MOD1 NPN
*
* SUPPLY, SIGNAL AND LOCAL OSCILLATOR
*
VCC 4 0 10
VS  2 1 0 SIN 0 1MV 1MEG
VLO 1 0 0 SIN 0.78 100MV 1.1MEG AC 1
*
.TRAN .2U 620U 600U 45N
.FOURIER 100K V(3)
*.TRAN 45N 601.8U 600U 45N
*.FOURIER 1.1M V(3)
.PLOT TRAN V(3)
.WIDTH OUT=80
.OPTIONS RELTOL=1E-4 ITL5=0
.END
```

Figure 8.8 SPICE2 input and DC bias point for mixer circuit.

```
****      BJT MODEL PARAMETERS                TEMPERATURE =   27.000 DEG C


                 MOD1
TYPE              NPN
IS              1.00D-16
BF              100.000
NF                1.000
BR                1.000
NR                1.000


****      SMALL SIGNAL BIAS SOLUTION        TEMPERATURE =   27.000 DEG C


   NODE    VOLTAGE       NODE   VOLTAGE       NODE    VOLTAGE

 (  2)     0.7800      (  3)   10.0000      (  4)   10.0000


      VOLTAGE SOURCE CURRENTS


      NAME        CURRENT


      VCC        -1.251D-03
      VLO        -1.251D-05


      TOTAL POWER DISSIPATION    1.25D-02   WATTS

****      OPERATING POINT INFORMATION        TEMPERATURE =   27.000 DEG C

**** BIPOLAR JUNCTION TRANSISTORS


               Q1
MODEL        MOD1
IB           1.25E-05
IC           1.25E-03
VBE             0.780
VBC            -9.220
VCE            10.000
BETADC        100.000
GM           4.84E-02
RPI          2.07E+03
RX           0.00E+00
RO           1.00E+12
CPI          0.00E+00
CMU          0.00E+00
CBX          0.00E+00
CCS          0.00E+00
BETAAC        100.000
FT           7.70E+17
```

Figure 8.8   *(continued)*

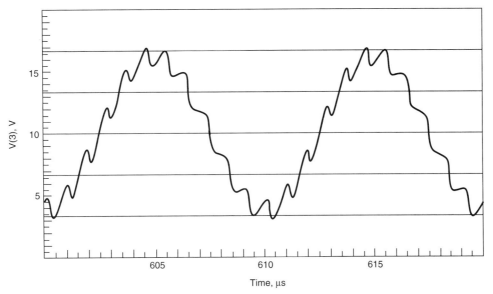

Figure 8.9   IF component waveform.

which request SPICE2 to plot the waveform of V(3), the collector voltage, for the last two periods ($T_{if} = 10\mu s$) after 60 cycles have been computed to assure that the circuit has reached steady state. The **.FOUR** statement requests the harmonics of the 100-kHz spectral component. The IF waveforms modulated by the LO signal

```
****     FOURIER ANALYSIS                    TEMPERATURE =    27.000 DEG C

FOURIER COMPONENTS OF TRANSIENT RESPONSE V(3)
DC COMPONENT =    1.000D+01
HARMONIC   FREQUENCY    FOURIER    NORMALIZED    PHASE     NORMALIZED
   NO        (HZ)      COMPONENT   COMPONENT     (DEG)     PHASE (DEG)

    1      1.000D+05   6.225D+00   1.000000    -89.950      0.000
    2      2.000D+05   5.624D-03   0.000903      5.330     95.280
    3      3.000D+05   1.593D-03   0.000256     42.211    132.161
    4      4.000D+05   1.347D-03   0.000216   -157.027    -67.077
    5      5.000D+05   1.059D-03   0.000170    -61.888     28.062
    6      6.000D+05   1.054D-03   0.000169     19.929    109.879
    7      7.000D+05   1.509D-03   0.000242    100.348    190.298
    8      8.000D+05   5.607D-04   0.000090     62.290    152.240
    9      9.000D+05   1.819D-03   0.000292    -75.714     14.236

    TOTAL HARMONIC DISTORTION =      0.106690  PERCENT
```

Figure 8.10   Fourier coefficients of the IF component.

are shown in Figure 8.9, the Fourier analysis results are listed in Figure 8.10. The amplitude of V_{oif} computed by SPICE2 is 6.225 V. This value can be verified by hand using Eq. 8.76:

$$V_{oif} = I_{cif}R_1 = 0.95I_{DC}\frac{V_s}{V_{th}}R_1$$

$$= 0.95 \cdot (1.25 \cdot 10^{-2}\text{A})\frac{10^{-3}}{0.0258} 15 \cdot 10^3 \ \Omega = 6.9 \text{ V} \qquad (8.81)$$

where

$$I_{DC} = I_C I_0(s) I_0(l) = (1.25 \cdot 10^{-3} \text{ A}) \cdot 10 = 1.25 \cdot 10^{-2} \text{ A}$$

with I_C given by SPICE2, and $I_0(l) = 10$, and $I_0(s) = 1$, from tables of Bessel functions.

The **.TRAN** and **.FOUR** statements for the next run must be changed in order to observe the 1.1-MHz LO spectral component:

```
.TRAN 45N 610.909U 609.090U 45N
.FOUR 1.1MEG V(3)
```

As in the IF signal plot, the last two periods of the 1.1-MHz LO signal waveform are shown in Figure 8.11; the resulting Fourier coefficients for the ω_{lo} signal are listed

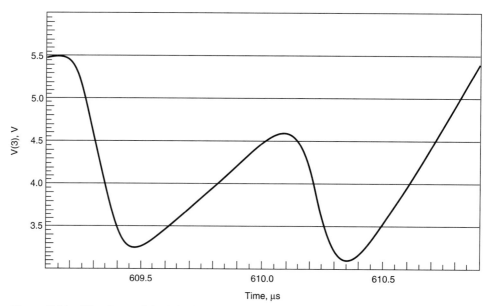

Figure 8.11   Waveform of the LO signal.

in Figure 8.12. The value computed by SPICE2 for V_{olo}, 0.89 V, is larger than that predicted by Eq. 8.80:

$$V_{olo} = \frac{I_{Clo}R_1}{Rej(\omega_{lo})} = \frac{1.9 \cdot 1.25 \cdot 10^{-2} \cdot 15 \cdot 10^3}{440} \text{ V} = 0.81 \text{ V} \qquad (8.82)$$

The approach of finding the high-frequency component of a modulated signal from a Fourier analysis with this component as fundamental is very inaccurate. The amplitude of the IF component can vary by 100% depending on which time interval of the IF signal's period the computation is performed. Note that the above **.TRAN** analyses were run with a maximum internal time step $TMAX = 45$ ns, which represents 1/20th of the highest-frequency component of interest in the signal. It is necessary to provide at least 20 points for one period of a sine wave for the calculation of the Fourier coefficients to be accurate.

Better accuracy can be obtained from a program that can compute more than nine harmonics. SPICE3 allows a user to define the number of Fourier components computed; in this example both the IF and the LO magnitudes result from a single analysis as the fundamental and the eleventh harmonic. The results of the SPICE3 Fourier analysis are listed in Figure 8.13.

The SPICE3 Fourier analysis verifies the ratio $V_{oif}/V_{olo} = 8.5$ predicted above. More details about Fourier and distortion analysis in SPICE3 can be found in the latest user's guide (Johnson, Quarles, Newton, Pederson, and Sangiovanni-Vincentelli 1992).

The same result can be arrived at by using the **.DISTO** small-signal analysis as long as the circuit behaves fairly linearly. Since V_{lo} is not much larger than V_{th}, it is

```
****      FOURIER ANALYSIS                 TEMPERATURE =   27.000 DEG C

FOURIER COMPONENTS OF TRANSIENT RESPONSE V(3)
DC COMPONENT =    4.099D+00
HARMONIC   FREQUENCY    FOURIER    NORMALIZED    PHASE      NORMALIZED
   NO        (HZ)      COMPONENT   COMPONENT    (DEG)      PHASE (DEG)

    1      1.100D+06    8.892D-01   1.000000    110.885       0.000
    2      2.200D+06    7.609D-02   0.085574     -0.298    -111.184
    3      3.300D+06    1.344D-01   0.151106   -140.456    -251.342
    4      4.400D+06    9.763D-02   0.109789   -169.624    -280.510
    5      5.500D+06    6.168D-02   0.069368   -165.533    -276.418
    6      6.600D+06    5.231D-02   0.058822   -156.038    -266.923
    7      7.700D+06    4.618D-02   0.051932   -151.451    -262.337
    8      8.800D+06    4.095D-02   0.046057   -146.867    -257.752
    9      9.900D+06    3.701D-02   0.041623   -142.171    -253.056

   TOTAL HARMONIC DISTORTION =     23.881368  PERCENT
```

Figure 8.12 Fourier components of the LO signal.

```
1> spice3d2
Spice 1 -> source 1xtor_mixer.ckt

Circuit: one transistor mixer circuit

Spice 2 -> run
Spice 3 -> set nfreqs=12
Spice 4 -> fourier 100k v(3)
Fourier analysis for v(3):
  No. Harmonics: 12, THD: 11.7802 %, Gridsize: 200, Interpolation Degree: 1


Harmonic Frequency   Magnitude   Phase       Norm. Mag   Norm.  Phase
-------- ---------   ---------   -----       ---------   -----------
   0      0          9.99887     0           0           0
   1      100000     6.224       -89.993     1           0
   2      200000     0.00663444  11.4065     0.00106594  101.4
   3      300000     0.00102807  71.4975     0.000165179 161.491
   4      400000     0.00222051  -168.16     0.000356766 -78.163
   5      500000     0.00106634  -79.296     0.000171327 10.6968
   6      600000     0.00140642  9.06029     0.000225968 99.0534
   7      700000     0.00115581  100.742     0.000185701 190.735
   8      800000     0.00095327  -158.17     0.00015316  -68.176
   9      900000     0.0016788   -71.311     0.00026973  18.6825
  10      1e+06      0.0181946   84.4315     0.0029233   174.425
  11      1.1e+06    0.732933    90.1382     0.117759    180.131


Spice 5 -> quit
Spice-3d2 done
```

Figure 8.13   Fourier components computed by SPICE3 for the mixer circuit.

instructive to verify the results of the **.DISTO** analysis. The IF component is the *DIM2* second-order difference intermodulation component. The SPICE2 input and the results of the **.AC** and **.DISTO** analyses are listed in Figure 8.14. It is important to provide the correct information in the **.DISTO** statement.

First, the reference power is calculated for an input amplitude $V_{lo} = 100$ mV

$$P_{ref} = \frac{1}{2}I_c^2 R_1 = \frac{1}{2}\left(4.84 \cdot 10^{-3}\right)^2 15 \cdot 10^3 \text{ mW} = 175 \text{ mW} \qquad (8.83)$$

where

$$I_c = g_m V_{lo} = 4.84 \times 10^{-3} \text{ A}$$

with $g_m = 4.84 \times 10^{-2}$ mho as computed by SPICE in Figure 8.8.

```
*******12/13/90 ********   SPICE 2G.6    3/15/83 ********18:44:38*****
ONE TRANSISTOR MIXER CIRCUIT
****      INPUT LISTING                    TEMPERATURE =   27.000 DEG C
*********************************************************************
Q1 3 2 0 MOD1
R1 3 4 15K
C1 3 4 4.244NF
L1 3 4 596.83UH
.MODEL MOD1 NPN
*
* SUPPLY, SIGNAL AND LOCAL OSCILLATOR
*
VCC 4 0 10
*VS 2 1 0 SIN 0 75.5UV 1MEG
VLO 2 5 SIN 0.78 100MV 1.1MEG AC 1
*
.DISTO R1 0 .90909 175M .01
.AC LIN 11 100K 1.1MEG
.PRINT AC VM(3)
.PRINT DISTO DIM2 SIM2 HD2 HD3 DIM3
.WIDTH OUT=80
.OPTIONS RELTOL=1E-4 ITL5=0
.END
****      AC ANALYSIS                      TEMPERATURE =   27.000 DEG C


     FREQ        VM(3)
  1.000E+05    7.255E+02
  2.000E+05    1.209E+01
  3.000E+05    6.802E+00
  4.000E+05    4.837E+00
  5.000E+05    3.779E+00
  6.000E+05    3.109E+00
  7.000E+05    2.645E+00
  8.000E+05    2.303E+00
  9.000E+05    2.041E+00
  1.000E+06    1.832E+00
  1.100E+06    1.663E+00


****      AC ANALYSIS                      TEMPERATURE =   27.000 DEG C

     FREQ        DIM2        SIM2        HD2         HD3         DIM3
  1.000E+05    4.424E-05   3.484E-04   1.609E-02   5.778E-03   2.647E-03
  2.000E+05    3.268E-01   4.887E-01   2.317E+01   5.753E+02   5.841E+01
  3.000E+05    1.618E+00   9.891E-01   4.707E+01   2.121E+03   1.906E+02
  4.000E+05    4.551E+00   1.447E+00   6.895E+01   4.399E+03   3.803E+02
  5.000E+05    1.019E+01   1.885E+00   8.986E+01   7.361E+03   6.256E+02
  6.000E+05    2.040E+01   2.312E+00   1.103E+02   1.099E+04   9.259E+02
  7.000E+05    3.882E+01   2.733E+00   1.303E+02   1.530E+04   1.281E+03
  8.000E+05    7.388E+01   3.149E+00   1.502E+02   2.026E+04   1.691E+03
  9.000E+05    1.507E+02   3.563E+00   1.700E+02   2.589E+04   2.155E+03
  1.000E+06    3.931E+02   3.976E+00   1.897E+02   3.218E+04   2.675E+03
  1.100E+06    3.676E+03   4.386E+00   2.093E+02   3.914E+04   3.249E+03
```

Figure 8.14 Small-signal distortion analysis results for mixer.   **275**

For the $\mathtt{.DISTO}$ specification the radio signal S is defined with respect to the LO signal; the amplitude V_s is .01 V_{lo} and the frequency ω_s is 0.909 ω_{lo}. The following changes must be made to the SPICE2 input used by the previous analyses (see Figure 8.8). First, the input signal source, V_s, is commented out, since only one AC source, V_{lo}, is needed to run the AC analysis. Second, the $\mathtt{.AC}$ and $\mathtt{.DISTO}$ lines replace the $\mathtt{.TRAN}$ and $\mathtt{.FOUR}$ lines; see the $\mathtt{INPUT\ LISTING}$ in Figure 8.14.

The ratio V_{oif}/V_{olo} can be obtained from the value of $DIM2$ computed by SPICE at 1.1 MHz. SPICE2 computes $DIM2$ according to

$$DIM2 = \frac{a_2 V_{lo} V_s}{a_1 V_{lo}} \cdot \frac{V_{on}}{a_1 V_{lo}} \tag{8.84}$$

where a_1 is the gain, a_2 is the second-order coefficient in the power series, and V_{on} is the normalized voltage amplitude at the output:

$$V_{on} = \sqrt{2 P_{ref} R_1} = 72.5 \text{ V} \tag{8.85}$$

The gain a_1 is frequency-dependent, and at 1.1 MHz it is found in the AC ANALYSIS listing for VM(3) equal to 1.66. a_2 can be evaluated by solving equations similar to Eqs. 8.23 and Eqs. 8.32 through 8.34:

$$a_2(j\omega_1, -j\omega_2) = g_{m2} Z_1(j\omega_1 - j\omega_2) \tag{8.86}$$

The value of $DIM2$ computed by SPICE2 is 3.68×10^3. From Eqs. 8.73, 8.80, and 8.84 the ratio of the IF to the LO signal at the output can be derived:

$$\frac{V_{oif}}{V_{olo}} = \frac{a_2 V_{lo} V_s}{a_1 V_{lo}} = DIM2 \frac{a_1 V_{lo}}{V_{on}} = 3.68 \cdot 10^3 \frac{1.66 \cdot 0.1}{72.5} = 8.4 \tag{8.87}$$

which is the same as the result obtained from large-signal Fourier analysis and hand calculations.

It is advantageous to use the $\mathtt{.DISTO}$ analysis whenever the circuit behaves linearly because the $\mathtt{.DISTO}$ and $\mathtt{.AC}$ analyses are much faster than the $\mathtt{.TRAN}$ and $\mathtt{.FOUR}$ analyses. The results obtained using $\mathtt{.DISTO}$ are correct for this example because the only distortion of the signal is due to the transistor nonlinearities.

8.4   SUMMARY

This chapter has described in detail how to evaluate distortion in electronic circuits using both small-signal AC and large-signal Fourier analysis in SPICE. The purpose has been to familiarize the user with the methodology of evaluating distortion, which can be applied with any SPICE program, regardless of whether the program supports small-signal distortion analysis.

All the capabilities of the small-signal .DISTO analysis of SPICE2 introduced in Chap. 5 have been presented here. Frequency-dependent elements were included in the general derivation of the spectral component calculation. The summary option of .DISTO was exercised in SPICE2 and the meanings of all the results were explained. Commonly used distortion measures were related to the results obtained from SPICE2 for a single-transistor amplifier.

In the second part of the chapter large-signal time-domain analysis and .FOUR analysis were applied to evaluate the same distortion measures derived with small-signal analysis. The Fourier components computed by SPICE correspond to the distortion components of the same order. Special consideration must be given to the accuracy of the .FOUR analysis, that is, to the number of time points computed by the simulator during the last period of the signal. The relationship between small-signal and large-signal analysis results was exemplified for a single-transistor mixer. This example also pointed to the advantages of the small-signal .DISTO analysis over the .FOUR analysis for circuits with two nearly linear input signals.

REFERENCES

Chisholm, S. H., and L. W. Nagel. 1973. Efficient computer simulation of distortion in Electronic circuits. *IEEE Transactions on Circuit Theory* (November).

Johnson, B., T. Quarles, A. R. Newton, D. O. Pederson, and A. Sangiovanni-Vincentelli. 1992 (April). *SPICE3 Version 3f User's Manual.* Berkeley: Dept. of Electrical Engineering and Computer Science, Univ. of California.

Meyer, R. G. 1979. Distortion analysis of solid-state circuits. In *Nonlinear Analog Circuits EECS 240 Class Notes.* Berkeley: Dept. of Electrical Engineering and Computer Science, Univ. of California.

Meyer, R. G., M. J. Shensa, and R. Eschenbach. 1972. Cross modulation and intermodulation in amplifiers at high frequencies. *IEEE Journal of Solid-State Circuits* SC-7 (February).

Narayanan, S. 1967. Transistor distortion analysis using Volterra series representation. *Bell System Technical Journal* (May–June).

Pederson, D. O., and K. Mayaram. 1990. *Integrated Circuits for Communication.* Boston: Kluwer Academic.

Nine

SPICE ALGORITHMS AND OPTIONS

9.1 OVERVIEW OF ALGORITHMS

A number of algorithms have proven well suited for the solution of the equations of electrical circuits and are implemented not only in SPICE simulators but in most circuit simulators in existence today. The various types of BCEs for electrical elements were described in Chaps. 2 and 3. The most general equations have been shown to be ordinary differential equations, ODE, with nonlinear BCEs, which must be solved in the time domain.

The solution process implemented in SPICE for the time-domain solution is shown in Figure 9.1. Generally the program first solves for a stable DC operating point. The solution starts with an initial guess of the operating point, which is followed by iterations for solving the DC nonlinear equations. The iterative process is represented by the inner loop in Figure 9.1. The solution the iterative process converges to represents either the small-signal bias solution (SSBS) or the initial transient solution (ITS); see Chaps. 4 and 6, respectively. This is the time zero solution. The iterative process is repeated for every time point at which the circuit equations are solved in transient analysis. The algorithms used in SPICE for the DC solutions of linear and nonlinear circuits are presented in Secs. 9.2 and 9.3, respectively.

The time-domain solution, which is represented as the outer loop in Figure 9.1, uses numerical integration to transform the set of ODEs into a set of nonlinear equations. The time-domain analysis is replaced by a sequence of quasi-static solutions. These techniques are described in Sec. 9.4.

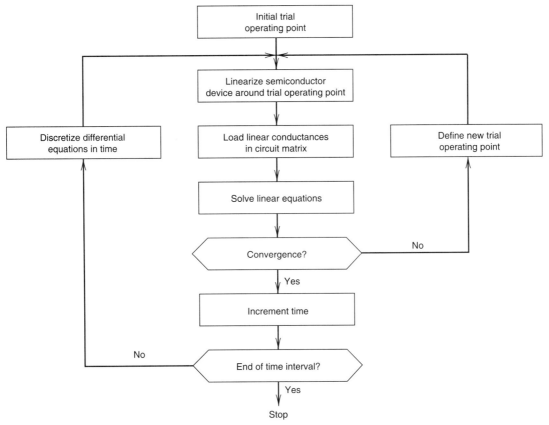

Figure 9.1   SPICE solution algorithm.

A circuit simulator is defined by the following sequence of specific algorithms; an implicit numerical integration method that transforms the nonlinear differential equations into nonlinear algebraic equations, linearization of these through a modified Newton-Raphson iterative algorithm, and finally Gaussian elimination and sparse matrix techniques that solve the linear equations. Simulators using these techniques, such as SPICE2, SPICE3, and their derivatives, are known as third-generation circuit simulators. The solution approach implemented in these simulators is also referred to as direct methods. These algorithms are described in detail by McCalla (1988) and Nagel (1975) and in overview papers, such as those by McCalla and Pederson (1971) and Hachtel and Sangiovanni-Vincentelli (1981).

An important characteristic of SPICE is the *analysis options*, which enable the user to select among several numerical methods and analysis tolerances. SPICE should not be treated as a black box that always provides the right answer to no matter what circuit. The choice of algorithms and tolerances is based on a large set of examples, and the majority of the circuits run well using the default settings. This chapter is intended to relate the analysis options accessible to the user with the solution algorithms. A good understanding of this chapter is important for overcoming the analysis failures commonly referred to as convergence problems.

The general form of the option statement is

.OPTIONS *OPT1=Name1/val1* <*OPT2=Name2/val2*> . . .

OPT1, OPT2, . . . are options recognized by the SPICE program. The option is followed either by a name, *Name1, Name2,* . . . or a number, *val1, val2,* . . . The most important options control the solution algorithms and tolerances and are introduced in this chapter. Each version of SPICE has a few options that differ from the ones in SPICE2, but the fundamental options for algorithms, tolerances, and iteration limits can be found in most versions. The analysis options introduced throughout this chapter are summarized at the end.

9.2   DC SOLUTION OF LINEAR CIRCUITS

This section describes the circuit equation formulation as well as the solution algorithms for linear systems. The linear-equation solution algorithms described here are used for the DC solution of linear circuits and for the the AC solution, which always involves linear circuits. The DC equations are formulated with real numbers, and the AC equations use complex numbers.

The matrix formulation for connectivity and nodal equations in SPICE is presented in Sec. 9.2.1 together with the solution algorithm. An extended version of nodal analysis (Dorf 1989; Nilsson 1990; Paul 1989), called modified nodal analysis (MNA), is used in SPICE to represent the circuit. Gaussian elimination and the associated factorization into lower triangular and upper triangular matrices, the LU factorization, are implemented in SPICE for the solution of a linear system of simultaneous equations. An important characteristic of the circuit admittance matrix, its sparsity, and its impact on the analysis is also described in this section.

Certain characteristics of the MNA formulation and Gaussian elimination associated with computer limitations of the representation of real numbers can lead to a loss of accuracy and a wrong solution. These pitfalls are detailed in Sec. 9.2.2. Two important issues are detailed there: the reordering of equations for accuracy and sparsity and the SPICE options available to the user for controlling the linear solution process.

9.2.1   Circuit Equation Formulation: Modified Nodal Equations

In Example 1.1 the DC solution of the bridge-T circuit (Figure 1.1) is derived from the nodal equations written for each node. In the evaluation of the node voltages, the value of the grounded voltage source, V_{BIAS}, was assigned, by inspection, to V_1, the voltage at node 1. Thus only two nodal equations in two unknowns must be solved; Eqs. 1.1

and 1.2 are reproduced here:

node2: $-G_1 \cdot V_1 + (G_1 + G_2 + G_3)V_2 \qquad -G_3 \cdot V_3 = 0$
node3: $-G_4 \cdot V_1 \qquad\qquad -G_3 \cdot V_2 + (G_3 + G_4)V_3 = 0$ $\qquad$ (9.1)

Eq. 9.1 can be expressed as a matrix equation:

$$\mathbf{GV} = \mathbf{I} \qquad\qquad (9.2)$$

where $\mathbf{G}$ is the conductance matrix of the circuit, $\mathbf{V}$ is the unknown node voltage vector, and $\mathbf{I}$ is the right-hand-side (RHS) current vector. Representation by matrices and vectors is well suited for programming and therefore is the methodology of choice in SPICE. The conductance matrix, $\mathbf{G}$, is easily set up by adding all conductances incident into a node to each diagonal term and subtracting the conductances connecting two nodes from the corresponding off-diagonal terms.

One problem in formulating the conductance matrix, $\mathbf{G}$, is that a voltage source cannot easily be included in the set of nodal equations: the conductance of an ideal voltage source is infinite, and its current is unknown. A voltage source connected between two circuit nodes complicates this issue even more and raises the need for a consistent formulation suited for programming. The problem is exemplified by the modified bridge-T circuit shown in Figure 9.2, which contains an additional voltage source, V_A, in series with R_4.

This problem led the developers of SPICE to extend the set of nodal equations to include voltage-source equations represented by currents in the unknown vector and by voltages in the RHS vector. This approach, modified nodal analysis, is therefore an extension of nodal analysis in that the node voltage equations are augmented by current equations for the voltage-defined elements (Nagel and Rohrer 1971; Ho, Ruehli, and Brennan 1975).

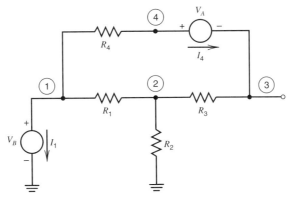

Figure 9.2   Modified bridge-T circuit.

The complete set of equations can now be written for the bridge-T circuit in Figure 9.2, including the voltage sources:

$$
\begin{aligned}
\text{node 1:} \quad & (G_1 + G_4)V_1 & -G_1V_2 & & -G_4V_4 & +I_1 & = 0 \\
\text{node 2:} \quad & -G_1V_1 & +(G_1 + G_2 + G_3)V_2 & -G_3V_3 & & & = 0 \\
\text{node 3:} \quad & & -G_3V_2 & +G_3V_3 & & -I_4 & = 0 \\
\text{node 4:} \quad & -G_4V_1 & & & +G_4V_4 & +I_4 & = 0 \\
V_B: \quad & V_1 & & & & & = V_B \\
V_A: \quad & & & -V_3 & +V_4 & & = V_A
\end{aligned}
\tag{9.3}
$$

In partitioned matrix form the equations become

$$
\left[
\begin{array}{cccc:cc}
G_1 + G_4 & -G_1 & 0 & -G_4 & 1 & 0 \\
-G_1 & G_1 + G_2 + G_3 & -G_3 & 0 & 0 & 0 \\
0 & -G_3 & G_3 & 0 & 0 & -1 \\
-G_4 & 0 & 0 & G_4 & 0 & 1 \\
\hdashline
1 & 0 & 0 & 0 & 0 & 0 \\
0 & 0 & -1 & 1 & 0 & 0
\end{array}
\right]
\left[
\begin{array}{c}
V_1 \\ V_2 \\ V_3 \\ V_4 \\ \hdashline I_1 \\ I_4
\end{array}
\right]
=
\left[
\begin{array}{c}
0 \\ 0 \\ 0 \\ 0 \\ \hdashline V_B \\ V_A
\end{array}
\right]
\tag{9.4}
$$

The above MNA equations can be rewritten in abbreviated form:

$$
\begin{bmatrix} \mathbf{G} & \mathbf{F} \\ \mathbf{B} & \mathbf{R} \end{bmatrix}
\begin{bmatrix} \mathbf{V} \\ \mathbf{I} \end{bmatrix}
=
\begin{bmatrix} \mathbf{C} \\ \mathbf{E} \end{bmatrix}
$$

where $\mathbf{C}$ and $\mathbf{E}$ are the vectors of the current and voltage sources, respectively.

So far the discussion has treated only voltage sources as being difficult to include into a set of nodal equations. Another circuit element that presents the same problem is the inductor, because it is a short in DC, and therefore the voltage across it is zero and the conductance is infinite. The inductor is also a voltage-defined element in SPICE and is included as a current equation in the MNA formulation. The total number of equations, N, used to represent a circuit in SPICE is

$$
N = n + n_v + n_l
\tag{9.5}
$$

where n is the number of circuit nodes excluding ground, n_v is the number of independent voltage sources, and n_l is the number of inductors. Note that all controlled sources except the voltage-controlled current source (VCCS), which is a transconductance, also introduce current equations. For complete details on the MNA matrix representation of different elements consult McCalla's book (1988).

The MNA set of equations that needs to be solved, Eq. 9.4, can be expressed as a matrix equation:

$$\mathbf{Ax} = \mathbf{b} \tag{9.6}$$

The solution vector, $\mathbf{x}$, can be computed by inverting matrix $\mathbf{A}$. This approach is very time-consuming, and Gaussian elimination (Forsythe and Moler 1967) is preferred for numerical solutions.

The Gaussian elimination procedure uses scaling of each equation followed by subtraction from the remaining equations in order to eliminate unknowns one by one until $\mathbf{A}$ is reduced to an upper triangular matrix. The solution can then be found by computing each element of vector $\mathbf{x}$ in reverse order (the back-substitution phase). For a 3×3 system of linear equations,

$$
\begin{array}{c}
e_1^{(0)}: \\
e_2^{(0)}: \\
e_3^{(0)}:
\end{array}
\begin{bmatrix}
a_{11}^{(0)} & a_{12}^{(0)} & a_{13}^{(0)} \\
a_{21}^{(0)} & a_{22}^{(0)} & a_{23}^{(0)} \\
a_{31}^{(0)} & a_{32}^{(0)} & a_{33}^{(0)}
\end{bmatrix}
\begin{bmatrix}
x_1 \\
x_2 \\
x_3
\end{bmatrix}
=
\begin{bmatrix}
b_1^{(0)} \\
b_2^{(0)} \\
b_3^{(0)}
\end{bmatrix}
\tag{9.7}
$$

the steps leading to the solution are outlined as follows. The equations are designated by $e_1^{(0)}, e_2^{(0)}$, and $e_3^{(0)}$. The superscripts, both in the equation designators and the matrix elements, represent the step of the elimination process. First, x_1 can be eliminated from $e_2^{(0)}$ and $e_3^{(0)}$ by subtracting $e_1^{(0)}$ multiplied by $a_{21}^{(0)}/a_{11}^{(0)}$ from $e_2^{(0)}$ and subtracting $e_1^{(0)}$ multiplied by $a_{31}^{(0)}/a_{11}^{(0)}$ from $e_3^{(0)}$:

$$
\begin{aligned}
e_1^{(1)} &= e_1^{(0)} \\
e_2^{(1)} &= e_2^{(0)} - (a_{21}^{(0)}/a_{11}^{(0)})e_1^{(0)} \\
e_3^{(1)} &= e_3^{(0)} - (a_{31}^{(0)}/a_{11}^{(0)})e_1^{(0)}
\end{aligned}
\tag{9.8}
$$

Second, x_2 is eliminated from $e_3^{(1)}$ by subtracting $e_2^{(1)}$ multiplied by $a_{32}^{(1)}/a_{22}^{(1)}$ from $e_3^{(1)}$:

$$
\begin{aligned}
e_1^{(2)} &= e_1^{(1)} \\
e_2^{(2)} &= e_2^{(1)} \\
e_3^{(2)} &= e_3^{(1)} - (a_{32}^{(1)}/a_{22}^{(1)})e_2^{(1)}
\end{aligned}
\tag{9.9}
$$

which yields an upper triangular equation system:

$$\begin{matrix} e_1^{(2)} : \\ e_2^{(2)} : \\ e_3^{(2)} : \end{matrix} \begin{bmatrix} a_{11}^{(0)} & a_{12}^{(0)} & a_{13}^{(0)} \\ 0 & a_{22}^{(1)} & a_{23}^{(1)} \\ 0 & 0 & a_{33}^{(2)} \end{bmatrix} \begin{bmatrix} x_1 \\ x_2 \\ x_3 \end{bmatrix} = \begin{bmatrix} b_1^{(0)} \\ b_2^{(1)} \\ b_3^{(2)} \end{bmatrix} \qquad (9.10)$$

Third, back-substitution leads to the following solution:

$$\begin{aligned} x_3 &= b_3^{(2)}/a_{33}^{(2)} \\ x_2 &= (b_2^{(1)} - a_{23}^{(1)}x_3)/a_{22}^{(1)} \\ x_1 &= (b_1^{(0)} - a_{13}^{(0)}x_3 - a_{12}^{(0)}x_2)/a_{11}^{(0)} \end{aligned} \qquad (9.11)$$

A variant of Gaussian elimination is *LU factorization*. This procedure transforms the circuit matrix **A** into a lower, **L**, and an upper, **U**, triangular matrix. Equation 9.6 can be rewritten as

$$\mathbf{LUx} = \mathbf{b} \qquad (9.12)$$

The first step of the method is to factorize **A** into **L** and **U**, which for the third-order system become

$$\mathbf{U} = \begin{bmatrix} a_{11} & a_{12} & a_{13} \\ 0 & a_{22} & a_{23} \\ 0 & 0 & a_{33} \end{bmatrix} \qquad \mathbf{L} = \begin{bmatrix} 1 & 0 & 0 \\ a_{21}/a_{11} & 1 & 0 \\ a_{31}/a_{11} & a_{32}/a_{22} & 1 \end{bmatrix} \qquad (9.13)$$

where **U** is the result of Gaussian elimination and **L** stores the scale factors at each elimination step. The second phase of the solution is the forward substitution, which results in a new RHS:

$$\mathbf{Ux} = \mathbf{L}^{-1}\mathbf{b} = \mathbf{b}' \qquad (9.14)$$

The last step is back-substitution, which computes the elements of the unknown vector, **x**:

$$\mathbf{x} = \mathbf{U}^{-1}\mathbf{b}' = \mathbf{U}^{-1}\mathbf{L}^{-1}\mathbf{b} \qquad (9.15)$$

Note that inverting **L** and **U** is trivial because they are triangular. The advantage of LU factorization over Gaussian elimination is that the circuit can be solved repeatedly for different excitation vectors, that is, different right-hand sides. This property is useful in certain SPICE analyses, such as sensitivity, noise, and distortion analyses.

An important observation about the conductance matrix **G** (Eq. 9.4) is that it is diagonally dominant and many off-diagonal terms are zero. The simple explanation of this is that off-diagonal terms are generated by conductances connected between pairs of nodes, and usually a node is connected to only two or three neighboring nodes. In the conductance matrix of a circuit having a few tens of nodes, only two or three out of a few tens of off-diagonal terms are nonzero; in other words the conductance matrix is *sparse*. The sparsity is maintained in the current-equation submatrix **R** (Eq. 9.4), where many diagonal elements are also zero. This issue is addressed in the following section on accuracy. The MNA matrix of the bridge-T circuit, Eq. 9.4, has 20 zero elements of a total of 36 elements in the matrix. The sparsity of the matrix can be defined as

$$sparsity = \frac{\text{number of elements equal to zero}}{\text{total number of elements in matrix}} \qquad (9.16)$$

The sparsity is 55.6% in this example. The number obtained from SPICE for the sparsity of this circuit in the accounting summary, option **ACCT** (Sec. 9.5), differs from the above number because SPICE includes the ground node in the computation, raising the total number of elements to 49. The zero-element count in SPICE is taken after reordering. Sparsity is a very useful feature, which can be exploited for reducing data storage and computation. The next section explains the need for careful reordering of the equations for maintaining accuracy and sparsity at the same time.

9.2.2 Accuracy and SPICE Options

Accuracy problems in the solution of a linear circuit can be classified as either topological or numerical. The voltage-defined elements, such as voltage sources and inductors, generate zero diagonal elements linked to the current equation (Ho, Ruehli, and Brennan 1975). This problem can be corrected in the setup phase based on a topological reordering, also referred to as *preordering*. In SPICE the row in the MNA matrix corresponding to the current equation of a voltage source is swapped with the node voltage equation corresponding to the positive terminal of the same voltage source (Cohen 1981).

A second problem, which is also topological in nature, is a cut set of voltage-defined elements. This leads to a cancellation of a diagonal element value during the factorization process. Earlier versions of SPICE2, up to version F, which used only topological reordering, could not correct this problem. A reordering algorithm has been proposed (Hajj, Yang, and Trick 1981) that finds an equation sequence free of topological problems. This additional topological reordering scheme is, however, not necessary once the numerical reordering known as *pivoting* is used. All SPICE2 version G releases use pivoting in the sparse matrix solution (Boyle 1978; Vladimirescu 1978).

A circuit that cannot be solved only through preordering is shown in Figure 9.3. The MNA matrix of the circuit is

$$
\begin{bmatrix}
1/5 & -1/5 & 0 & 0 & 0 & 0 \\
-1/5 & 1/5 & 0 & 0 & 1 & 0 \\
0 & 0 & 1/2 & -1/2 & -1 & 0 \\
0 & 0 & -1/2 & 1/2 & 0 & 1 \\
0 & 1 & -1 & 0 & 0 & 0 \\
0 & 0 & 0 & 1 & 0 & 0
\end{bmatrix}
\begin{bmatrix}
V_1 \\ V_2 \\ V_3 \\ V_4 \\ I_L \\ I_V
\end{bmatrix}
=
\begin{bmatrix}
1 \\ 0 \\ 0 \\ 0 \\ 0 \\ 3
\end{bmatrix}
\tag{9.17}
$$

After preordering, which swaps row I_L with node 2 and row I_V with node 4, and reordering to maintain sparsity, described below, the circuit matrix becomes

$$
\begin{bmatrix}
1 & 0 & 0 & 0 & 0 & 0 \\
1/2 & 1 & 0 & 0 & 0 & -1/2 \\
0 & 0 & 1/5 & -1/5 & 0 & 0 \\
0 & 0 & -1/5 & 1/5 & 1 & 0 \\
-1/2 & 0 & 0 & 0 & -1 & 1/2 \\
0 & 0 & 0 & 1 & 0 & -1
\end{bmatrix}
$$

Matrix equations 3 and 4 form a diagonal block

$$
\begin{bmatrix}
G_1 & -G_1 \\
-G_1 & G_1
\end{bmatrix}
$$

where $G_1 = \frac{1}{5}$; during the LU factorization a zero is created on the diagonal at row 4. This example shows the need of a reordering scheme based on the matrix entries at each step of the LU decomposition.

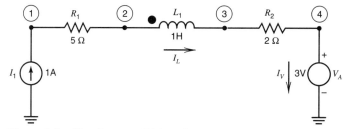

Figure 9.3   Circuit exemplifying diagonal cancellation.

Not all problems associated with the solution of the MNA matrix are topological, however. Another problem is that computers only have finite precision. The limit of the number of digits in the mantissa of a floating-point number, up to 15 decimal digits for a double-precision, or 64-bit, real number, in IEEE floating-point format, can lead to the loss of significance of a matrix term relative to another during the solution of the linear equations.

The simple circuit (Freret 1976) shown in Figure 9.4 demonstrates this point. It is assumed that the computer can represent only four digits of floating-point accuracy. It is easy to imagine that having a switch element (see Chap. 2) A circuit can have resistors with a range from 1 mΩ to 1 GΩ, that is, 12 orders of magnitude. The equations of the circuit are

$$\begin{bmatrix} 1 & -1 \\ -1 & 1.0001 \end{bmatrix}\begin{bmatrix} V_1 \\ V_2 \end{bmatrix} = \begin{bmatrix} 1 \\ 0 \end{bmatrix} \tag{9.18}$$

The hypothetical computer in this case, however, rounds off element a_{22} to

$$a_{22} = G_1 + G_2 = 1.000$$

due to the four-digit limitation. As in the previous example, a zero is created on the diagonal during Gaussian elimination, resulting in erroneous values of infinity for V_1 and V_2. This problem is due to the insufficient accuracy of the computer and cannot be corrected by reordering. The conductance matrix of the circuit is singular.

Another accuracy problem, also due to the limited number of digits, can be caused by the Gaussian elimination process. This case is exemplified by the circuit in Figure 9.5. The nodal equations for this circuit are

$$G_1 V_1 + g_2 V_2 = I$$
$$-g_1 V_1 + G_2 V_2 = 0 \tag{9.19}$$

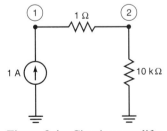

Figure 9.4   Circuit exemplifying limited floating-point range.

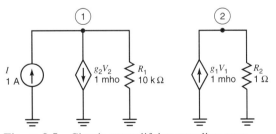

Figure 9.5   Circuit exemplifying rounding error.

Substitution of the values of the conductances, G_1 and G_2, and the transconductances, g_1 and g_2, yields the following system of equations:

$$\begin{bmatrix} 0.0001 & 1 \\ -1 & 1 \end{bmatrix} \begin{bmatrix} V_1 \\ V_2 \end{bmatrix} = \begin{bmatrix} 1 \\ 0 \end{bmatrix}$$

After one elimination step the system becomes

$$\begin{bmatrix} 0.0001 & 1 \\ 0 & 10,000 \end{bmatrix} \begin{bmatrix} V_1 \\ V_2 \end{bmatrix} = \begin{bmatrix} 1 \\ 10,000 \end{bmatrix}$$

with the following solution on a computer with four-digit accuracy:

$$V_2 = 1$$
$$V_1 = 0$$

which is obviously incorrect. If the rows of the equation are swapped before factorization in order to bring the largest element onto the diagonal, the system of equations becomes

$$\begin{bmatrix} -1 & 1 \\ 0.0001 & 1 \end{bmatrix} \begin{bmatrix} V_1 \\ V_2 \end{bmatrix} = \begin{bmatrix} 0 \\ 1 \end{bmatrix}$$

with the more realistic solution

$$V_2 = 1$$
$$V_1 = 1$$

The accuracy of this solution is still affected by the limited number of digits; the solution is correct, however, for the available number representation. The exact solution is

$$V_2 = 0.9999$$
$$V_1 = 0.9999$$

This type of accuracy loss can also be observed in the case of a series of high-gain stages connected in a feedback loop, such as a ring oscillator; the off-diagonal elements are transconductances, which grow as a power-law function of the gain factors during factorization and can eventually swamp out the diagonal term. A ring oscillator similar to that in Figure 6.9 but implemented with bipolar transistors is shown in Figure 9.6. Replacement of the transistors with a linearized model during each iteration, as

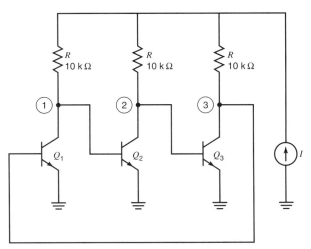

Figure 9.6   BJT ring oscillator exemplifying rounding error.

described in Sec. 9.3, leads to the following system of equations:

$$
\begin{bmatrix}
G + g_{\pi 2} & 0 & g_{m1} \\
g_{m2} & G + g_{\pi 3} & 0 \\
0 & g_{m3} & G + g_{\pi 1}
\end{bmatrix}
\begin{bmatrix}
V_1 \\
V_2 \\
V_3
\end{bmatrix}
=
\begin{bmatrix}
I_{eq1} \\
I_{eq2} \\
I_{eq3}
\end{bmatrix}
\tag{9.20}
$$

The solutions of the node voltages, V_1, V_2, and V_3, should be identical, but they differ if the above system is solved as is, because the self-conductance of each node, or diagonal term, loses its contribution during the elimination process.

Exercise
Assume that $g_m = 3.607 \times 10^{-2}$ mho, $g_\pi = 3.607 \times 10^{-4}$ mho, $R = 10$ kΩ, and $I_{eq} = 2.192 \times 10^{-2}$ A. Carry out the Gaussian elimination steps to find the solution for V_1, V_2, and V_3 with four-digit accuracy.

Another numerical problem occurs when the matrix entries at a certain step of the elimination process become very small, with the ratio between the largest value in the remainder matrix and the largest value in the initial matrix being of more orders of magnitude than can be represented by the computer.

A parameter, **PIVTOL**, is used in SPICE that defines the lowest threshold for accepting an element as a diagonal element, or a *pivot*. If an element larger than this value is not found in the remainder matrix at any step, then the matrix is declared singular and SPICE aborts the analysis. The default for this parameter, $PIVTOL = 10^{-13}$, was chosen under the assumption that for typical circuits the maximum conductance is 1 mho and that, at least 13 digits of accuracy are used by computers to store the matrix entries. Note that SPICE lists the value of the largest element in the remainder matrix in the *ERROR*: MAXIMUM ENTRY IN THIS COLUMN AT STEP *number value* IS LESS THAN PIVTOL message (see Appendix B for error messages). If this value is nonzero, *PIVTOL* can be reset using the .OPTION statement to accommodate it. The solution obtained afterward can be correct, but the user must double-check the circuit for possible high-impedance nodes, very high ratios of conductance values, and so on.

The above examples demonstrate that a topological reordering is not sufficient and that the order of the MNA sparse matrix has to be based on actual values generated by the circuit as well. Sometimes it is even necessary to reorder during the iterative process. Reordering based on selecting the largest element for the diagonal, called *pivoting*, is essential for an accurate solution to a set of equations if the original values can be altered significantly by the factorization process. The larger the circuit, the more important pivoting becomes, because of the accumulation of rounding error in the solution process. A detailed presentation of numerical accuracy issues can also be found in the thesis by Cohen (1981).

As part of numerical reordering, a partial pivoting strategy, which picks the largest element in the column or row, or a full pivoting strategy, which selects the largest element in the remainder of the matrix, can be chosen. The second constraint mentioned in the previous section is preservation of the sparsity of the matrix. The Markowitz algorithm is used in SPICE to select, among a number of acceptable pivots, the one that introduces the fewest *fill-in* terms. Fill-ins are the matrix terms that are zero at the beginning of the factorization process and become nonzero during LU decomposition. According to the Markowitz algorithm, the best element to be picked as the next pivot is the one that has the minimum number of off-diagonal entries in the row and column as measured by

$$m = (r - 1)(c - 1) \qquad (9.21)$$

where r and c are the numbers of nonzero entries of a row and a column, respectively. In SPICE2 the topological aspects are considered in the setup phase when the sparse-matrix pointers are defined. The numerical reordering is based on two criteria: partial pivoting for accuracy and the Markowitz algorithm for minimum fill-in. Reordering is performed at the very first iteration after the actual MNA values have been loaded.

The selection of a pivot in SPICE proceeds as follows. Pivoting is performed on the diagonal elements; if no pivot can be found on the diagonal the rest of the submatrix is searched. First, the largest element is found in the remainder matrix, and then the diagonal element with the best Markowitz number is checked as to whether it satisfies the following magnitude test:

$$a_{ii} \geq PIVREL \cdot a_{iMax} \qquad (9.22)$$

where a_{iMax} is the maximum entry at the ith elimination step and **PIVREL** is a SPICE option parameter, which defaults to 10^{-3}. Therefore SPICE accepts as pivot the element that introduces the fewest fill-ins as long as it is not *PIVREL* orders of magnitude smaller than the largest element at that elimination step. An increase of *PIVREL* forces a better-conditioned matrix at the expense of introducing fill-in terms. **PIVREL** is the other option parameter that a user can modify if SPICE aborts because of singular matrix problems.

Once an optimal order has been found, it is used throughout the analysis unless at some point a diagonal element is less than *PIVTOL*. In this case *pivoting on the fly* is performed for the remainder of the matrix.

In summary, two option parameters control the linear equation solution in SPICE, **PIVREL** and **PIVTOL**. The user is not advised to modify these parameters unless SPICE cannot find a solution because of a singular matrix problem. In order to modify one or both linear equation option parameters, one adds the following line to the SPICE input file:

.OPTIONS PIVTOL=*value1* PIVREL=*value2*

PIVTOL should be reset to a smaller value than the maximum entry listed in the SPICE message as long as the value is nonzero. If the largest entry is zero, one should carefully check the circuit first, try increasing *PIVREL*, or skip the DC solution and run a transient analysis with the **UIC** flag on. The time-domain admittance matrix adds the contributions of the charge storage elements, possibly leading to a well-conditioned matrix.

9.3   DC SOLUTION OF NONLINEAR CIRCUITS

In Chap. 3 semiconductor devices were shown to be described by nonlinear I-V characteristics that are exponential or quadratic functions. Controlled sources can also be described by nonlinear BCEs that are limited to polynomials in SPICE2 but can be any arbitrary function in SPICE3 and other commercial versions. The modified nodal equations for a circuit with transistors are a set of nonlinear simultaneous equations. The iterative loop marked in Figure 9.1 for the DC solution of nonlinear equations is implemented in SPICE using the Newton-Raphson algorithm. This algorithm is described in Sec. 9.3.1.

The iterative solution process continues until the values of the unknown voltages and currents *converge*, or in other words, until the solutions of two consecutive iterations are the same. The criteria for convergence and the options available to the user to control convergence are presented in Sec. 9.3.2.

9.3.1   Newton-Raphson Iteration

The Newton-Raphson algorithm for the solution of the equations of nonlinear circuits is introduced with the simple circuit consisting of one diode, one conductance, G, and one current source, I_A, shown in Figure 9.7a. The graphical representation of the BCEs for the two elements,

$$I_D = I_S(e^{V/V_{th}} - 1) \tag{9.23a}$$

$$I_G = GV \tag{9.23b}$$

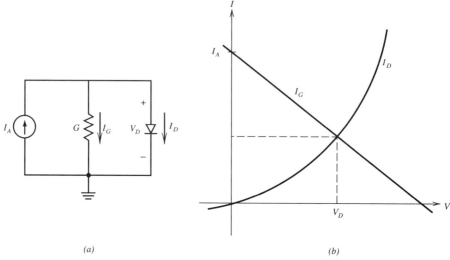

(a)   (b)

Figure 9.7   Diode circuit: (*a*) circuit diagram; (*b*) graphical solution.

is shown in Figure 9.7*b*; the solution V_D is located at the intersection of the two functions I_D and I_G, which must satisfy Kirchhoff's current law.

$$I_A = I_D + I_G \tag{9.24}$$

Nonlinear equations of this type are generally solved through an iterative approach, such as Newton's method, also known as the tangent method (Ortega and Rheinholdt 1970). The current I_D in Eq. 9.23*a* can be approximated by a Taylor series expansion around a trial solution, or operating point, V_{D0}:

$$I_D = I_{D0} + \left.\frac{dI_D}{dV_D}\right|_{V_{D0}} (V_D - V_{D0}) = I_{D0} + G_{D0}(V_D - V_{D0}) = I_{DN0} + G_{D0}V_D \tag{9.25}$$

where only the first-order term is considered. Equation 9.25 represents the linearized BCE of the diode around a trial operating point, V_{D0}. The diode can therefore be represented by a *companion model*, a Norton equivalent with current I_{DN0} and conductance G_{D0}. The new circuit and the graphical representation of the linearized BCE of the diode are shown in Figure 9.8. After Eq. 9.25 is substituted into Eq. 9.24, the nodal equation for this circuit becomes

$$(G + G_{D0})V = I_A - I_{DN0} \tag{9.26}$$

The solution of the above equation is V_{D1}, which becomes the new trial operating point for the diode. Equation 9.26 is solved repeatedly, or iteratively, with new values for G_D and I_{DN} at each iteration $(i + 1)$, which are based on the voltage at the previous iteration

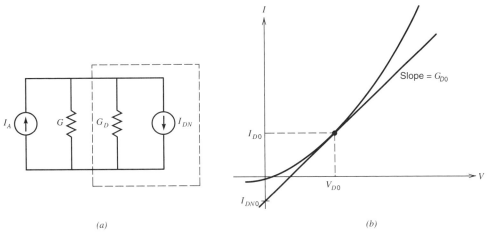

Figure 9.8 Linearized companion model of the diode circuit: (*a*) circuit diagram; (*b*) graphical solution.

(i). The iterative process is described by the nodal equation

$$(G + G_D^{(i)})V^{(i+1)} = I_A - I_{DN}^{(i)} \tag{9.27}$$

which represents for this simple case the Newton iteration. The graphical representation of the iterative process for this circuit is given in Figure 9.9. Intersection of the linearized diode conductance with the load line G define the values V_{Di} at successive iterations. This process converges to the solution, $\hat{V}$.

The general Newton iteration applied to a nonlinear function $g(x)$ of a single variable is

$$x^{(i+1)} = x^{(i)} - g^{(i)}/g'^{(i)} \tag{9.28}$$

where for the diode equation

$$g(x) = I_D(V) = I_S(e^{V/V_{th}} - 1)$$

and at the solution $\hat{x}$, $I_D^{(i+1)} = I_D^{(i)}$.

The generalized linearization approach for a set of nonlinear equations $\mathbf{g}(\mathbf{x}) = 0$ consists in computing the partial derivatives with respect to the controlling voltages, according to Eq. 9.25. These values form the conductance contribution of the nonlinear BCEs to the MNA of the circuit, called the Jacobian, $\mathbf{J}$. For an arbitrary circuit the set of all nonlinear equations is replaced at each iteration by the following set of linear equations:

$$\mathbf{J}(\mathbf{x}^{(i)}) \cdot \mathbf{x}^{(i+1)} = \mathbf{J}(\mathbf{x}^{(i)}) \cdot \mathbf{x}^{(i)} - \mathbf{g}(\mathbf{x}^{(i)}) \tag{9.29}$$

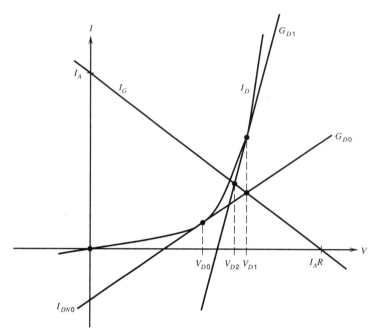

Figure 9.9   Newton iteration for the diode circuit.

where $\mathbf{J}(\mathbf{x}^{(i)})$ is the Jacobian computed with $\mathbf{x}^{(i)}$, the solution at the previous iteration. The conductance matrix $\mathbf{A}$ and RHS $\mathbf{b}$ for a nonlinear circuit have the following matrix representation:

$$\mathbf{A} = \mathbf{J}(\mathbf{x}^{(i)}) + \mathbf{G} \tag{9.30}$$

$$\mathbf{b} = \mathbf{J}(\mathbf{x}^{(i)}) \cdot \mathbf{x}^{(i)} - \mathbf{g}(\mathbf{x}^{(i)}) + \mathbf{C} \tag{9.31}$$

Equation 9.31 includes the contributions of the linear elements and independent current sources in the circuit, $\mathbf{G}$ and $\mathbf{C}$, respectively. At each iteration the circuit is described as a linear system, Eqs. 9.6, 9.30, and 9.31, which is solved using the LU factorization described in Sec. 9.2.1.

An important question is whether the above equations converge to a solution and how many iterations it takes. The nonlinear I-V characteristics of semiconductor devices are exponential or quadratic functions. As exemplified by the diode characteristic in Figure 9.8, the equivalent conductance can vary from 0 in the reverse region to infinity in the forward bias region. With the limited range of floating-point numbers available on a computer and the unboundedness of solutions provided by the Newton algorithm according to Eq. 9.28, the iterative scheme can fail to converge.

An algorithm that controls the changes in the state variables of the nonlinear elements from iteration to iteration is very important if the simulated circuit is to converge to the correct solution. This algorithm is known as a *limiting algorithm* and determines the convergence features of the simulator. A limiting algorithm selectively accepts the solution unchanged or, when large changes in the value of the nonlinear function would occur, limits it by correcting the new value of the Newton iteration (Calahan 1972) from Eq. 9.28 to the following:

$$x^{(i+1)} = x^{(i)} - \alpha g^{(i)} / g'^{(i)} \tag{9.32}$$

The parameter $\alpha(0 < \alpha \le 1)$ indicates that only a fraction of the change is accepted at each iteration. Equation 9.32 defines the Newton-Raphson iterative algorithm. The choice of α is implemented through the limiting algorithm, which is tailored to the different nonlinear characteristics of each semiconductor device. This parameter assumes a different value at each iteration and for each device.

A practical implementation of the above modified Newton-Raphson iteration, Eq. 9.32, is shown in Figure 9.10. The scheme proposed by Colon is successfully used in SPICE and its derivatives to limit the new junction voltage on diodes and BJTs. At each iteration the new solution of the junction voltage, V_1, if larger than the previous value, V_0, is used to derive a new current, I'_1, based on the last linearization of the diode characteristic. The voltage corresponding to I'_1 on the nonlinear diode characteristic is V'_1, which is smaller than the original solution, V_1; V'_1 becomes the new trial operating point for the diode in the current iteration. If V_1 is smaller than V_0, it is accepted directly

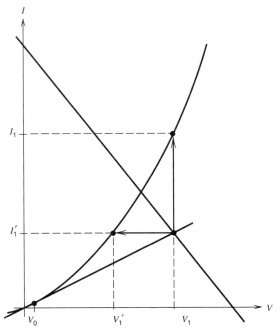

Figure 9.10 Newton-Raphson iteration with current limiting.

as the new trial operating point because there is no danger of a runaway solution. The danger in the absence of any limiting is that a trial solution can generate very large values of the exponential function, which cannot be corrected in subsequent iterations, and the process fails to converge.

The Newton-Raphson algorithm has quadratic convergence properties if the initial guess, V_0 in Figure 9.10, is close to the solution. Therefore, it is important in a circuit simulator to provide an initial guess as a set either of node voltages or of terminal voltages for the nonlinear semiconductor devices. Circuit simulators start out usually with all unknowns set to zero. SPICE2 initializes the semiconductor devices in a starting operating point such that nonzero conductances can be loaded into the MNA circuit matrix at the first iteration. An additional protection built into SPICE is a parallel minimum conductance, *GMIN*, with a default value of 10^{-12} mho, connected in parallel to every pn junction, such as BE and BC for BJTs and BD and BS for MOSFETs, to prevent zero-valued conductances from being loaded into the circuit admittance matrix when these junctions are reverse biased.

9.3.2   Convergence and SPICE Options

This section addresses the issues of how and when convergence is achieved. Equation 9.32 defines the Newton-Raphson iteration; the iterative process is finished when the following two conditions are met:

1. All the voltages and currents of the unknown vector are within a prescribed tolerance for two consecutive iterations.

2. The values of the nonlinear functions and those of the linear approximations are within a prescribed tolerance.

Next, the tolerances used in SPICE to establish convergence are introduced. Each tolerance, for voltage or current variables, is formed from a relative term and an absolute term. The voltage tolerance for node n, ϵ_{Vn}, is defined as

$$\epsilon_{Vn} = RELTOL \cdot \max\left(|V_n^{(i+1)}|, |V_n^{(i)}|\right) + VNTOL \qquad (9.33)$$

The values of node voltages at two consecutive iterations have to satisfy the following inequality for convergence:

$$\left|V_n^{(i+1)} - V_n^{(i)}\right| \leq \epsilon_{Vn} \qquad (9.34)$$

RELTOL and **VNTOL** are SPICE option parameters representing the relative and the absolute voltage tolerance, respectively. The default values are set to

$$RELTOL = 10^{-3}$$
$$VNTOL = 1 \; \mu V$$

The absolute tolerance defines the minimum value for which a given variable is still accurate; with the SPICE defaults, voltages are accurate to 1 part in 1000 down to 1 μV of resolution. This means that a node with 100 V is accurate to 100 mV and a voltage of 10 μV is accurate to only 1 μV.

Convergence is based not only on the circuit variables but also on the values of the nonlinear functions that define the BCEs of the nonlinear elements. In the case of semiconductor devices, the nonlinear functions are the currents, for example, I_D for a diode, I_C and I_B for a BJT, and I_{DS} for all FETs. In the following derivation the nonlinear function will be referred to for simplicity as a current.

SPICE defines the difference between the nonlinear expression, I_D, evaluated for the last junction voltage solution $V_J^{(i)}$ and the linear approximation $\hat{I}_D$, using the present voltage solution, $V_J^{(i+1)}$, as the test for convergence:

$$\hat{I}_D = G_D V_J^{(i+1)} - I_{DN}$$
$$I_D = I_S\left(e^{V_J^{(i)}/V_{th}} - 1\right)$$

(9.35)

The tolerance is defined as

$$\epsilon_I = RELTOL \cdot \max\left(\hat{I}_D, I_D\right) + ABSTOL$$

(9.36)

and the iteration process converges when

$$\left|\hat{I}_D - I_D\right| \le \epsilon_I$$

(9.37)

$$\left|V_J^{(i+1)} - V_J^{(i)}\right| \le \epsilon_{VJ}$$

(9.38)

ABSTOL is the absolute current tolerance and defaults to 10^{-12} A in SPICE2 and SPICE3. This default is very accurate for the base current of BJTs but may be too restrictive for other devices, such as FETs, whose operation is controlled by the gate voltage and for which I_{DS} is in most applications 1 μA and higher.

There is a limit to the number of iterations computed by SPICE. This limit is set by the option parameter **ITL1**, which defaults to 100. If the inequalities of Eqs. 9.34, 9.37, and 9.38 are not satisfied in *ITL1* iterations, SPICE2 prints the message *ERROR*: NO CONVERGENCE IN DC ANALYSIS and the last node voltages.

When SPICE computes DC transfer curves, the number of allowed iterations is *ITL1* only for the first value of the sweeping variable and then is reduced to *ITL2*, which defaults to 50. The advantage of subsequent analyses is that the unknown vector is initialized with the values from the last point.

Convergence failure can also occur during a transient analysis, which consists of quasi-static iterative solutions at a discrete set of time points. For each time point only **ITL4** iterations are allowed before a failure of convergence is decreed. *ITL4* defaults to 10. In transient analysis failure to converge at a time does not result in the abortion of the analysis but causes a reduction of the time step, as explained in Sec. 9.4.

When SPICE fails to find a DC solution, an additional option can be used to achieve convergence, called *source ramping*. This method consists in finding the DC solution by ramping all independent voltage and current sources from zero to the actual values. This is equivalent to an .OP analysis using a .DC transfer curve approach. Source ramping can be viewed as a variation of the general modified Newton-Raphson solution algorithm. SPICE2 does not automatically use source ramping if it fails to converge. Option parameter ITL6 must be set to the number of iterations to be performed for each stepped value of the source, similar to ITL2 for DC transfer curves. SPICE3 and PSpice perform source ramping automatically when the regular iterative process fails to converge. A more detailed presentation of convergence problems along with examples and solutions can be found in Chap. 10.

All *ITLx* options can be reset by the user, although this should be rarely necessary. A detailed presentation on the situations requiring the change of these options is provided in the following chapter.

During the solution process the linear equivalent of each nonlinear element must be evaluated. This is a very time-consuming process and may not always be necessary. In time-domain analysis, during the iterations performed at any one time point each nonlinear element is checked for a change in the terminal voltages and the output current from the last time point. If the controlling variables and resulting function of a device have not changed, a new linearized model is not computed for this device; it is *bypassed*. In other words, the linearized conductances obtained at the previous time point are used again in the circuit matrix.

The bypass operation results for a majority of circuits in analysis time savings without affecting the end result. There are, however, cases when bypass can cause nonconvergence at a later time point. The check for bypassing a device is based on Eqs. 9.37 and 9.38. This check can limit the devices being bypassed by reducing the tolerances; however, this reduction can negatively affect the convergence test for the overall circuit.

9.4 TIME-DOMAIN SOLUTION

The time-domain solution is the most complex analysis in a circuit simulator, because it involves all the algorithms presented so far in this chapter, as shown graphically in the flowchart of Figure 9.1. As described in the introductory overview of algorithms (Sec. 9.1), transient analysis is divided into a sequence of quasi-static solutions.

This section outlines the numerical techniques used in SPICE to transform a set of differential equations, such as the ones representing the BCEs of capacitors and inductors, into a set of algebraic equations. The presentation of the numerical methods is simplified for the purpose of providing the SPICE user with insight into the workings of the program; a rigorous description of the integration algorithm can be found in the book by Chua and Lin (1975) and L. Nagel's Ph.D. thesis (1975). The specific implementation in SPICE is described here; furthermore, the accuracy of these algorithms is analyzed along with the options available to the user regarding numerical methods and error bounds.

9.4.1 Numerical Integration

The different numerical integration algorithms and their properties are best introduced with an example.

EXAMPLE 9.1

Consider the series RC circuit shown in Figure 9.11a; at $t = 0$ a voltage step of magnitude V_i is applied at the input. Find the voltage, $v_R(t)$, across the resistor and the voltage, $v_C(t)$, across the capacitor over time.

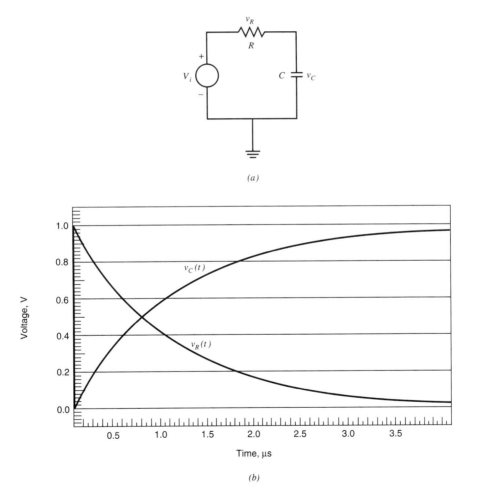

(a)

(b)

Figure 9.11   RC circuit: (a) circuit diagram; (b) plots of solutions $v_R(t)$ and $v_C(t)$.

Solution

From the BCEs of the resistor and capacitor the following equation is obtained:

$$v_R(t) = Ri_R(t) = RC\frac{dv_C}{dt} \tag{9.39}$$

KVL applied to the circuit allows the following substitution of v_C:

$$v_C(t) = V_i - v_R(t) \tag{9.40}$$

yielding the following differential equations for $v_R(t)$ and $v_C(t)$:

$$\dot{v}_R(t) = -\frac{1}{RC}v_R = -\frac{1}{\tau}v_R = \lambda v_R = f(v_R)$$

$$\dot{v}_C(t) = \frac{1}{RC}(V_i - v_C) = \frac{1}{\tau}(V_i - v_C) = f(v_C) \tag{9.41}$$

The solutions $v_R(t)$ and $v_C(t)$ are

$$v_R(t) = V_i e^{-t/\tau}$$

$$v_C(t) = V_i(1 - e^{-t/\tau}) \tag{9.42}$$

and are presented in Figure 9.11*b*.

In SPICE the solution of the above equation in the interval 0 to *TSTOP* (see also Sec. 6.2) is performed at a number of discrete time points, where the differential equation is replaced by an algebraic equation. For simplicity, let x be the time function to be solved for and x_n the values at the discrete time points t_n:

$$x_n = x(t_n)$$

The solution at t_{n+1}, x_{n+1}, can be expressed by a Taylor series expansion around x_n:

$$x_{n+1} = x_n + h\dot{x}_n \tag{9.43}$$

where h is the time step, assumed equal for all time points for simplicity. This is identical to the finite-difference approximation of the derivative of x and represents the *forward-Euler* (FE) integration formula. Substitution of Eqs. 9.41 into 9.43 yields the following recursive solution for v_C at t_{n+1}:

$$v_C(t_{n+1}) = v_C(t_n) + h\dot{v}_C(t_n) = v_C(t_n) + hf(v_{Cn}) \tag{9.44}$$

This is represented graphically in Figure 9.12*a*. A rather sizeable error can be noticed for $v_C(t_{n+1})$ computed with Eq. 9.44.

A different solution is obtained for v_C if in Eq. 9.43 x_{n+1} is expressed in terms of the derivative at $t_{n+1}, \dot{x}_{n+1}$,

$$x_{n+1} = x_n + h\dot{x}_{n+1} \tag{9.45}$$

This represents the *backward-Euler* (BE) integration formula. Because Eq. 9.45 must be solved simultaneously for x as well as for its derivative, this formula is known as an *implicit method*, whereas the FE is an explicit method. The graphical interpretation of this solution is shown in Figure 9.12*b*. It can be seen that v_{n+1} as given by the BE formula is less sensitive to the size of the time step h than that given by the FE formula.

An important measure of the accuracy of a numerical integration method is the *local truncation error, LTE*, evaluated at each time point. For the FE and BE methods *LTE* can be approximated by the first discarded term in the Taylor expansion:

$$x_{n+1} = x_n + h\dot{x}_n + \frac{h^2}{2}\ddot{x}_n$$

$$LTE = \left| \frac{h^2}{2}\ddot{x}_n \right| \tag{9.46}$$

Algorithms for automatic time-step control such as the one used in SPICE are based on checking whether the LTE of each time-dependent BCE is within prescribed bounds. The actual algorithm is detailed in the following section.

The two methods introduced so far are known as first-order methods, because higher-order terms are neglected in the series. Based on the above definition of *LTE*, it can be thought that using higher-order terms of the series in the solution of x_{n+1} can lead to smaller *LTE*. The trapezoidal integration is a second-order method that can be derived based on the observation that a more accurate solution $v_C(t_{n+1})$ can be obtained if in Eqs. 9.43 and 9.45 the average of the slopes at t_n and t_{n+1} are used as compared to either one or the other:

$$x_{n+1} = x_n + \frac{1}{2}h(\dot{x}_n + \dot{x}_{n+1}) \tag{9.47}$$

The higher accuracy of this method is obvious from the graphical solution of $v_C(t_{n+1})$ shown in Figure 9.13.

The LTE of the trapezoidal integration formula can be derived by first substituting $\dot{x}_{n+1}$ in Eq. 9.47 with a Taylor expansion:

$$\dot{x}_{n+1} = \dot{x}_n + h\ddot{x}_n + \frac{h^2}{2}\dddot{x}_n \tag{9.48}$$

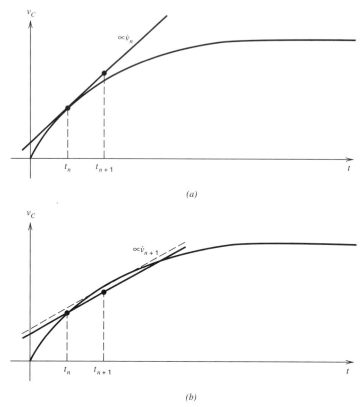

(a)

(b)

Figure 9.12   Solutions of $v_C(t)$ between t_n and t_{n+1}: (*a*) FE solution; (*b*) BE solution.

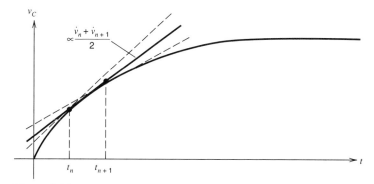

Figure 9.13   Trapezoidal solution of $v_C(t)$ between t_n and t_{n+1}.

and then subtracting the trapezoidal solution, Eq. 9.47, from the exact solution given by the first three terms of the Taylor series:

$$x_{n+1} = x_n + h\dot{x}_n + \frac{h^2}{2}\ddot{x}_n + \frac{h^3}{6}\dddot{x}_n \tag{9.49}$$

The resulting LTE of the trapezoidal integration method for x_{n+1} is

$$LTE = \left|\frac{h^3}{12}\dddot{x}_n\right| \tag{9.50}$$

The LTE for $\dot{x}_{n+1}$ is obtained by first substituting x_{n+1} in Eq. 9.47 by Eq. 9.49 and then subtracting the resulting equation in $\dot{x}_{n+1}$ from Eq. 9.48:

$$LTE = \left|\frac{h^2}{6}\dddot{x}_n\right| \tag{9.51}$$

EXAMPLE 9.2

Apply the trapezoidal integration method to the BCE of a capacitor. Use the result to develop a companion model of the capacitor to be used in nodal equations.

Solution
The BCE of a capacitor, Eqs. 2.4 and 2.5, can be rewritten for a nodal interpretation:

$$\int_0^t i\,dt = \int_0^v C\,dv_C \tag{9.52}$$

The trapezoidal integration formula, Eq. 9.47, is applied to the above equation and results in

$$i_{n+1} = -i_n + \frac{2C}{h}(v_{n+1} - v_n) \tag{9.53}$$

Equation 9.53 can be rewritten as a nodal equation at t_{n+1}:

$$\frac{2C}{h}v_{n+1} = i_n + \frac{2C}{h}v_n \tag{9.54}$$

or

$$G_{eq}v_{n+1} = I_{eq}$$

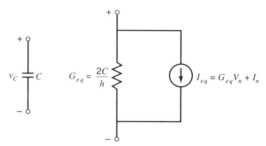

Figure 9.14 Companion model for a capacitor.

The companion model of the capacitor for nodal analysis is shown in Figure 9.14. It is formed of the parallel combination of the equivalent conductance G_{eq} and the equivalent current source I_{eq}. In SPICE Eq. 9.54 is updated at each time point and the contributions are loaded into the circuit matrix and RHS vector. The companion model for an inductor can be derived similarly.

An important property of an integration method is its *stability* or convergence feature. Whereas the LTE is a local measure of accuracy at each time point, stability is a global measure of how the solution computed by a given method approaches the exact solution as time proceeds to infinity. Stability is also a function of the specific circuit. A quantitative analysis of the stability of the integration methods introduced so far can be performed for the RC circuit in Figure 9.11. The exact solution for $v_R(t)$, Eqs. 9.42, can be compared with the FE, BE, and TR solutions computed after n time steps:

$$\text{(FE)} \quad V_i(1 - h/\tau)^n \tag{9.55}$$

$$\text{(BE)} \quad \frac{V_i}{(1 + h/\tau)^n} \tag{9.56}$$

$$\text{(TR)} \quad V_i\frac{(1 - h/2\tau)^n}{(1 + h/2\tau)^n} \tag{9.57}$$

The FE solution can be seen to lead to the wrong solution if the step-size, h, is greater than 2τ. The BE solution, by contrast, decreases to zero as does the exact solution $v_R(t)$ in Eqs. 9.42 as time increases. An interesting result is offered by the TR method, which converges toward zero but does so in an oscillatory manner if $h > 2\tau$. This behavior of the TR method can be observed in SPICE especially when the solution goes through discontinuities.

Electronic circuits have time constants that can differ by several orders of magnitude; the equations representing these circuits constitute *stiff systems*. The integration methods used to solve such systems must be *stiffly stable*; in other words, these methods must provide the correct solution without constraining the time step to the smallest time constant in the circuit. The implicit methods introduced so far, BE and TR, are stiffly stable, but explicit methods, such as the FE method, are not. The TR and BE integration methods are the default in the majority of SPICE versions.

Additional integration formulas have been developed that fall in the general category of polynomial integration methods defined by

$$x_{n+1} = \sum_{i=0}^{n} a_i x_{n-i} + \sum_{i=-1}^{n} b_i \dot{x}_{n-i} \tag{9.58}$$

If b_{-1} is zero, the method is explicit, and if b_{-1} is nonzero, the method is implicit. The algorithm is a multistep algorithm if $i > 1$, that is, if more than one time point from the past is needed to compute x_{n+1}. The *Gear integration* (Gear 1967) formulas of order 2 to 6 have proven to have good stability properties. The Gear formulas of varying order for x_{n+1} are listed in Appendix D. The Gear integration formulas order 2 to 6 are implemented in SPICE2, SPICE3, and most commercial SPICE versions as an alternative to the default TR method. PSpice uses only the Gear algorithms.

The time-domain response of a circuit can differ depending on the integration method used. Although in the vast majority of cases both the TR and the Gear methods lead to the same solution, the two have different characteristics. The TR method converges to a solution in an oscillatory manner (Eq. 9.57) when the time step is larger than a certain limit. The Gear formula of order 2 has an opposite behavior, which leads to a damped response. This difference between the two methods is demonstrated by the following example.

EXAMPLE 9.3

Find the time response of the LC circuit shown in Figure. 9.15 assuming that at $t = 0$ the switch is opened. Use both the **TRAP** and the **GEAR** options, the latter with a *MAXORD* of 2, if running SPICE2 or SPICE3, and compare the results.

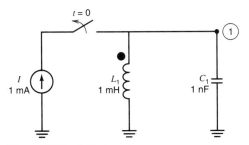

Figure 9.15   LC circuit.

```
LC CIRCUIT
*
L1 1 0 1M IC=-1M
C1 1 0 1N
*
*.OPTION METHOD=GEAR MAXORD=2
*
.TRAN 1U 400U 0 UIC
.PLOT TRAN V(1)
.END
```

The SPICE input is listed above. Note that the current source must not appear in the circuit description, which is valid for $t \geq 0$; its effect, however, must be taken into account by setting the appropriate initial condition for the inductor current. In order to start the analysis from the initial condition at the time the switch is opened, the **UIC** keyword must be specified in the **.TRAN** statement.

The result of trapezoidal integration is free oscillations with an amplitude of 1 V at the resonant frequency of 10^6 rad/s, or 159 kHz, as seen in the upper trace of Figure 9.16.

The inclusion of the **.OPTIONS** statement by removal of the asterisk at the beginning of the line leads to the solutions computation by the Gear formula. The result is shown in the lower trace of Figure 9.16 to be a decaying oscillation, which is wrong. Recent versions of PSpice produce the decaying waveform, probably because the Gear 2 method is used as default.

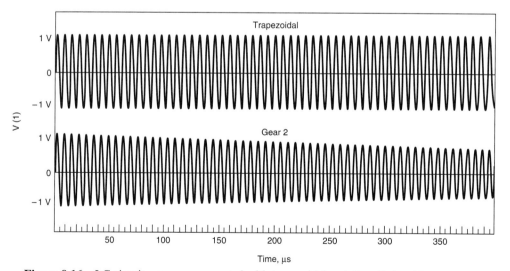

Figure 9.16 LC circuit response computed with trapezoidal and Gear 2 algorithms.

The stability of integration methods is presented in more detail in the works by McCalla (1988) and by Nagel (1975). Several conclusions can be drawn based on the first-order analysis presented in this section and the more thorough analysis found in the references.

First, the LTE of a numerical method diminishes for higher-order methods; by contrast, the stability deteriorates as the order of the method increases. Second, the equations of electronic circuits must be solved by stiffly stable integration methods for which the time step is determined by LTE and not by stability constraints.

9.4.2   Integration Algorithms in SPICE, Accuracy, and Options

This section describes the implementation in SPICE of the integration algorithms introduced in the previous section and the options available to the user to improve the accuracy of a solution.

As mentioned above, most SPICE programs support two integration algorithms, second-order trapezoidal and Gear order 2 to 6. The default method is trapezoidal, but a user can select the Gear algorithm with the options **METHOD** and **MAXORD**. The choices for **METHOD** are **TRAP** or **GEAR**, and for **MAXORD** a number between 2 and 6 is required. SPICE implements the Gear algorithm as a variable-order, multistep method. **MAXORD** limits the order of the integration formula used for the variable-order Gear method and is therefore relevant only when METHOD=GEAR.

EXAMPLE 9.4

Change the SPICE2 default integration method to Gear and limit the order of the integration formula to 3.

Solution
This can be achieved by adding the following **.OPTIONS** line to the SPICE circuit description:

```
.OPTIONS METHOD=GEAR MAXORD=3
```

The variable-order algorithm in SPICE selects at each time point the order that allows for the maximum time step. A higher-order method has a smaller LTE, as shown in the previous section, enabling the variable time step algorithm in SPICE to select a larger time step. However, caution must be exercised with higher orders because inaccuracy is introduced in the computation of the LTE and of the resulting time step.

The truncation-error-based time-step control algorithm is described next for trape-zoidal integration. This algorithm is common to most SPICE programs. An upper bound is calculated for the truncation error at each time point based on the computation of charges or currents of capacitors and fluxes or voltages of inductors. This error is similar to the one defined for nonlinear equations (Eqs. 9.33 and 9.36) and consists of a relative and an absolute error:

$$\epsilon_{\dot{x}} = \epsilon_r \cdot \max(|\dot{x}_{n+1}|, |\dot{x}_n|) + \epsilon_a \tag{9.59}$$

$\dot{x}_{n+1}$ in the above equation represents the current of capacitors or the voltage across inductors; SPICE defines also a charge or flux error:

$$\epsilon_x = \epsilon_r \cdot \max \frac{(|x_{n+1}|, |x_n|, |\epsilon_{qa}|)}{h_n} \tag{9.60}$$

The LTE at each time point is taken as the maximum of the two errors:

$$E = \max(\epsilon_x, \epsilon_{\dot{x}}) \tag{9.61}$$

A new SPICE option, **CHGTOL**, is introduced for the absolute charge or flux error, ϵ_{qa}. The exact SPICE implementation of the truncation errors is

$$\epsilon_{\dot{x}} = RELTOL \cdot \max(|\dot{x}_{n+1}|, |\dot{x}_n|) + ABSTOL \tag{9.62}$$

$$\epsilon_x = RELTOL \cdot \max(|x_{n+1}|, |x_n|, CHGTOL)/h_n \tag{9.63}$$

The default value for $CHGTOL$ is 10^{-14} C.

Based on the upper bound E for the LTE at each time point, the next time step, h_{n+1}, is given by the following inequality:

$$h_{n+1} \le \sqrt{\frac{6E}{\left|\dfrac{d^3 x_n}{dt^3}\right|}} \tag{9.64}$$

which results from the definition of the LTE of the trapezoidal method for $\dot{x}_{n+1}$, Eq. 9.51.

It is important to get an accurate estimate of the third derivative of the charge, $d^3 x/dt^3$, in the above inequality. The high-order derivatives are approximated in SPICE by divided differences using the following definition:

$$\frac{d^k x}{dt^k} = k! DD_k \qquad (9.65)$$

which sets the relation between the kth derivative, $d^k x / dt^k$, and the divided difference of order k, DD_k. The recursive formula for divided differences is

$$DD_k = \frac{DD_{k-1}(t_{n+1}) - DD_{k-1}(t_n)}{\sum_{i=1}^{k} h_{n+1-i}} \qquad (9.66)$$

DD_1 is the numerical approximation of the derivative of x between t_n and t_{n+1}:

$$DD_1 = \frac{x_{n+1} - x_n}{h_n}$$

With these formulas for divided differences, the time step computation in SPICE, Eq. 9.64, becomes

$$h_{n+1} = \sqrt{\frac{E}{DD_3(t_{n+1})}} \qquad (9.67)$$

A value for the maximum time step given by Eq. 9.67 is computed for every linear or nonlinear charge- or flux-defined element in the circuit. Comparisons between the exact LTE for the circuit in Figure 9.11, Eq. 9.50, and the one approximated by divided differences have shown that the divided difference overestimates the LTE several times (Nagel 1975). This observation has led to the conclusion that a larger time step can be used than the one defined by Eq. 9.67. An option parameter has been introduced in SPICE, **TRTOL**, which scales down the divided difference and therefore the LTE. With the new factor the predicted time step becomes

$$h_{n+1} = \sqrt{\frac{TRTOL \cdot E}{\max\left(\dfrac{DD_3}{12}, \epsilon_a\right)}} \qquad (9.68)$$

The default value of 7 for *TRTOL* has proven to provide a good compromise between accuracy and speed for a large number of circuits.

The automatic time-step control algorithm in SPICE selects h_{n+1} based on the minimum value resulted from evaluating Eq. 9.68 for all capacitors and inductors in the circuit. The SPICE time-selection algorithm is outlined below.

$$t_{n+1} = t_n + h_n$$
solve at t_{n+1}
if $iter\_num < ITL4$
      compute $h_{n+1} = f(LTE)$
      if $(h_{n+1} < 0.9 \cdot h_n)$ then
            reject t_{n+1}
            $h_n = h_{n+1}$
            recompute at new t_{n+1}
    else
            accept t_{n+1}
            $h_{n+1} = \min(h_{n+1}, 2 \cdot h_n, TMAX)$
            proceed with t_{n+2}
else
    reject t_{n+1}
    $h_n = h_n/8$
    reduce integration order to 1 (BE)
    if $(h_n > h_{min})$ then
            recompute at new t_{n+1}
    else
            `TIME STEP TOO SMALL`; abort

Assume that the solution at t_n has been accepted and h_n has been selected as the new time step. After removing the time dependency at t_{n+1} using transformations of the type given by Eq. 9.53, the set of nonlinear equations is solved as described in Sec. 9.3.

First, the program checks whether the nonlinear solution has converged in less than *ITL4* iterations, where **ITL4** is an option parameter defaulting to 10. If a solution could not be obtained in *ITL4* iterations, a new value is defined for h_n that is $\frac{1}{8}$ of the previous value. If this value for the time step is larger than the minimum acceptable time step, h_{min}, which is approximately eight orders of magnitude smaller than the print step, *TSTEP* (see Chap. 6), a new t_{n+1} is defined,

$$\text{new } t_{n+1} = t_n + \frac{h_n}{8}$$

and the solution is repeated. The solution at the newly defined t_{n+1} is performed with the first-order BE method. The method is changed back to the second-order TR only if the new t_{n+1} is accepted. If the new h_n is not larger than h_{min}, SPICE aborts the analysis and issues the message `*ERROR*: INTERNAL TIMESTEP TOO SMALL IN TRANSIENT ANALYSIS` followed by the value of t_n and h_n.

If the solution at t_{n+1} is obtained in less than *ITL4* iterations, h_{n+1} is computed based on the prescribed LTE, Eq. 9.51. The value of h_{n+1} is accepted if it is at least $0.9h_n$ which implies that the LTE is within bounds; if larger than h_n, h_{n+1} is allowed only to double at each time point. Also, the time step can only increase up to the lesser of $2 \cdot TSTEP$ or $TMAX$. If h_{n+1} evaluated according to Eq. 9.51 is less than $0.9h_n$, the solution at t_{n+1} is rejected and the smaller value obtained for the time step is assigned to h_n, which defines in turn the new t_{n+1}.

There are a few additional details in the integration implementation that a knowledgeable user should be aware of.

First, there are certain time points, called *breakpoints*, that are treated differently. The time-step control algorithm is based on an estimate of the LTE, which in turn is a function of the divided difference approximation of the derivative of a voltage or current waveform. The approximation of the truncation error is therefore based on an often untrue assumption that $x(t)$ is continuously differentiable to order k in the time interval of interest. Large inaccuracies are avoided by introducing the concept of a *breakpoint*, which is a time point where an abrupt change in the waveform is anticipated based on the shape of independent source signals. At a breakpoint a solution of the circuit differential equations is enforced and a first-order method, the backward Euler method, is used to solve for the time point immediately following the breakpoint. The method applied at the first time point following a breakpoint is known as a *start-up method*. The LTE of the first-order method is used at the new time point to minimize the error of the approximation. Breakpoints are important for an accurate solution because they prevent the evaluation of the LTE based on time points preceding the discontinuity.

A second accuracy-enhancing technique is the prediction of the circuit-variable values for a new time point. In the first iteration of a new time point SPICE employs a linear prediction step for the voltages and charges of nonlinear branches.

A third implementation detail is the *bypass* option. At every new time point after the first iteration the nonlinear branch voltages, such as V_{BE} and V_{BC} of a BJT, and the resulting nonlinear function, such as I_C, are compared to the corresponding values at the previous time point. If they have not changed more than the tolerance errors ϵ_V and ϵ_I (Eqs. 9.37 and 9.38), the computation of the linear equivalent is bypassed and the conductance values from the last time point are used. Bypassing the reevaluation of certain devices can result in 20% savings in analysis time, but it also can cause convergence problems for slowly moving variables that are within the prescribed tolerances compared to the last time point but may differ more than the error tolerance if compared to k time points previously.

In SPICE2 and PSpice the only way to reduce bypassing is to tighten the relative error, **RELTOL**. This must be done carefully since it can have the adverse effect of nonconvergence depending on the nonlinear equations. SPICE3 provides an option at compile time that prohibits bypass.

In SPICE2 and other commercial SPICE programs there is an alternate time-step controlling mechanism, based on the Newton-Raphson iteration count used at each time point. The time-step selection algorithm is defined by the **LVLTIM** option. The default value is 2, which represents the LTE-based algorithm; if $LVLTIM = 1$, the iteration-count time-step algorithm is used, which doubles the time step for any time point where a solution is obtained in less than $ITL3$ iterations and reduces the time step by 8 when more than $ITL4$ iterations are required. The iteration time-step control proves more conservative but more foolproof for cases when no charge storage parameters are specified in the model definition of a nonlinear device or when the default error bounds are not suitable. This time-step control algorithm is not available in SPICE3 or PSpice.

So far the time-step control mechanism has been presented only for the TR method. As mentioned previously, the Gear algorithms order 2 to 6 are also implemented in SPICE. The time-step is also controlled by the LTE in the same manner as described above. An additional feature for the Gear method is that the order of the method can also be modified dynamically. The order leading to the largest time step within the prescribed LTE is chosen. The highest order of the Gear method to be used is limited by *MAXORD* and the number of previous time points available since the last breakpoint. The time step following a time point where the method of order k has been used is evaluated for orders $k - 1, k$, and $k + 1$. If the value of h_{n+1} for higher or lower order method is 1.05 times larger than the value obtained for the current order, an order change is performed. Nagel (1975) has found that for a set of benchmarks using the default tolerances, the Gear algorithm used order 2 most of the time. Only for tolerances reduced by 2 to 3 orders of magnitude are the higher-order Gear algorithms being exercised. The problem with the fifth- and sixth-order Gear algorithms is that the divided difference introduces a sizable error. In general it has been found that most problems where Gear is beneficial, such as where there is numerical ringing or convergence problems, a *MAXORD* of 2 or 3 is sufficient. Although use of the Gear method can result in a reduction of computed time points and iterations, the savings in execution time over the TR method is small, if there is any, because of the overhead of computing the coefficients of the method at each time step (Nagel 1975).

9.5 SUMMARY OF OPTIONS

The various options available to a user for controlling solution algorithms and tolerances have been introduced throughout this chapter. Although options differ among SPICE versions, the majority of the options introduced in this text can be found in most versions. This section summarizes the **.OPTIONS** parameters by function. This summary covers most common SPICE options; for a complete list of options available in a specific SPICE version the reader is advised to consult the corresponding user's guide.

9.5.1 Analysis Summary

Before reviewing the analysis options it is instructive to introduce a SPICE option that provides insight into the analysis: the accounting option, **ACCT**. A sample output of the information printed by the **ACCT** request in SPICE2 is shown in Figure 9.17; the information is for the slew-rate transient analysis of the complete μA741 circuit, whose SPICE input is listed in Figure 7.28 and whose schematic is shown in Figure 7.29. The list starts with circuit statistics; NUNODS is the number of nodes defined at the top

```
******* 03/23/92 ******** SPICE 2G.6 3/15/83 ******** 17:55:34 *****

UA741 FULL-MODEL SLEW RATE

****      JOB STATISTICS SUMMARY        TEMPERATURE = 27.000 DEG C


**********************************************************************

NUNODS   NCNODS   NUMNOD   NUMEL   DIODES   BJTS   JFETS   MFETS
    6       27       79      42        0     26       0       0

NUMTEM   ICVFLG   JTRFLG   JACFLG   INOISE   IDIST    NOGO
    1        0      201        0        0       0       0

NSTOP    NTTBR    NTTAR    IFILL     IOPS    PERSPA
  82.     370.     472.     102.    1059.    92.980

NUMTTP   NUMRTP   NUMNIT   MAXMEM   MEMUSE   COPYKNT
 284.      56.    1231.    400000    15484    41543.

             READIN            0.60
             SETUP             0.15
             TRCURV            0.00                    0.
             DCAN              3.68                   84.
             DCDCMP            9.783                   3.
             DCSOL             4.283
             ACAN              0.00                    0.
             TRANAN           51.53                 1231.
             OUTPUT            0.45
             LOAD             35.067
             CODGEN            0.000                   0.
             CODEXC            0.000
             MACINS            0.000
             OVERHEAD          0.02
             TOTAL JOB TIME   56.43
```

Figure 9.17 SPICE statistics for the μA741 slew-rate analysis in Sec. 7.5.

level of the circuit, NCNODS is the number of the actual circuit nodes resulting after expansion of subcircuits, and NUMNOD is the number of nodes after adding the internal nodes generated because of the parasitic series resistances of semiconductor devices. These numbers are followed by NUMEL, the total number of elements, which is broken down into the different types of semiconductor devices. These numbers can be verified from the data in Chap. 7.

The second set of data summarizes the analyses requested. NUMTEM is the number of temperatures, ICVFLG is the requested points in a DC transfer curve, JTRFLG is the number of transient print/plot points, $\texttt{JACFLG}$ is the number of frequency points in the AC analysis, and $\texttt{INOISE}$ and $\texttt{IDIST}$ indicate whether a small-signal noise or distortion analysis has been performed. Last in the analysis category is the variable $\texttt{NOGO}$, which is 0, or FALSE, if the analysis has finished, and 1, or TRUE, if the analysis has been aborted because of an error.

The third set of data contains the linear system matrix statistics. $\texttt{NSTOP}$ is the number of MNA equations, $\texttt{NTTBR}$ is the total number of nonzero terms before reordering, $\texttt{NTTAR}$ is the total number of nonzero terms after reordering, $\texttt{IFILL}$ is the number of fill-ins created during the LU decomposition process and is equal to $\texttt{NTTAR-NTTBR}$, $\texttt{IOPS}$ is the number of floating-point multiplications and divisions required for each solution of the linear system, and $\texttt{PERSPA}$ is the sparsity of the MNA matrix expressed as a percentage.

The fourth set summarizes information about the transient solution and memory use. $\texttt{NUMTTP}$ is the number of time points at which the circuit has been solved, $\texttt{NUMRTP}$ is the number of rejected time points, where the time step h_{n+1} had to be rejected and the analysis restarted at t_n, and $\texttt{NUMNIT}$ is the total number of iterations performed for the transient analysis. $\texttt{MAXMEM}$ is the amount of memory available, $\texttt{MEMUSE}$ is the memory used by the present circuit, and $\texttt{CPYKNT}$ is the number of memory transfers.

The last part of the summary lists the times in seconds for the different analyses and solutions and the number of iterations. None of the above information has any bearing on the actual solution, but it may be of interest to users who want to relate the knowledge about algorithms acquired in this chapter to a specific circuit as well as to learn about the ease or difficulty of convergence and time-domain solution. Note that the outputs produced by different SPICE versions for this option vary but relate some of the same statistics.

9.5.2   Linear Equation Options

The following options are related to the linear equation solution:

$\texttt{GMIN}$=*value* defines the minimum conductance connected in parallel to a pn junction; the default is 10^{-12} mho.

$\texttt{PIVTOL}$=*value* (SPICE2 and SPICE3) sets the smallest MNA matrix entry that can be accepted as a pivot. If no entry is larger than *value* at any LU decomposition step, the circuit matrix is declared singular. The default is 10^{-13}.

$\texttt{PIVREL}$=*value* (SPICE2 and SPICE3) represents the ratio between the smallest acceptable pivot and the maximum entry in the respective column; a larger value for this option can lead to a better-conditioned matrix at the expense of more fill-ins. The default is 10^{-3}.

These options should not be modified unless convergence failure is caused by singular matrix problems. Only increasing $\texttt{GMIN}$ can lead in some cases to better convergence.

9.5.3   Nonlinear Solution Options

The options that control the Newton-Raphson solution can be grouped as convergence tolerances and iteration count limits.

The convergence tolerances are the following:

RELTOL=*value* defines the relative error tolerance within which voltages and device currents are required to converge as set forth by Eqs. 9.33 and 9.36. The default is 10^{-3}. This option can have direct impact on convergence, time-step control, and bypass.

ABSTOL=*value* represents the absolute current tolerance as defined by Eq. 9.36. The smallest current that can be monitored is equal to *value*, which defaults to 10^{-12} A.

VNTOL=*value* is the absolute voltage tolerance defined by Eq. 9.33. It represents the smallest observable voltage and defaults to 10^{-6} V.

A few options control the number of iterations allowed in the nonlinear equation solution. These options are the following:

ITL1=*value* sets the maximum number of iterations used for the DC solution; this is also the number of iterations used for a first solution in the time-domain when UIC is present on the .**TRAN** line. The default is 100. A higher *value* may lead to a solution.

ITL2=*value* sets the number of iterations allowed for any new source value in a .**DC** transfer curve analysis. The default is 50. In PSpice *value* is also used as the maximum number of iterations at each source value during source ramping.

ITL3=*value* is meaningful only in SPICE2, in connection with the LVLTIM=1 option where it defines the lower iteration limit at a time point; the time step is doubled when the circuit converges in fewer iterations.

ITL4=*value* sets an upper limit to the number of iterations performed at a time point before it is rejected and the time step reduced by 8; the default is 10.

ITL5=*value* is the total number of iterations allowed in a transient analysis; it defaults to 5000. This option is a protection against very long simulations and can be turned off by setting *value* to zero.

ITL6=*value* (SPICE2 and SPICE3) represents both a flag for source ramping in a DC solution and the maximum number of iterations allowed for each stepped value of the supplies.

9.5.4   Numerical Integration

The options for the time-domain solution can set integration methods as well as tolerances specific to this analysis.

A user can select from two integration methods, several integration formula orders, and two time-step control mechanisms as follows:

METHOD=**TRAP**/**GEAR** (SPICE2 and SPICE3) selects the numerical integration formula; the default is the second-order trapezoidal method, **TRAP**, as defined by Eq. 9.47.

MAXORD=*value* (SPICE2 and SPICE3) sets the maximum order of the Gear method when selected by the **METHOD** option; the default is 2. SPICE implements a variable-order Gear integration formula contained in Appendix D.

LVLTIM=*value* (SPICE2) selects whether the time-step is controlled by the local truncation error, LTE, of the method (*value* = 2) or by the iteration count needed at each time point for convergence (*value* = 1). The default value is 2.

The tolerances that can be modified in the transient analysis are the following:

TRTOL=*value* is a scale factor for LTE as defined in Eq. 9.67; it defaults to 7.

CHGTOL=*value* is the absolute charge tolerance at any time point according to Eq. 9.62; it defaults to 10^{-14} C.

9.5.5   Miscellaneous Options

A number of options in SPICE control the analysis environment, the analysis time, global device properties, and which information is output.

The following option modifies the analysis environment:

TNOM=*value* sets the reference temperature at which all device parameters are assumed to be measured; the default is 27°C. Note that this option effects the analysis results at temperatures specified in the **.TEMP** statement; see Sec. 4.1.3. In SPICE3 the analysis temperature is also an option, **TEMP**=*value*, rather than a command line as in SPICE2 and PSpice.

The information and its format saved by SPICE in the output file is controlled by the following options:

LIST generates a comprehensive summary of all elements in the circuit with connectivity and values; the default is **NOLIST**.

NOMOD suppresses the listing of device model parameters; by default the model parameters are printed.

NOPAGE suppresses new pages for different analyses and header printing.

NODE requests the output of a node table, which lists the elements connected at every node; the default is **NONODE**.

OPTS causes a complete list of all options parameter settings.

NUMDGT=*value* selects the number of digits to be printed after the decimal point for results; *value* is an integer number between 0 and 8 and defaults to 4. Note that this option does not affect the computation of the results but only how many digits are printed. More than 4 digits may be meaningless unless **RELTOL** is reduced.

LIMPTS=*value* sets the number of points to be saved for a .**PRINT** or a .**PLOT**. *value* must be larger than the number of data points resulting from the analysis; it defaults to 201.

Global geometric dimensions can be defined for MOSFETs as option parameters:

DEFW=*value* sets the global, or default, device channel width, *W*; the SPICE built-in default is 1 meter. The channel width, **W**, on the device line (see Chap. 3) overrides the *DEFW* value.

DEFL=*value* sets the global, or default, device channel length, *L*; the SPICE built-in default is 1 meter. The channel length, **L**, on the device line (see Chap. 3) overrides the *DEFL* value.

DEFAD=*value* sets the global, or default, drain area, *AD*; the SPICE built-in default is 1 m^2. The drain area, **AD**, on the device line (see Chap. 3) overrides the *DEFAD* value.

DEFAS=*value* sets the global, or default, source area, *AS*; the SPICE built-in default is 1 m^2. The source area, **AS**, on the device line (see Chap. 3) overrides the *DEFAS* value.

PSpice and specific implementations of SPICE2 allow the user to limit the analysis time through the following options parameter:

CPTIME=*value* sets the maximum CPU time for the analysis.

There is an additional command belonging in the options category that controls the line length in the SPICE2 output file:

.**WIDTH OUT**=*value*

where *value* equals the number of characters per line. By default the SPICE2 output line is 120 characters long.

REFERENCES

Boyle, G. R. 1978. Personal communication.

Calahan, D. A. 1972. *Computer-Aided Network Design.* New York: McGraw-Hill.

Chua, L. O., and P. M Lin. 1975. *Computer-Aided Analysis of Electronic Circuits: Algorithms and Computational Techniques.* Englewood Cliffs, NJ: Prentice Hall.

Cohen, E. 1981. Performance limits of integrated circuit simulation on a dedicated minicomputer system. Univ. of California, Berkeley, ERL Memo UCB/ERL M81/29 (May).

Dorf, R. C. 1989. *Introduction to Electric Circuits.* New York: John Wiley & Sons.

Forsythe, G. E., and C. B. Moler. 1967. *Computer Solution of Linear Algebraic Systems.* Englewood Cliffs, NJ: Prentice Hall.

Freret, J. P. 1976. *Minicomputer Calculation of the DC Operating Point of Bipolar Circuits.* Technical Report No. 5015-1, Stanford Electronics Labs., Stanford Univ., Stanford, CA. (May).

Gear, C. W. 1967. Numerical integration of stiff ordinary equations. Report 221, Dept. of Computer Science, Univ. of Illinois, Urbana.

Hachtel, G., and A. L. Sangiovanni-Vincentelli. 1981. A survey of third-generation simulation techniques. *IEEE Proceedings* 69 (October).

Hajj, I. N., P. Yang, and T. N. Trick. 1981. Avoiding zero pivots in the modified nodal approach. *IEEE Transactions on Circuits and Systems* CAS-28 (April): 271–278.

Ho, C. W., A. E. Ruehli, and P. A. Brennan. 1975. The modified nodal approach to network analysis. *IEEE Transactions on Circuits and Systems* CAS-22 (June): 504–509.

McCalla, W. J. 1988. *Fundamentals of Computer-Aided Circuit Simulation.* Boston: Kluwer Academic.

McCalla, W. J., and D. O. Pederson. 1971. Elements of computer-aided circuit analysis. *IEEE Transactions on Circuit Theory* CT-18 (January).

Nagel, L. W. 1975. SPICE2: A computer program to simulate semiconductor circuits. Univ. of California, Berkeley, ERL Memo ERL M520 (May).

Nagel, L. W., and R. Rohrer. 1971. Computer analysis of nonlinear circuits, excluding radiation (CANCER). *IEEE Journal of Solid-State Circuits* SC-6 (August): 166–182.

Nilsson, J. W. 1990. *Electric Circuits.* 3d ed. Reading, MA: Addison-Wesley.

Ortega, J. M., and W. R. Rheinholdt. 1970. *Iterative Solution of Non-Linear Equations in Several Variables.* New York: Academic Press.

Paul, C. R. 1989. *Analysis of Linear Circuits.* New York: McGraw-Hill.

Vladimirescu, A. 1978. "Sparse Matrix Solution with Pivoting in SPICE2," EECS 290H project, Univ. of California, Berkeley.

Ten

CONVERGENCE ADVICE

10.1 INTRODUCTION

Generally, SPICE finds a solution to most circuit problems. However, because of the nonlinearity of the circuit equations and a few imperfections in the analytical device models a solution is not always guaranteed when the circuit and its specification are otherwise correct.

In the majority of the cases when a solution failure occurs it is due to a circuit problem, either its specification or its inoperability. A convergence problem can be categorized as either failure to compute a DC operating point or abortion of the transient analysis because of the reduction of the time step below a certain limit without finding a solution. The two most common messages that SPICE2 prints when it fails to find a solution are

```
*ERROR*: NO CONVERGENCE IN DC ANALYSIS
```

and

```
*ERROR*: INTERNAL TIME STEP TOO SMALL IN TRANSIENT ANALYSIS
```

This chapter describes the most common causes of convergence failure and the appropriate remedies.

From the perspective of the previous chapter, failure to find a solution can occur at the level of the linear equation, the Newton-Raphson iteration, or the numerical integration. Rather than present the convergence issues based on the algorithm causing the problem, it has been deemed beneficial to describe the causes for failure from a user's perspective. Section 10.2 contains the most common remedies for SPICE solution failure.

Specific procedures can be followed when SPICE fails to find a DC solution of the circuit. The prescribed remedies include redefinition of analysis options, use of built-in convergence-enhancing algorithms, and DC operating point solution with a different analysis. These approaches are presented and exemplified in Sec. 10.3.

Time-domain analysis can provide an inaccurate solution or fail because of a number of reasons related either to the integration method and associated time-step control or the iterative solution of nonlinear equations. Section 10.4 describes some of the problems and several approaches that can lead to a solution in these cases.

Knowledge of the specifics of different types of electronic circuits can assist the user in finding an accurate solution by specifying appropriate analysis modes, options, tolerances, and suitable model parameters. Thus, oscillators require certain initializations not necessary for amplifiers, and bipolar circuits may need different convergence tolerances than do MOS circuits. An overview of circuit-specific analyses and issues is provided in Sec. 10.5.

All convergence issues described in this chapter are illustrated by small circuits that can be easily understood by a new user. Although convergence failure is more common for large circuits, the problems and their remedies are the same as for the smaller circuits described in this chapter.

Note that the convergence problems described are specific to the simulators mentioned in the text. Because of differences between SPICE simulators, the problems presented below can be duplicated only in the specified simulator. Sometimes the same simulator can succeed or fail to converge depending on the platform; different computers may use different floating-point representations and different mathematical libraries of elementary functions. The results presented in this chapter are obtained from SPICE2 and SPICE3 running on SUN workstations and PSpice running on 386/486 PCs.

10.2   COMMON CAUSES OF SOLUTION FAILURE

10.2.1   Circuit Description

The first thing a user should do after a convergence error occurs is to check the circuit description carefully; the SPICE input should be compared to the schematic for correct connectivity. Additional SPICE information can be helpful for this verification. A list of every node and the elements connected to it can be obtained by adding the **NODE** option to the input description:

```
.OPTIONS NODE
```

The user should specifically look for and identify nodes that are floating or undefined in DC.

SPICE checks every circuit for topological and component value correctness. Common error messages related to circuit topology are

```
*ERROR*: LESS THAN 2 CONNECTIONS AT NODE number
*ERROR*: NO DC PATH TO GROUND FROM NODE number
*ERROR*: INDUCTOR/VOLTAGE SOURCE LOOP FOUND, CONTAINING
   Vname
```

The first message identifies any node that has only one terminal of one element connected to it. The second message points to the nodes that are floating in DC and cannot be solved. Most often such nodes are connected to ground through capacitors, which are open circuits in DC. The third error message records a violation of Kirchhoff's voltage law. As described in the previous chapter, inductors are equivalent to zero-valued voltage sources in DC analysis and therefore cannot form a mesh or loop with voltage sources or other inductors.

Although SPICE identifies and reports most topology errors, some evade the scrutiny. One such example is the gate terminals of MOSFETs which need to be connected properly for DC biasing. The gate terminal of a MOSFET, unlike the base of a BJT, has no DC connection to the other terminals of the device (Grove 1967), drain, source, or bulk, and therefore must be properly biased outside the device. The following example illustrates an analysis failure due to the improper connection of a MOSFET which goes undetected by SPICE.

EXAMPLE 10.1

Find the time response of the circuit shown in Figure 10.1 to the input signals V_A and V_B, defined in the SPICE deck listed in Figure 10.2.

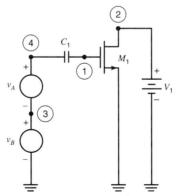

Figure 10.1   Floating-gate MOS circuit.

```
*******01/16/92  ******** SPICE  2G.6 3/15/83*********11:48:25*****

FLOATING GATE OF MOSFET ERROR

****      INPUT LISTING                TEMPERATURE =  27.000 DEG C

*************************************************************************
*
M1 2 1 0 0 NMOS L=100U W=100U
C1 4 1 10P IC=0
VA 4 3 PULSE 0 5 70U 2N 2N 10U 100U
VB 3 0 PULSE 0 5 10U 20U 20U 10U 100U
V1 2 0 6
*
.MODEL NMOS NMOS VTO=1.3
*
.TRAN 1U 100U 0
.PRINT TRAN I(V1) I(VA) I(VB)
.OPTIONS NODE
.WIDTH OUT=80
.END
*******01/16/92  ******** SPICE  2G.6 3/15/83*********11:48:25*****

FLOATING GATE OF MOSFET ERROR

****      ELEMENT NODE TABLE           TEMPERATURE = 27.000 DEG C

*************************************************************************

      0    VB       V1      M1        M1
      1    C1       M1
      2    V1       M1
      3    VA       VB
      4    C1       VA
*******01/16/92  ******** SPICE  2G.6    3/15/83 *********11:48:25*****

FLOATING GATE OF MOSFET ERROR

****      MOSFET MODEL PARAMETERS        TEMPERATURE =   27.000 DEG C

*************************************************************************

          NMOS
TYPE      NMOS
LEVEL     1.000
VTO       1.300
KP        2.00D-05

*ERROR*: MAXIMUM ENTRY IN THIS COLUMN AT STEP 2(0.000000D+00)IS LESS THAN PIVTOL
*ERROR*: NO CONVERGENCE IN DC ANALYSIS
LAST NODE VOLTAGES:

NODE   VOLTAGE  NODE   VOLTAGE  NODE   VOLTAGE  NODE    VOLTAGE

( 1)    0.0000  ( 2)    0.0000  ( 3)    0.0000  ( 5)    0.0000

          ***** JOB ABORTED
```

Figure 10.2 SPICE2 input and output with node table for circuit in Figure 10.1.

Solution

The SPICE2 output, including the input circuit, is shown in Figure 10.2. The initial transient solution, that is, the DC analysis, fails with

```
*ERROR*: NO CONVERGENCE IN DC ANALYSIS
```

A closer look at the output file reveals that the circuit matrix is singular within the SPICE tolerances; the following singular matrix message is printed by SPICE2:

```
*ERROR*: MAXIMUM ENTRY IN THIS COLUMN AT STEP 2 (0.000E0) IS LESS THAN
    PIVTOL
```

The submatrix at the second elimination step is singular because the circuit is open at node 1; the capacitor, which connects the signals V_A and V_B during the transient response, leaves the gate terminal floating in DC. This problem goes undetected by the SPICE topology checker, and it is up to the user to understand the problem.

If the singular matrix problem occurs but the value of the maximum element at a certain elimination step is nonzero, the actual value is listed in parentheses in the error message; *PIVTOL* can then be lowered to a value less than the maximum entry and the analysis rerun, as described in Example 10.2. The node table for this circuit, which is printed by SPICE when the **NODE** option is set (see Figure 10.2), shows that only C_1 and M_1 are connected at node 1. Unfortunately, the information about which terminal of M_1 is connected at a given node is not provided.

There are several remedies once the cause of this type of failure is understood. If only the time-domain response is of interest, the DC solution should be bypassed by specifying the **UIC** keyword in the **.TRAN** statement. Alternatively, if a DC solution is needed, proper biasing should be provided to the gate. It is left as an exercise for the reader to experiment with the suggested workarounds to find whether they result in the completion of the analysis.

Another element that can cause singular matrix problems is a current source that voluntarily or involuntarily is set to zero. The current source in this case is an open circuit. It can be used to defeat the topology checker, as described in Examples 4.6 and 5.3, but care must be exercised.

EXAMPLE 10.2

The setup in Figure 10.3 can be used to measure the collector cutoff current, I_{CBO}, a common data sheet parameter, by setting the emitter current supplied by the current

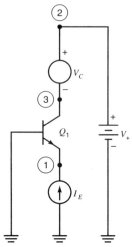

Figure 10.3 Cut-off current measurement circuit.

source I_E to zero and measuring I_C of the transistor. This is equivalent to leaving the emitter of Q_1 open, which is not permitted in SPICE.

Solution

The input is listed in Figure 10.4 together with the SPICE analysis results for $I_E = 0$. The same message of a singular matrix is printed as in Example 10.1 but the maximum entry is a finite number, 5.77×10^{-14}. This problem can be overcome by lowering the value of *PIVTOL*; see also Sec. 9.2.

The cause for this error is that the circuit equations that SPICE solves constitute an underdetermined system and the emitter voltage, $V_E = V_1$, can be set to any value. Although PSpice finds a solution without flagging a singular matrix, the problem remains that random numbers are generated during the solution process for V_E. Note that for $I_E \neq 0$ SPICE might not find a solution because of the erroneous circuit setup: Q_1 cannot conduct the driving current.

In the case of correctly defined circuits, lowering *PIVTOL* below the value of the maximum entry can cure the problem when a singular matrix is encountered. The following .**OPTIONS** line added to the circuit description in Figure 10.4,

```
.OPTIONS PIVTOL=1E-14
```

causes SPICE2 to finish the analysis. Lowering of *PIVTOL* allows SPICE to accept smaller values as pivots in the linear equation solution. The default is $PIVTOL = 10^{-13}$.

```
*******01/16/92 ******** SPICE  2G.6   3/15/83*********11:41:02*****

COLLECTOR CUT-OFF CURRENT MEASUREMENT CIRCUIT

****      INPUT LISTING                    TEMPERATURE =  27.000 DEG C

**********************************************************************
*
Q1 3 0 1 M2N2501 AREA=1
.MODEL M2N2501 NPN
+ BF=166.9 NR=1.038 MJE=0.3 VAF=49.25
+ VAR=10 CJE=5.9P RE=741.8M VJC=800M NC=1.526
+ VJE=757.1M NE=1.38 IKR=2.294 NF=1 BR=744.4M CJC=3.5P RC=696.7M
+ TF=61.52P VTF=1.188K TR=29.31N XTF=100 ITF=317.8M PTF=0
+ CJS=0 MJS=0 VJS=750M XCJC=1 XTB=1.5 IS=6.534F
+ IRB=100U RB=0 RBM=0 XTI=3 ISC=100P EG=1.11 ISE=23.42F
+ IKF=113.8M MJC=0.2 AF=1 KF=0 FC=0.5
V+ 2 0 15
VC 2 3 0.0
IE 0 1 0
*
*.OPTION PIVTOL=1.0E-14
*
.OPT ACCT
.WIDTH OUT=80
.OP
.END

PIVOT CHANGE ON FLY: N= 7 NXTI= 7 NXTJ= 7 ITERNO= 23 TIME= 0.00000D+00
*ERROR*: MAXIMUM ENTRY IN THIS COLUMN AT STEP 8 (5.773160D-14) IS LESS THAN PIVTOL
*ERROR*: NO CONVERGENCE IN DC ANALYSIS
LAST NODE VOLTAGES:

NODE    VOLTAGE  NODE    VOLTAGE  NODE    VOLTAGE

( 1)    0.1474  ( 2)   15.0000  ( 3)   15.0000

        ***** JOB ABORTED
```

Figure 10.4 SPICE2 input and output for circuit in Figure 10.3.

The conclusion of this example is that a singular matrix message can be caused by an error in the circuit specification and that if the circuit is correct and no floating nodes are found, lowering *PIVTOL* below the smallest matrix entry indicated in the error message enables SPICE to compute the solution.

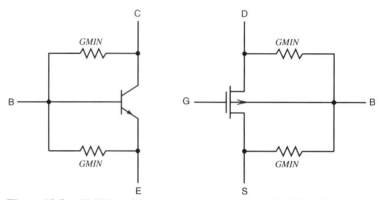

Figure 10.5   *GMIN* conductance across pn junctions for BJT and MOSFET.

A high-impedance node can result in the same failure mode as above, because numerically it can become an open circuit, similar to the gate node of transistor M_1 in Example 10.1. As protection against this problem the internal SPICE models of the pn junctions in diodes, BJTs, MOSFETs, and JFETs have a very small conductance, **GMIN**, connected in parallel, which can be set as an option. *GMIN* prevents the occurrence of floating nodes in a transistor circuit. The schematics for the BJT and the MOSFET including *GMIN* are shown in Figure 10.5. Large resistances, smaller than 1/*GMIN*, can be added to high-impedance nodes if they do not disturb the operation; an alternate way is to use the **.NODESET** command to initialize the voltages at these nodes.

10.2.2   Component Values

In the case of an analysis failure or erroneous results, after the topology has been verified, the component values and model parameters should be double-checked. The **LIST** option can be introduced in the SPICE input to obtain a comprehensive summary of all elements and their values. The semiconductor model parameters are printed by default in the output file obtained from SPICE2 and PSpice. Ideal elements and unrealistic component values and semiconductor device model parameters can lead to voltages and currents that combined with transcendental equations generate numbers outside the computer range. The simple example presented below leads to a solution failure or erroneous result depending on which SPICE version is used.

EXAMPLE 10.3

Use SPICE to find the operating point and the time-domain response of the half-wave diode rectifier shown in Figure 10.6.

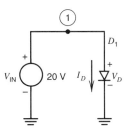

Figure 10.6 Half-wave diode rectifier circuit.

Solution

An uninformed user may try to simulate this circuit with the default diode model parameters, that is, with an ideal diode according to the SPICE input listed in Figure 10.7. The simulation fails because the absence of current limiting leads to numerical range error in computing the value of

$$I_D = IS(e^{V_D/V_{th}} - 1) = 10^{-14}e^{20/0.0258} = 10^{324} \text{ A} \qquad (10.1)$$

This number is greater than the largest value, 10^{308}, that can be represented in double precision. For smaller values of the voltage source V_{IN} SPICE may find a solution as long as I_D is in the range of floating-point numbers. Some versions of SPICE, such as older PSpice releases, limit the value of the exponent to 80 in the internal computation which combined with automatic source ramping leads to a solution; this solution is wrong, however, because the current is limited at

$$I_D = 10^{-14}e^{80} = 10^{22} \text{ A}$$

The error in this example is due to the lack of a parasitic resistance in the diode model. In the absence of an external resistor the role of the diode parasitic series resistance is to provide current limiting in the half-wave rectifier circuit. In reality the

```
DIODE CIRCUIT WITHOUT CURRENT LIMITING
*
VIN 1 0 20
D1 1 0 DMOD
.MODEL DMOD D
.OP
.WIDTH OUT=80
.END
```

Figure 10.7 SPICE description of half-wave diode rectifier.

pn junction of the diode would melt if no proper current-limiting resistor is added in the circuit. This error can also occur for JFETs and MOSFETs when improperly biased because of the presence of an ideal pn junction between gate and drain or source or between bulk and drain or source, respectively.

Other model parameters that can cause numerical problems for bipolar devices are the emission coefficients, such as **N** for a diode, **NF** and **NR** for a BJT (see Chapter 3), as well as **FC**, which approximates the junction capacitance, C_J, for a forward-biased junction. The value of C_J increases toward infinity as the junction voltage V_D approaches the built-in voltage **VJ** according to Eq. 3.3. In SPICE the characteristic described by Eq. 3.3 is replaced for pn junctions in all semiconductor models with the tangent to the curve at a junction voltage determined by FC and the built-in voltage, VJ:

$$V_x = FC \cdot VJ \quad \text{where} \quad 0 \le FC < 1. \tag{10.2}$$

The approximation is shown graphically in Figure 10.8. The default for FC is 0.5; that is, Eq. 3.3 is replaced by the tangent to the curve it describes when the junction voltage reaches half the value of the built-in voltage.

For an accurate solution it is important to observe certain limits on the element values. The maximum conductance cannot be more than 14 to 15 orders of magnitude larger than the smallest conductance in order to satisfy the constraint set by the limited accuracy of number representation in a computer (see Sec. 9.2.2). It was shown above that the smallest conductance used by SPICE is $GMIN = 10^{-12}$. This constraint limits the smallest resistance to 1 mΩ. Power electronics is the one area where resistances

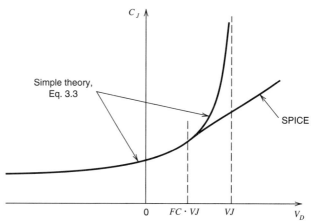

Figure 10.8   Junction capacitance approximation in forward bias.

smaller than 1 Ω need to be modeled, but the lower limit defined above should still be observed.

The use of ideal switches also leads to extreme values for the ON and OFF resistances. As explained in Sec. 2.3.3, it is not necessary to set these two parameters, *RON* and *ROFF*, more than six orders of magnitude apart as long as they are virtual shorts and open circuits by comparison with the other conductances in the circuit. The OFF resistance need not be as large as 1/*GMIN*.

The same rules on the ratio between the largest and smallest values of a component must be observed also for capacitors and inductors. The value of the equivalent conductance for these elements is also a function of the integration time step, as explained in Sec. 9.4. For practical circuits the largest capacitances should not surpass 1 μF; the smallest values are dictated by semiconductor models and are of the order of femtofarads (10^{-15} F).

Convergence is related not only to specific values of model parameters but also to the complexity of the models. The presence of a parameter causes the inclusion of a specific physical effect in the behavior of a semiconductor device, whereas the absence of the parameter eliminates modeling of the corresponding effect. For example, the presence of *VAF* in a BJT model results in a finite output conductance of the transistor; similarly, a nonzero value for *LAMBDA* in a FET model introduces a finite output conductance in the saturation region. MOSFETs can be described by more complex and accurate models than the Schichman-Hodges model presented in Chap. 3; these models are accessed by the **LEVEL** parameter, which can go to 3 in SPICE2 (Vladimirescu and Liu 1981), to 6 in SPICE3 (Sheu, Scharfetter, and Ko 1985), and to 4 in PSpice. The higher-level models include such second-order effects as subthreshold current, velocity-limited saturation, and small-size effects.

As mentioned earlier, the equations implemented in various versions of SPICE are not perfect. For a specific combination of arguments the function describing the conductances can become discontinuous. The simpler the model, the lower the chance for such an occurrence. Therefore, when a very complex model is used for a device and the simulation of the circuit fails, a selective omission of second-order effects, such as base-resistance modulation, parameters **RBM** and **IRB**, in a BJT (Appendix A) or small-size effects in a MOSFET, can help convergence. This approach is exemplified by the BiCMOS voltage reference circuit in the following section. Once a first solution is obtained, convergence with the initial, more complex model can be achieved by initializing key nodes of the circuit.

Convergence problems also occur when the most elementary or ideal models are used, as demonstrated by Example 10.3. In this situation the user is advised to add those parameters that increase accuracy to the simulation, such as finite output conductance and charge storage. Also, subthreshold conduction in a MOSFET can help convergence, as described in Sec. 10.5.2.

Once all the above guidelines on circuit correctness and component values limits have been observed and SPICE still does not converge, it is necessary to use initialization of critical nodes and adjust some options parameters. These two issues are exemplified in the following two sections, on convergence improvement in DC and transient analysis.

10.3 DC CONVERGENCE

This section describes several approaches that can be followed when a circuit that has passed the scrutiny of the previous section fails to converge in DC analysis. Note that various SPICE versions handle nonconvergence in different ways. Thus, among the three versions used in this text, PSpice and the latest versions of SPICE3 automatically run a built-in convergence-enhancing algorithm after failing to find a solution in the first *ITL1* iterations; SPICE2, however, aborts the run if no convergence is reached in *ITL1* iterations, and the user must specifically request the source-ramping convergence algorithm by specifying the **ITL6** option.

The first step to be taken after a convergence failure is to rerun the circuit for more than the default *ITL1* iterations. *ITL1* defaults to 100 in SPICE2 and SPICE3 and to 40 in PSpice. It is questionable whether it makes sense to use a higher *ITL1* in PSpice, which automatically ramps the supplies after *ITL1* iterations and may find a solution faster through this approach. It definitely makes sense in SPICE2, especially for large circuits, which may need more than 100 iterations to converge.

EXAMPLE 10.4

Compute the operating point of the CMOS operational amplifier shown in Figure 10.9 using SPICE2. This circuit is extracted from the collection of circuits with convergence problems in SPICE prepared by the Micro-electronics Center of North Carolina. The following **LEVEL**=2 model parameters should be used for the two types of MOS transistors:

```
NMOS: VTO=0.71 GAMMA=0.29 TOX=225E-10 NSUB=3.5E16 UO=411 LAMBDA=0.02
      CGSO=2.89E-10 CGDO=2.89E-10 CJ=3.74E-4 MJ=0.4
PMOS: VTO=-0.76 GAMMA=0.6 TOX=225E-10 NSUB=1.6E16 UO=139 LAMBDA=0.02
      XJ=0.2E-6 CGSO=3.35E-10 CGDO=3.35E-10 CJ=4.75E-4 MJ=0.4
```

Solution
The circuit is a conventional three-stage CMOS operational amplifier (Gray and Meyer 1985) in a closed-loop unity-gain feedback configuration. The function of the various transistors is documented in the input file, listed in Figure 10.10. The analysis of this circuit produces a DC convergence error in SPICE2. The iteration count is increased by adding the line

```
.OPTION ITL1=300
```

to the input file; the DC operating point is found by SPICE2 in 108 iterations, and the results are as listed in Figure 10.11. This circuit does not converge in PSpice or SPICE3, which use automatic ramping algorithms.

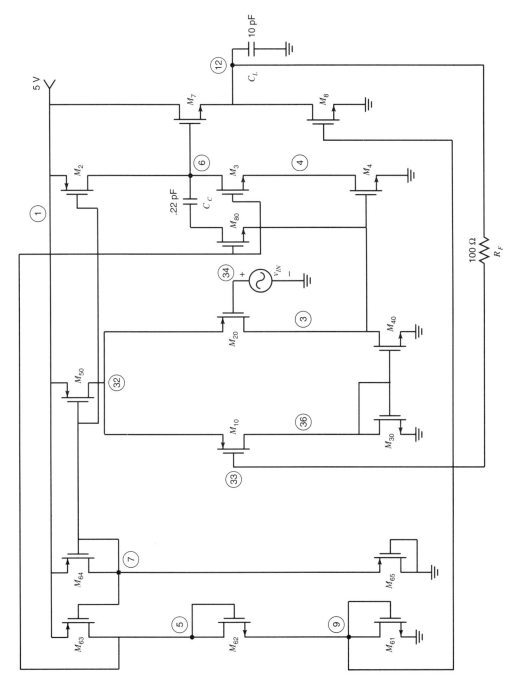

Figure 10.9 CMOS opamp circuit diagram.

```
CMOS OPAMP
*
* This opamp is a conventional 3-stage, internally compensated, CMOS opamp
* from the MCNC SPICE test circuits; The simulation file is set up for closed
* loop (unity-gain feedback) analysis of transient and ac performance.
*
* SUPPLY VOLTAGES
*
VDD 1 0 DC 5
VAP 34 99 PULSE ( 0.0 0.5 0 1E-9 1E-9 50E-9 100E-9 )
VIN 99 0 DC 2.5
*
* ANALOG INPUT
*
*
* BIAS CIRCUIT
*
M65 0 0 7 1 PCH W=4.5U L=40U
M64 7 7 1 1 PCH W=71U  L=10U
M63 5 7 1 1 PCH W=69U  L=10U
M62 5 5 9 0 NCH W=35U  L=10U
M61 9 9 0 0 NCH W=12U  L=10U
*
* DIFFERENTIAL AMPLIFIER STAGE
*
M10 36 33 32 1 PCH W=11U L=2U AD=24P  AS=24P
M20  3 34 32 1 PCH W=11U L=2U AD=24P  AS=24P
M30 36 36  0 0 NCH W=6U  L=3U AD=136P AS=136P
M40  3 36  0 0 NCH W=6U  L=3U AD=136P AS=136P
M50 32  7  1 1 PCH W=14U L=2U AD=24P  AS=24P
*
* FOLDED CASCODE STAGE WITH COMPENSATION
*
M2  6 7 1 1 PCH W=80U L=2U AD=24P  AS=24P
M3  6 5 4 0 NCH W=24U L=2U AD=136P AS=136P
M4  4 3 0 0 NCH W=46U L=2U AD=136P AS=136P
M80 11 5 3 0 NCH W=4U L=3U
CC  6 11 .22PF
*
* COMMON DRAIN OUTPUT STAGE
*
M7  1 6 12 12 NCH W=100U L=2U AD=136P AS=136P
M8 12 9  0  0 NCH W=63U  L=2U AD=136P AS=136P
```

Figure 10.10 SPICE input for CMOS opamp.

```
*
* LOAD
CL 12 0 10PF
*
* FEEDBACK CONNECTION
RF 12 33 100
*
* MOSFET PROCESS MODELS
*
.MODEL NCH NMOS LEVEL=2 CGSO=2.89E-10 VTO=0.71 GAMMA=0.29
+ CGDO=2.89E-10 CJ=3.74E-4 MJ=0.4 TOX=225E-10 NSUB=3.5E16
+ UO=411 LAMBDA=0.02
*
.MODEL PCH PMOS LEVEL=2 VTO=-0.76 GAMMA=0.6 CGSO=3.35E-10
+ CGDO=3.35E-10 CJ=4.75E-4 MJ=0.4 TOX=225E-10 NSUB=1.6E16
+ XJ=0.2E-6 UO=139 LAMBDA=0.02
*
* ANALYSES
*
.OP
.OPTIONS ACCT ITL1=300
.END
```

Figure 10.10 *(continued)*

```
*******01/21/92  ******** SPICE 2G.6    3/15/83********15:44:54*****

CMOS OPAMP

****     SMALL SIGNAL BIAS SOLUTION        TEMPERATURE =   27.000 DEG C

*************************************************************************

NODE   VOLTAGE    NODE   VOLTAGE    NODE   VOLTAGE    NODE   VOLTAGE

( 1)   5.0000   (   3)   1.0208   (   4)   1.1081   (   5)   2.3756
( 6)   3.6144   (   7)   3.8663   (   9)   1.2190   ( 11)   1.0208
(12)   2.5000   (  32)   3.7841   ( 33)   2.5000   ( 34)   2.5000
(36)   1.0207   (  99)   2.5000

    VOLTAGE SOURCE CURRENTS

    NAME        CURRENT

    VDD        -3.250D-04
    VAP         0.000D+00
    VIN         0.000D+00

    TOTAL POWER DISSIPATION 1.62D-03 WATTS
```

Figure 10.11   CMOS opamp DC solution.

CMOS OPAMP

**** OPERATING POINT INFORMATION   TEMPERATURE = 27.000 DEG C

****

**** MOSFETS

	M65	M64	M63	M62	M61	M10	M20
MODEL	PCH	PCH	PCH	NCH	NCH	PCH	PCH
ID	-8.62E-06	-8.63E-06	-8.72E-06	8.72E-06	8.72E-06	-5.37E-06	-5.37E-06
VGS	-3.866	-1.134	-1.134	1.157	1.219	-1.284	-1.284
VDS	-3.866	-1.134	-2.624	1.157	1.219	-2.763	-2.763
VBS	1.134	0.000	0.000	-1.219	0.000	1.216	1.216
VTH	-1.062	-0.750	-0.748	0.865	0.710	-0.965	-0.965
VDSAT	-2.389	-0.292	-0.293	0.265	0.444	-0.271	-0.271
GM	6.21E-06	4.52E-05	4.55E-05	5.99E-05	3.44E-05	3.37E-05	3.37E-05
GDS	1.89E-07	2.36E-07	2.24E-07	1.79E-07	1.79E-07	2.92E-07	2.92E-07
GMB	1.08E-06	1.42E-05	1.42E-05	5.98E-06	5.07E-06	5.49E-06	5.49E-06
CBD	0.00E+00	0.00E+00	0.00E+00	0.00E+00	0.00E+00	5.58E-15	5.58E-15
CBS	0.00E+00	0.00E+00	0.00E+00	0.00E+00	0.00E+00	7.88E-15	7.88E-15
CGSOVL	1.51E-15	2.38E-14	2.31E-14	1.01E-14	3.47E-15	3.68E-15	3.68E-15
CGDOVL	1.51E-15	2.38E-14	2.31E-14	1.01E-14	3.47E-15	3.68E-15	3.68E-15
CGBOVL	0.00E+00	0.00E+00	0.00E+00	0.00E+00	0.00E+00	0.00E+00	0.00E+00
CGS	1.84E-13	7.26E-13	7.06E-13	3.58E-13	1.23E-13	2.25E-14	2.25E-14
CGD	0.00E+00	0.00E+00	0.00E+00	0.00E+00	0.00E+00	0.00E+00	0.00E+00
CGB	0.00E+00	0.00E+00	0.00E+00	0.00E+00	0.00E+00	0.00E+00	0.00E+00

	M30	M40	M50	M2	M3	M4	M80
MODEL	NCH	NCH	PCH	PCH	NCH	NCH	NCH
ID	5.37E-06	5.37E-06	-1.07E-05	-6.19E-05	6.19E-05	6.19E-05	-1.25E-12
VGS	1.021	1.021	-1.134	-1.134	1.267	1.021	1.355
VDS	1.021	1.021	-1.216	-1.386	2.506	1.108	0.000
VBS	0.000	0.000	0.000	0.000	-1.108	0.000	-1.021
VTH	0.710	0.710	-0.707	-0.706	0.854	0.710	0.844
VDSAT	0.269	0.269	-0.331	-0.332	0.376	0.269	0.463
GM	3.47E-05	3.47E-05	5.07E-05	2.92E-04	2.99E-04	4.00E-04	3.23E-12
GDS	1.10E-07	1.10E-07	5.48E-07	3.05E-06	1.30E-06	1.27E-06	4.29E-05
GMB	5.33E-06	5.33E-06	1.36E-05	7.80E-05	3.03E-05	6.14E-05	3.51E-13
CBD	3.66E-14	3.66E 14	7.88E 15	7.63E-15	2.57E-14	3.59E-14	0.00E+00
CBS	5.09E-14	5.09E-14	1.14E-14	1.14E-14	3.59E-14	5.09E-14	0.00E+00
CGSOVL	1.73E-15	1.73E-15	4.69E-15	2.68E-14	6.94E-15	1.33E-14	1.16E-15
CGDOVL	1.73E-15	1.73E-15	4.69E-15	2.68E-14	6.94E-15	1.33E-14	1.16E-15
CGBOVL	0.00E+00	0.00E+00	0.00E+00	0.00E+00	0.00E+00	0.00E+00	0.00E+00
CGS	1.84E-14	1.84E-14	2.86E-14	1.64E-13	4.91E-14	9.41E-14	5.15E-15
CGD	0.00E+00	0.00E+00	0.00E+00	0.00E+00	0.00E+00	0.00E+00	1.16E-14
CGB	0.00E+00	0.00E+00	0.00E+00	0.00E+00	0.00E+00	0.00E+00	0.00E+00

Figure 10.11   *(continued)*

	M7	M8
MODEL	NCH	NCH
ID	2.35E-04	2.35E-04
VGS	1.114	1.219
VDS	2.500	2.500
VBS	0.000	0.000
VTH	0.710	0.710
VDSAT	0.352	0.444
GM	1.17E-03	9.28E-04
GDS	4.95E-06	4.95E-06
GMB	1.76E-04	1.37E-04
CBD	2.89E-14	2.89E-14
CBS	5.09E-14	5.09E-14
CGSOVL	2.89E-14	1.82E-14
CGDOVL	2.89E-14	1.82E-14
CGBOVL	0.00E+00	0.00E+00
CGS	2.05E-13	1.29E-13
CGD	0.00E+00	0.00E+00
CGB	0.00E+00	0.00E+00

Figure 10.11   *(continued)*

The next step in overcoming a convergence error is to relax the two key tolerances *ABSTOL*, the absolute current tolerance, which defaults to 1 pA, and *RELTOL,* the relative convergence tolerance, which defaults to 10^{-3}. Especially for MOS circuits the default *ABSTOL* can be too small.

When the above options do not lead to a solution, the source-ramping mechanism must be invoked in SPICE2, or some of the iteration options must be changed in PSpice and SPICE3. If the ramping methods fail, it is recommended to use initialization either by setting the values of key nodes with **.NODESET** and **.IC** (see Sec. 4.6 and Sec. 6.3) or by identifying cutoff devices with the **OFF** keyword (see Chap. 3).

Next an operational amplifier that fails DC convergence will be considered, and the steps that lead to a solution will be outlined.

EXAMPLE 10.5

Find the DC bias point of the μA741 operational amplifier with an external emitter-follower circuit shown in Figure 10.12 using SPICE2. The external output stage is formed of two emitter-follower stages built for higher current capability with discrete transistors 2N2222 for Q_{21} and 2N3055 for Q_{22}. The opamp is connected in a unity-feedback loop. The SPICE input file is listed in Figure 10.13.

Solution
The simulation of this circuit results in a convergence failure. Neither a higher number of iterations, *ITL1*, nor looser tolerances help the solution of this circuit. The next

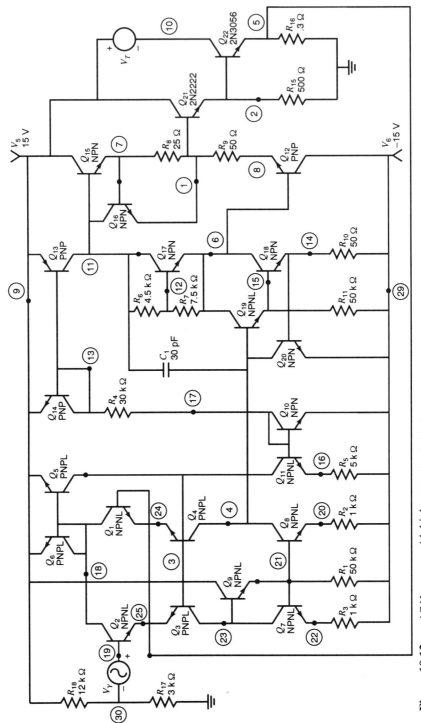

Figure 10.12 μA741 opamp with high-current output stage.

336

```
UA741 W/ POWER OUTPUT STAGE
*
* THIS CKT FAILS DC CONVERGENCE
*
Q1 18 5 24 NPNL AREA=1
Q2 18 19 25 NPNL AREA=1
Q3 23 3 25 PNPL AREA=4
Q4 4 3 24 PNPL AREA=4
Q5 3 18 9 PNPL AREA=5
Q6 18 18 9 PNPL AREA=5
Q7 23 21 22 NPNL AREA=1
Q8 4 21 20 NPNL AREA=1
Q9 9 23 21 NPNL AREA=0.5
Q10 17 17 29 NPN AREA=2
Q11 3 17 16 NPNL AREA=2
Q12 29 6 8 PNP AREA=120
Q13 11 13 9 PNP AREA=30
Q14 13 13 9 PNP AREA=12
Q15 9 11 7 NPN AREA=60
Q16 11 7 1 NPN AREA=3 OFF
* Q16 11 7 1 NPN AREA=3
Q17 11 12 6 NPN AREA=7
Q18 6 15 14 NPN AREA=7
Q19 6 4 15 NPNL AREA=5
Q20 4 14 29 NPN AREA=4 OFF
* Q20 4 14 29 NPN AREA=4
Q21 9 1 2 2N2222 AREA=1
Q22 10 2 5 2N3055 AREA=1
*
R1 29 21 50K
R2 29 20 1K
R3 29 22 1K
R4 17 13 30K
R5 29 16 5K
R6 12 11 4.5K
R7 6 12 7.5K
R8 1 7 25
R9 8 1 50
R10 29 14 50
R11 29 15 50K
R15 0 2 500
R16 0 5 300M
R17 30 0 3K
R18 9 30 12K
```

(continued on next page)

Figure 10.13   SPICE input file for circuit of Figure 10.12.

```
*
C1 11 4 30P
*
VY 19 30
VT 9 10 0.0
V6 29 0 -15
V5 9 0 15
*
.MODEL 2N3055 NPN IS=10.3P BF=120 NF=1.02167 VAF=50 IKF=3 ISE=
+ 500P NE=2 BR=8.1 NR=1.02167 VAR=500 IKR=1 ISC=0 NC=2 RB=1.8
+ IRB=80M RBM=100M RE=5M RC=50M CJE=711P VJE=530M MJE=530M
+ TF=20N XTF=5 VTF=10 ITF=10 PTF=0 CJC=650P VJC=580M MJC=400M
+ XCJC=0.5 TR=400N CJS=0 VJS=0.7 MJS=0.5 XTB=2.1 EG=1.11 XTI=3
+ KF=0 AF=1 FC=0.5
.MODEL 2N2222 NPN IS=166.78F BF=150 NF=1.074 VAF=78 IKF=500M
+ ISE=3.92P NE=1.776 BR=2.394 NR=1.074 VAR=500 IKR=0
+ ISC=0 NC=1 RB=676M IRB=0 RBM=676M RE=100M RC=654M
+ CJE=22.25P VJE=1.333 MJE=0.522 TF=454.4P
+ XTF=13.24 VTF=4.83 ITF=216.3M PTF=0 CJC=8.37P VJC=1.333
+ MJC=0.518 XCJC=0.5 TR=117.5N CJS=0 VJS=0.7 MJS=0.5
+ XTB=2.34 EG=1.11 XTI=3 KF=0 AF=1 FC=0.5
.MODEL NPNL NPN IS=4.479F BF=260 NF=1.07 VAF=260 IKF=1M
+ ISE=3.471P NE=3.66 BR=1 NR=1.07 VAR=500 IKR=0 ISC=0 NC=1 RB=36
+ IRB=0 RBM=36 RE=500M RC=1 CJE=910F VJE=661M
+ MJE=294M TF=112P XTF=120 VTF=0 ITF=0 PTF=0
+ CJC=835F VJC=1 MJC=280M XCJC=0.5 TR=1N
+ CJS=0 VJS=700M MJS=0.5 XTB=2.3 EG=1.11 XTI=3 KF=0 AF=1 FC=0.5
.MODEL PNPL PNP IS=218.88F BF=150 NF=1.221 VAF=150 IKF=
+ 1.50M ISE=23.64F NE=2.155 BR=4.68 NR=1.221 VAR=500 IKR=0
+ ISC=0 NC=1 RB=12 IRB=0 RBM=12 RE=100M RC=1.5 CJE=6.25P
+ VJE=916M MJE=389.8M TF=361.72P XTF=21 VTF=4.7 ITF=
+ 260M PTF=0 CJC=6.6P VJC=1.67 MJC=406.5M XCJC=0.5 TR=19.92N
+ CJS=0 VJS=700M MJS=0.5 XTB=1.7 EG=1.11 XTI=3 KF=0 AF=1 FC=0.5
.MODEL NPN NPN IS=4.479F BF=260 NF=1.07 VAF=260 IKF=100M
+ ISE=347.1P NE=3.66 BR=1 NR=1.07 VAR=500 IKR=0 ISC=0 NC=1 RB=36
+ IRB=0 RBM=36 RE=500M RC=1 CJE=910F VJE=661M
+ MJE=294M TF=112P XTF=120M VTF=0 ITF=0 PTF=0
+ CJC=835F VJC=1 MJC=280M XCJC=0.5 TR=1N
+ CJS=0 VJS=700M MJS=0.5 XTB=2.3 EG=1.11 XTI=3 KF=0 AF=1 FC=0.5
.MODEL PNP PNP IS=218.88F BF=150 NF=1.221 VAF=150 IKF=
+ 150M ISE=2.364P NE=2.155 BR=4.68 NR=1.221 VAR=500 IKR=0
+ ISC=0 NC=1 RB=12 IRB=0 RBM=12 RE=100M RC=1.5 CJE=6.25P
+ VJE=916M MJE=389.8M TF=361.72P XTF=21 VTF=4.7 ITF=
+ 260M PTF=0 CJC=6.6P VJC=1.67 MJC=406.5M XCJC=0.5 TR=19.92N
+ CJS=0 VJS=700M MJS=0.5 XTB=1.7 EG=1.11 XTI=3 KF=0 AF=1 FC=0.5
*
.OP
*.OPT ITL6=40 ACCT
.WIDTH OUT=80
.END
```

Figure 10.13 *(continued)*

step is to use the built-in source-ramping algorithm. This convergence method is exercised by adding the following line to the above circuit file:

```
.OPTIONS ITL6=40
```

The simulation results using this approach are shown in Figure 10.14. As can be verified, the results are correct because of the correct biasing of the output, which is node 5 and the emitter of Q_{22}, at 3 V, close to the value of node 19, the positive input of the opamp. A quick inspection of the operating point of the devices shows that the current through the external transistor Q_{22} is higher than the current in the class AB output stage of the opamp, Q_{12} and Q_{15}. The source ramping that is invoked automatically in PSpice fails to find a solution for this circuit; in such a situation it is suggested that the user change *ITL2*, which is the number of iterations taken at each source value. For source ramping the function of **ITL2** in PSpice is similar to that of **ITL6** in SPICE2.

Another way of finding the bias point of this circuit is by understanding the role of each transistor. As a result of this inspection one can see that transistors Q_{20} and Q_{16} have the role of limiting the current through the gain stage $Q_{18} - Q_{19}$ and the output

```
*******01/21/92 ******** SPICE 2G.6    3/15/83********11:58:49*****

UA741 W/ POWER OUTPUT STAGE
****      SMALL SIGNAL BIAS SOLUTION         TEMPERATURE =  27.000 DEG C

***********************************************************************

NODE   VOLTAGE   NODE   VOLTAGE    NODE   VOLTAGE    NODE VOLTAGE

(  1)   4.7906  (  2)    3.9773  (    3)    1.8958  (    4) -13.6087
(  5)   3.0004  (  6)    4.3167  (    7)    4.8498  (    8)   4.7862
(  9)  15.0000  ( 10)   15.0000  (   11)    5.4823  (   12)   5.0097
( 13)  14.3784  ( 14)  -14.8792  (   15)  -14.1871  (   16) -14.8930
( 17) -14.2969  ( 18)   14.4726  (   19)    2.9999  (   20) -14.9902
( 21) -14.3947  ( 22)  -14.9902  (   23)  -13.7767  (   24)   2.4057
( 25)   2.4054  ( 29)  -15.0000  (   30)    2.9999

     VOLTAGE SOURCE CURRENTS

     NAME         CURRENT

     VY         -3.859D-08
     VT          9.753D+00
     V6          3.526D-03
     V5         -1.001D+01

     TOTAL POWER DISSIPATION   1.50D+02 WATTS
```
                                                    (continued on next page)

Figure 10.14 SPICE DC solution for the circuit of Figure 10.12.

```
*******01/21/92  ******** SPICE  2G.6  3/15/83*********11:58:49*****
```

UA741 W/ POWER OUTPUT STAGE
****   OPERATING POINT INFORMATION   TEMPERATURE = 27.000 DEG C

```
**************************************************************************
```

**** BIPOLAR JUNCTION TRANSISTORS

	Q1	Q2	Q3	Q4	Q5	Q6	Q7
MODEL	NPNL	NPNL	PNPL	PNPL	PNPL	PNPL	NPNL
IB	3.89E-08	3.86E-08	-6.05E-08	-6.10E-08	-1.32E-07	-1.32E-07	3.99E-08
IC	9.94E-06	9.87E-06	-9.85E-06	-9.92E-06	-2.12E-05	-1.95E-05	9.81E-06
VBE	0.595	0.594	-0.510	-0.510	-0.527	-0.527	0.595
VBC	-11.472	-11.473	15.672	15.504	12.577	0.000	-0.618
VCE	12.067	12.067	-16.182	-16.014	-13.104	-0.527	1.213
BETADC	255.873	255.829	162.865	162.706	160.137	147.733	245.722
GM	3.56E-04	3.53E-04	3.11E-04	3.14E-04	6.69E-04	6.17E-04	3.51E-04
RPI	7.37E+05	7.42E+05	5.26E+05	5.21E+05	2.40E+05	2.40E+05	7.17E+05
RX	3.60E+01	3.60E+01	3.00E+00	3.00E+00	2.40E+00	2.40E+00	3.60E+01
RO	2.73E+07	2.75E+07	1.68E+07	1.67E+07	7.66E+06	7.66E+06	2.65E+07
CPI	6.20E-12	6.17E-12	3.43E-11	3.43E-11	4.36E-11	4.36E-11	6.13E-12
CMU	2.06E-13	2.06E-13	5.10E-12	5.12E-12	6.90E-12	1.65E-11	3.65E-13
CBX	2.06E-13	2.06E-13	5.10E-12	5.12E-12	6.90E-12	1.65E-11	3.65E-13
CCS	0.00E+00	0.00E+00	0.00E+00	0.00E+00	0.00E+00	0.00E+00	0.00E+00
BETAAC	262.203	262.219	163.611	163.445	160.476	148.047	251.646
FT	8.56E+06	8.55E+06	1.11E+06	1.12E+06	1.85E+06	1.28E+06	8.14E+06

	Q8	Q9	Q10	Q11	Q12	Q13	Q14
MODEL	NPNL	NPNL	NPN	NPNL	PNP	PNP	PNP
IB	3.99E-08	4.42E-08	4.85E-06	8.18E-08	-1.80E-06	-2.04E-05	-8.18E-06
IC	9.81E-06	1.21E-05	9.52E-04	2.13E-05	-8.50E-05	-2.46E-03	-9.28E-04
VBE	0.595	0.618	0.703	0.596	-0.470	-0.622	-0.622
VBC	-0.786	-28.777	0.000	-16.193	19.317	8.896	0.000
VCE	1.381	29.395	0.703	16.789	-19.786	-9.518	-0.622
BETADC	245.881	274.882	196.394	260.623	47.341	120.222	113.489
GM	3.51E-04	4.29E-04	3.42E-02	7.62E-04	2.69E-03	7.78E-02	2.94E-02
RPI	7.17E+05	6.38E+05	6.88E+03	3.50E+05	2.56E+04	1.73E+03	4.31E+03
RX	3.60E+01	7.20E+01	1.80E+01	1.80E+01	1.00E-01	4.00E-01	1.00E+00
RO	2.65E+07	2.38E+07	2.73E+05	1.29E+07	1.99E+06	6.46E+04	1.61E+05
CPI	6.14E-12	6.52E 12	7.26E-12	1.31E-11	9.93E-10	3.08E-10	1.23E-10
CMU	3.55E-13	8.07E-14	8.35E-13	3.77E-13	1.42E-10	4.68E-11	3.96E-11
CBX	3.55E-13	8.07E-14	8.35E-13	3.77E-13	1.42E-10	4.68E-11	3.96E-11
CCS	0.00E+00	0.00E+00	0.00E+00	0.00E+00	0.00E+00	0.00E+00	0.00E+00
BETAAC	251.809	274.104	235.400	266.642	68.796	134.210	126.691
FT	8.16E+06	1.02E+07	6.10E+08	8.76E+06	3.36E+05	3.08E+07	2.32E+07

Figure 10.14   *(continued)*

	Q15	Q16	Q17	Q18	Q19	Q20	Q21
MODEL	NPN	NPN	NPN	NPN	NPNL	NPN	2N2222
IB	2.53E-05	9.04E-10	1.26E-05	1.23E-05	1.11E-07	3.58E-09	2.28E-03
IC	2.34E-03	1.45E-12	2.33E-03	2.40E-03	2.84E-05	4.10E-12	2.54E-01
VBE	0.633	0.059	0.693	0.692	0.578	0.121	0.813
VBC	-9.518	-0.633	-0.473	-18.504	-17.925	-1.271	-10.209
VCE	10.150	0.692	1.166	19.196	18.504	1.391	11.023
BETADC	92.615	0.002	184.109	195.513	256.745	0.001	111.186
GM	8.46E-02	4.11E-12	8.37E-02	8.65E-02	1.02E-03	5.11E-11	6.96E+00
RPI	2.04E+03	4.86E+07	2.76E+03	2.84E+03	2.63E+05	1.90E+07	1.24E+01
RX	6.00E-01	1.20E+01	5.14E+00	5.14E+00	7.20E+00	9.00E+00	6.76E-01
RO	1.15E+05	9.74E+11	1.12E+05	1.16E+05	9.76E+06	9.71E+11	3.47E+02
CPI	9.55E-11	2.81E-12	2.08E-11	2.12E-11	2.07E-11	3.86E-12	1.05E-08
CMU	1.30E-11	1.09E-12	2.62E-12	1.27E-12	9.16E-13	1.33E-12	1.38E-12
CBX	1.30E-11	1.09E-12	2.62E-12	1.27E-12	9.16E-13	1.33E-12	1.38E-12
CCS	0.00E+00	0.00E+00	0.00E+00	0.00E+00	0.00E+00	0.00E+00	0.00E+00
BETAAC	172.648	0.000	230.797	246.215	268.691	0.001	86.406
FT	1.11E+08	1.31E-01	5.11E+08	5.81E+08	7.23E+06	1.25E+00	1.05E+08

	Q22
MODEL	2N3055
IB	2.48E-01
IC	9.75E+00
VBE	0.977
VBC	-11.023
VCE	12.000
BETADC	39.344
GM	2.13E+02
RPI	1.07E-01
RX	6.81E-01
RO	6.22E+00
CPI	1.49E-05
CMU	9.91E-11
CBX	9.97E-11
CCS	0.00E+00
BETAAC	22.797
FT	2.28E+06

Figure 10.14 *(continued)*

transistor Q_{15}, respectively. In normal operation these two transistors should be turned off. Adding the keyword **OFF** at the end of the lines corresponding to these transistors and commenting out the line defining *ITL6* result in the quick convergence of the circuit, in only 11 iterations in SPICE2. PSpice and SPICE3 also converge easily to a solution.

The above example has illustrated how both source ramping and the **OFF** initialization of transistors can lead to a solution. Another circuit that demonstrates the need of initialization is the bistable circuit in Figure 10.15, implemented with two μA741 opamps as the gain stages.

EXAMPLE 10.6

Find a DC solution to the bistable circuit shown in Fig. 10.15 prior to computing the response to a sawtooth input voltage V_A.

Solution

The input description of the circuit is listed in Figure 10.16. The macro-model UA741MAC, given in Sec. 7.5.1 (Figure 7.28), is used for the μA741. The sub-circuit definition of UA741MAC has to be copied from Figure 7.28 into the deck of Figure 10.16. All the approaches mentioned above, including source ramping, may fail to lead to a solution for this circuit depending on the model used for the opamp. This circuit, containing high-gain stages and high-impedance nodes, is one that requires

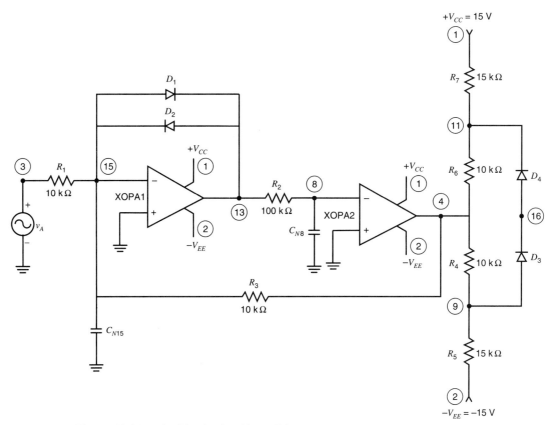

Figure 10.15 Bistable circuit with μA741 opamps.

```
BI-STABLE CIRCUIT W/ UA741 OPAMPS
*
*VA 3 0 -15
VA 3 0 DC -15 PWL 0 -15 1M 15 2M -15 2.5M 0
VCC 1 0 15
VEE 2 0 -15
*
D1 15 13 M1M1N914 AREA=1
D2 13 15 M1M1N914 AREA=1
D3 9 16 M1M1N914 AREA=1
D4 16 11 M1M1N914 AREA=1
*
.MODEL M1M1N914 D IS=1N RS=500M N=1.55 TT=5N CJO=1.85P
+ VJ=650M M=180M EG=1.11 XTI=3 KF=0 AF=1 FC=0.5 BV=120 IBV=1
*
R7 1 11 15K
R6 11 4 10K
R5 9 2 15K
R4 4 9 10K
R2 8 13 100K
R1 15 3 10K
*
* FEEDBACK RESISTOR
*
R3 4 15 10K
*
CN8 8 0 1P
CN15 15 0 1P
*
XOPA1 0 15 13 1 2 UA741MAC
XOPA2 0 8 4 1 2 UA741MAC
*
.SUBCKT UA741MAC 3 2 6 7 4
* NODES: IN+ IN- OUT VCC VEE
*
* INSERT UA741 MACRO-MODEL DEFINITION FROM FIG. 7.28
*
.ENDS
*
.OP
.TRAN .01M 2M 0
.PLOT TRAN V(3) V(15) V(13) V(8) V(4)
*
* INITIALIZATION FOR VIN=-15
*
.NODESET V(4)=-15 V(13)=.6 V(15)=0 V(8)=.6
*.OPTIONS ITL1=300 ABSTOL=1U
.OPTIONS ACCT
.WIDTH OUT=80
.END
```

Figure 10.16 SPICE input for bistable circuit with μA741 opamps.

the initialization mentioned in the previous section. The high-impedance opamp input, nodes 8 and 15, as well as the opamp output, nodes 4 and 13, must be initialized.

As described in Example 4.8, for an NMOS flip-flop SPICE finds the metastable state as solution. This state corresponds to all inputs and outputs of the opamps being at 0 V. The output of XOPA1, node 13, is limited by the two diodes D_1 and D_2 to between -0.6V and 0.6 V, with the input at a very low value. Therefore, one of the states of this circuit can be specified with the following **.NODESET** statement:

```
.NODESET V(15)=0 V(13)=0.6 V(8)=0.6 V(4)=-15
```

With this line added to the circuit description of Figure 10.16 the following refined solution is obtained:

$$V(15) = 0.0003$$
$$V(13) = 0.5964$$
$$V(8) = 0.5452$$
$$V(4) = -12.6983$$

Note that a DC solution is a prerequisite for a transient analysis unless the **UIC** keyword is used on the **.TRAN** line; in this case an **.IC** line should be used to define the initial state of the circuit in order to avoid a convergence failure at the first time point.

The waveform at node 4 together with the triangular input voltage is shown in Figure 10.17.

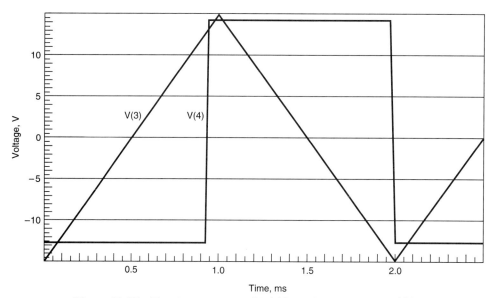

Figure 10.17   Transient response of $V(4)$ to triangular input $V(3)$.

Circuits with high-impedance nodes are often difficult convergence cases for circuit simulators. For such circuits a combination of ramping methods, initialization and variations in model parameters and complexity can help convergence. High-impedance nodes are common in CMOS and BiCMOS circuits using cascode configurations. A good example for a difficult convergence case is the BiCMOS reference circuit shown in Figure 10.18.

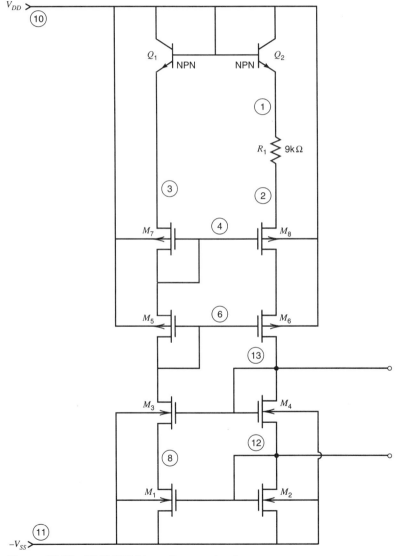

Figure 10.18 BiCMOS bias reference circuit.

EXAMPLE 10.7

Find the currents in the two branches of the thermal-voltage-referenced current source in Figure 10.18.

Solution

The areas of the MOS transistors in the left branch are five times the areas of the MOS-FETs in the right branch of the circuit; therefore we expect the current in the left branch to be approximately five times the current in the right branch, which can be estimated from the V_{BE} difference of the two identical BJTs (Gray and Meyer 1985):

$$I = \frac{V_{th}}{R} \ln 5 = 4.6 \ \mu A \tag{10.3}$$

Because of the cascode current-source configuration the drain connections between transistors M_3 and M_5 and between M_4 and M_6 are high-impedance nodes, which make convergence to the DC solution difficult. The transistors with the larger area are modeled as they are implemented on the layout, namely, as five transistors connected in parallel. For the input specification of this circuit shown in Figure 10.19, PSpice finds a solution after source ramping, and SPICE2 requires *ITL6* to be set to 40 for convergence.

Note that a model of higher accuracy is used for the MOSFETs than the one described in Chap. 3. The **LEVEL**=2 model includes such second-order effects as subthreshold conduction, set by parameter **NFS**; saturation due to carrier velocity limitation, represented by parameters **VMAX** and **NEFF**; and mobility modulation by the gate voltage, represented by parameters **UCRIT** and **UEXP**. The complete model equations for **LEVEL**=2 can be found in Appendix A or in the text by Antognetti and Massobrio (1988). For more details on the semiconductor device physics see the works by Muller and Kamins (1977) and by Sze (1981).

Two approaches are suggested for solving this convergence problem. First, set the **ITL6** option in SPICE2 to invoke the source-ramping method. The solution to the SPICE input of Figure 10.19 is the DC operating point shown in Figure 10.20. The currents in the right and left branches are 4.66 μA and 23.3 μA, respectively, consistent with our estimates.

The second approach recommended in convergence cases of circuits using complex models is to eliminate some of the second-order effects. A first approximation is to neglect for all MOSFETs the small-size effects, such as narrow-width modulation of the threshold voltage, represented by parameter **DELTA**, and velocity-limited saturation, represented by parameters **VMAX** and **NEFF**. This is achieved by deleting the **DELTA** parameter from the .**MODEL** statement. A rerun of the circuit results in the desired solution in SPICE2 in 96 iterations without the need of ramping. Deletion of **VMAX** and **NEFF** as well leads to a SPICE2 solution in only 25 iterations.

```
BICMOS BIAS REFERENCE
*
Q1 10 10 3 NPNMOD 400
Q2 10 10 1 NPNMOD 400
R1 1 2 9K
M1A 8 12 11 11 N
M1B 8 12 11 11 N
M1C 8 12 11 11 N
M1D 8 12 11 11 N
M1E 8 12 11 11 N
M2 12 12 11 11 N
M3A 6 13  8 11 N
M3B 6 13  8 11 N
M3C 6 13  8 11 N
M3D 6 13  8 11 N
M3E 6 13  8 11 N
M4 13 13 12 11 N
M5A 6 6 4 10 P W=60U
M5B 6 6 4 10 P W=60U
M5C 6 6 4 10 P W=60U
M5D 6 6 4 10 P W=60U
M5E 6 6 4 10 P W=60U
M6 13 6 5 10 P W=60U
M7A 4 4 3 10 P W=60U
M7B 4 4 3 10 P W=60U
M7C 4 4 3 10 P W=60U
M7D 4 4 3 10 P W=60U
M7E 4 4 3 10 P W=60U
M8  5 4 2 10 P W=60U
*
VDD 10 0 5
VSS 11 0 -5
*
.MODEL NPNMOD NPN
+ IS=2E-17 BR=.4 BF=100 ISE=6E-17 ISC=26E-17 IKF=3E-3 IKR=1E-3
+ VAF=100 VAR=30 RC=100K RB=200K RE=1K
.MODEL N NMOS LEVEL=2
+ VTO=0.8 TOX=500E-10 NSUB=1.3E16
+ XJ=0.5E-6 LD=.5E-6 UO=640 UCRIT=6E4 UEXP=0.1
+ VMAX=5E4 NEFF=4 DELTA=4 NFS=4E11
.MODEL P PMOS LEVEL=2
+ VTO=-0.8 TOX=500E-10 NSUB=2E15
+ XJ=0.5E-6 LD=.5E-6 UO=220 UCRIT=5.8E4 UEXP=0.18
+ VMAX=3E4 NEFF=3.5 DELTA=2.5 NFS=3E11
*
.OPTION ACCT DEFL=10U DEFW=20U
*.OPTION ABSTOL=1N
.WIDTH OUT=80
.OP
.END
```

Figure 10.19 SPICE description of BiCMOS bias reference circuit.

```
*******02/11/92 ******** SPICE  2G.6    3/15/83********15:57:31*****

BICMOS BIAS REFERENCE

****     SMALL SIGNAL BIAS SOLUTION          TEMPERATURE = 27.000 DEG C

*********************************************************************

   NODE   VOLTAGE    NODE   VOLTAGE    NODE   VOLTAGE    NODE   VOLTAGE

(   1)    4.4778  (   2)    4.4359  (   3)    4.4360  (   4)    3.2322
(   5)    3.2169  (   6)    1.8759  (   8)   -3.8354  (  10)    5.0000
(  11)   -5.0000  (  12)   -3.8484  (  13)   -2.6967

       VOLTAGE SOURCE CURRENTS

       NAME        CURRENT

       VDD      -2.798D-05
       VSS       2.798D-05

     TOTAL POWER DISSIPATION   2.80D-04  WATTS
*******02/11/92 ******** SPICE  2G.6    3/15/83********15:57:31*****

BICMOS BIAS REFERENCE

****     OPERATING POINT INFORMATION     TEMPERATURE = 27.000 DEG C

*********************************************************************

**** BIPOLAR JUNCTION TRANSISTORS

              Q1         Q2
MODEL      NPNMOD     NPNMOD
IB         2.84E-07   6.36E-08
IC         2.30E-05   4.60E-06
VBE          0.564      0.522
VBC          0.000      0.000
VCE          0.564      0.522
BETADC      81.116     72.318
GM         8.90E-04   1.78E-04
RPI        9.67E+04   4.46E+05
RX         5.00E+02   5.00E+02
RO         4.26E+06   2.14E+07

**** MOSFETS

              M1A        M1B        M1C        M1D        M1E        M2         M3A
MODEL       N          N          N          N          N          N          N
ID         4.66E-06   4.66E-06   4.66E-06   4.66E-06   4.66E-06   4.66E-06   4.66E-06
VGS          1.152      1.152      1.152      1.152      1.152      1.152      1.139
VDS          1.165      1.165      1.165      1.165      1.165      1.152      5.711
VBS          0.000      0.000      0.000      0.000      0.000      0.000      0.000
VTH          0.841      0.841      0.841      0.841      0.841      0.841      0.833
VDSAT        0.248      0.248      0.248      0.248      0.248      0.248      0.246
GM         2.51E-05   2.51E-05   2.51E-05   2.51E-05   2.51E-05   2.51E-05   2.54E-05
GDS        1.01E-07   1.01E-07   1.01E-07   1.01E-07   1.01E-07   1.01E-07   5.60E-08
GMB        1.25E-05   1.25E-05   1.25E-05   1.25E-05   1.25E-05   1.25E-05   1.26E-05
```

Figure 10.20   DC operating point of BiCMOS reference circuit.

	M3B	M3C	M3D	M3E	M4	M5A	M5B
MODEL	N	N	N	N	N	P	P
ID	4.66E-06	4.66E-06	4.66E-06	4.66E-06	4.66E-06	-4.66E-06	-4.66E-06
VGS	1.139	1.139	1.139	1.139	1.152	-1.356	-1.356
VDS	5.711	5.711	5.711	5.711	1.152	-1.356	-1.356
VBS	0.000	0.000	0.000	0.000	0.000	1.768	1.768
VTH	0.833	0.833	0.833	0.833	0.841	-1.097	-1.097
VDSAT	0.246	0.246	0.246	0.246	0.248	-0.275	-0.275
GM	2.54E-05	2.54E-05	2.54E-05	2.54E-05	2.51E-05	3.04E-05	3.04E-05
GDS	5.60E-08	5.60E-08	5.60E-08	5.60E-08	1.01E-07	1.31E-07	1.31E-07
GMB	1.26E-05	1.26E-05	1.26E-05	1.26E-05	1.25E-05	3.29E-06	3.29E-06

	M5C	M5D	M5E	M6	M7A	M7B	M7C
MODEL	P	P	P	P	P	P	P
ID	-4.66E-06	-4.66E-06	-4.66E-06	-4.66E-06	-4.66E-06	-4.66E-06	-4.66E-06
VGS	-1.356	-1.356	-1.356	-1.341	-1.204	-1.204	-1.204
VDS	-1.356	-1.356	-1.356	-5.914	-1.204	-1.204	-1.204
VBS	1.768	1.768	1.768	1.783	0.564	0.564	0.564
VTH	-1.097	-1.097	-1.097	-1.091	-0.939	-0.939	-0.939
VDSAT	-0.275	-0.275	-0.275	-0.267	-0.270	-0.270	-0.270
GM	3.04E-05	3.04E-05	3.04E-05	3.14E-05	2.98E-05	2.98E-05	2.98E-05
GDS	1.31E-07	1.31E-07	1.31E-07	1.00E-07	1.29E-07	1.29E-07	1.29E-07
GMB	3.29E-06	3.29E-06	3.29E-06	3.37E-06	4.58E-06	4.58E-06	4.58E-06

	M7D	M7E	M8
MODEL	P	P	P
ID	-4.66E-06	-4.66E-06	-4.66E-06
VGS	-1.204	-1.204	-1.204
VDS	-1.204	-1.204	-1.219
VBS	0.564	0.564	0.564
VTH	-0.939	-0.939	-0.939
VDSAT	-0.270	-0.270	-0.270
GM	2.98E-05	2.98E-05	2.98E-05
GDS	1.29E-07	1.29E-07	1.29E-07
GMB	4.58E-06	4.58E-06	4.58E-06

Figure 10.20   *(continued)*

This example shows that both source ramping and selective deletion of some second-order effects can lead to convergence for CMOS and BiCMOS circuits. Another possible approach is to increase the values of **ABSTOL** and **RELTOL**. Once a solution is obtained, initialization of critical nodes can be used to obtain convergence of the circuit using the complex models and default tolerances.

Another approach to finding the DC bias point of a circuit is to ramp up the supplies in a transient analysis. This method is applied in the next example to obtain the hysteresis curve of a CMOS Schmitt trigger, which displays very abrupt switching at the two thresholds. The advantage of time-domain analysis for solving circuits with positive feedback and high loop gain resides in the presence of charge storage elements, which do not allow instantaneous switching to take place.

EXAMPLE 10.8

Compute the hysteresis characteristic of the CMOS Schmitt trigger (Jorgensen 1976) shown in Figure 10.21. The n- and p-channel transistors have the following **LEVEL**=1 model parameters:

```
NMOS: VTO=2  KP=20U RD=100 CGSO=1P
PMOS: VTO=-2 KP=10U RD=100 CGSO=1P
```

If a finite output conductance in saturation needs to be modeled, use $LAMBDA = 0.05$ V^{-1} for both transistors. All transistors have the same geometry, with $W = 10$ μm and $L = 10$ μm. Compare the results from PSpice and SPICE2.

Solution
The most appropriate analysis for computing the transfer characteristic is **.DC**, the DC sweep analysis, presented in Sec. 4.3. The output voltage, V_o can be computed while V_i first increases from 0 to V_{CC} and then decreases from V_{CC} to 0. The positive and negative thresholds of the hysteresis curve are set by the voltages at nodes 5

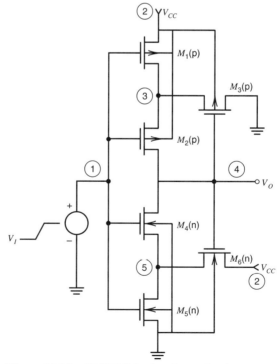

Figure 10.21   CMOS Schmitt trigger circuit.

and 3, respectively. The transistors M_2 and M_4 form a comparator, while the p-channel transistor pair M_1-M_3 and the n-channel transistor pair M_5-M_6 are voltage dividers, which define the two thresholds. When V_i is 0, p-channel transistors M_1 and M_2 are turned on and n-channel transistors M_4 and M_5 are off, with V_o at V_{CC}. This value of V_o biases M_3 off and M_6 on. When $V_i = V_{TOn}$, or 2 V, M_5 turns on and the voltage at node 5 is set to approximately $\frac{1}{2}V_{CC}$. When $V_i = V_5 + V_{TOn}$, M_4 starts turning on, triggering regenerative switching and resulting in V_o going to 0 V and transistors M_2 and M_6 turning off. This point is the positive switching threshold, V_{st+} of the Schmitt trigger and is equal to

$$V_{st+} = \frac{V_{CC}}{2} + V_{TOn} \tag{10.4}$$

The same process takes place when V_i is varied from V_{CC} to 0, except that the switching point occurs at

$$V_i = V_{st-} = \frac{V_{CC}}{2} - |V_{TOp}| \tag{10.5}$$

```
CMOS SCHMITT TRIGGER
*
* THIS CIRCUIT DIES IN .DC AROUND VT+ AND VT-;
* PSPICE SOURCE RAMPING CANNOT SAVE IT; ONLY A .TRAN CAN PROVIDE THE HYSTERESIS;
* FOR IMPROVED CONVERGENCE LAMBDA > 0, CAPS IN MODEL.
*
M1 2 1 3 2 P
M2 3 1 4 2 P
M3 0 4 3 2 P
M4 4 1 5 0 N
M5 5 1 0 0 N
M6 2 4 5 0 N
.OPTION DEFL=10U DEFW=10U
.MODEL P PMOS VTO=-2 KP=10U RD=100
*+ LAMBDA=0.05 CGSO=1P
.MODEL N NMOS VTO=2 KP=20U RD=100
*+ LAMBDA=0.05 CGSO=1P
VCC 2 0 12
VI  1 0 0
*VI 1 0 6 PWL 0 0 100N 12 200N 0
.DC VI 0 12 0.1
*.TRAN 1N 200N
.OPTION ABSTOL=100N
*.PROBE
.END
```

Figure 10.22   SPICE input for CMOS Schmitt trigger.

The SPICE input circuit is listed in Figure 10.22. The analysis of this circuit fails both in SPICE2 and PSpice. A basic rule when simulating stacked CMOS transistors is to provide a finite output conductance in saturation for all transistors. In the basic **LEVEL**=1 model, finite output conductance is modeled when a value is specified for **LAMBDA** in the **.MODEL** statement. Models of higher **LEVEL** compute a finite conductance internally from process parameters, and specifying **LAMBDA** is necessary only when the internal value must be overridden.

After the addition of **LAMBDA**, SPICE2 completes the DC transfer characteristic. PSpice however fails to converge when V_i reaches the positive switching threshold, V_{st+}. In general, any SPICE simulator can fail to converge in DC analysis when strong regenerative feedback is present in the circuit.

The approach to curing this problem is to find a time-domain solution rather than a DC solution, by sweeping V_i over time from 0 V to V_{CC} and then back to 0 V. VI must be changed in the input circuit to include the ramping in time, and the **.TRAN** line must be added in Figure 10.22:

```
VI 1 0 0 PWL 0 0 100N 12 200N 0
.TRAN 1N 200N
```

The solution is obtained over 200 ns, with V_i rising to V_{CC} at 100 ns and falling back to 0 at the end of the analysis interval, at 200 ns.

Every SPICE program can complete this analysis. The hysteresis of the Schmitt trigger is shown in Figure 10.23: it is the plot of V_o as a function of V_i. The two switching points agree between the two programs and analyses.

In order to demonstrate the importance of the charge storage in the circuit it is left for the reader as an exercise to remove CGSO from both **.MODEL** definitions in Figure 10.22 and observe the result of the analysis.

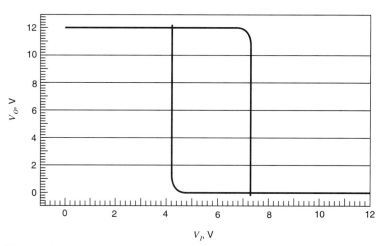

Figure 10.23 Hysteresis curve of the Schmitt trigger.

10.4 TIME-DOMAIN CONVERGENCE

This section addresses the convergence problems that can occur during a transient analysis and the ways to overcome these problems. As in the case of DC analysis, the three SPICE versions used in this text differ slightly in the implementation of the integration algorithms and the control a user can exert on the time-domain solution.

Through the **METHOD** option SPICE2 and SPICE3 offer the user the choice between trapezoidal and Gear variable-order integration, whereas PSpice does not offer this choice. SPICE2 and SPICE3 use trapezoidal integration by default, and PSpice exclusively uses the Gear algorithm. In addition, SPICE2 differs from SPICE3 by offering the user the choice of the time-step control method through the **LVLTIM** option. By default the time step is controlled by the truncation error in both programs, corresponding to $LVLTIM = 2$ in SPICE2. If $LVLTIM = 1$, an iteration-count time-step control is used in SPICE2, which doubles the time step at any time point when the program does not need more than 3 iterations to converge, and cuts it by 8 when convergence is not reached within $ITL4$, equal to 10, iterations.

In the following examples both potential traps of time-domain solution, convergence failure and accuracy, are addressed. The most important causes for convergence failure are incomplete semiconductor model descriptions and lack of charge storage elements. Ideal representations of semiconductor devices, that is, models described only by the default parameters are often the cause of a transient analysis failure.

The following example illustrates a `TIME STEP TOO SMALL` (TSTS) failure due to an incomplete model used to represent a diode in a full-wave rectifier circuit.

EXAMPLE 10.9

Use SPICE to compute the time-domain response of the circuit shown in Figure 10.24. The model parameters of the diodes are $IS = 10^{-14}$ A and $CJO = 10$ pF. Then perform a second analysis including the breakdown voltage for the two diodes, $BV = 100$ V.

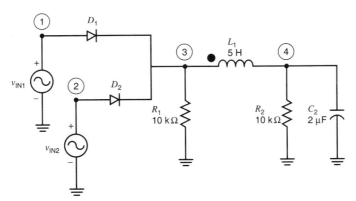

Figure 10.24 Diode rectifier circuit.

```
CHOKE CKT - FULL WAVE CHOKE INPUT
VIN1 1 0 SIN(0 100 50)
VIN2 2 0 SIN(0 -100 50)
D1 1 3 DIO
D2 2 3 DIO
R1 3 0 10K
L1 3 4 5.0
R2 4 0 10K
C2 4 0 2UF IC=80
.MODEL DIO D IS=1.0E-14 CJO=10PF
.TRAN 0.2MS 200MS UIC
.PLOT TRAN V(1) V(2) V(3) V(4)
.OPT ACCT
.WIDTH OUT=80
.END
```

Figure 10.25   SPICE input for diode rectifier.

Solution

In the circuit of Figure 10.24 the two diodes, D_1 and D_2, driven by the two sinusoidal sources represent a full-wave rectifier, which is followed by an RLC low-pass filter. The SPICE input file for this circuit is listed in Figure 10.25, and the voltage waveforms at nodes 3 and 4 are plotted in Figure 10.26. The output voltage on capacitor C_2 is initialized at 80 V, which is the peak value and is reached when the voltage at node 3 reaches its lowest point.

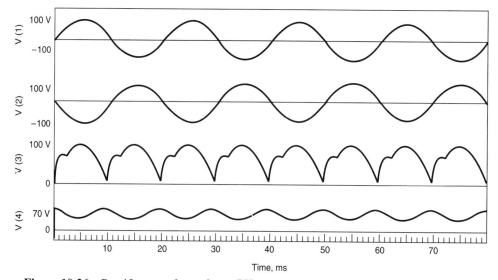

Figure 10.26   Rectifier waveforms for no **BV**.

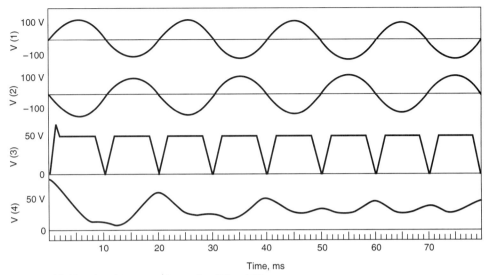

Figure 10.27 Rectifier waveforms for $BV = 100$.

In order to model the breakdown characteristic of the diode in the reverse region, the value of parameter **BV** is added to the **.MODEL** line. Resimulation of the circuit results in a TSTS error in PSpice and a numerical exception in SPICE2. Because of the ideality of the diodes in the previous analysis only one is conducting at any one time and proper current limiting is provided by the load circuit. When a breakdown characteristic is added to the diodes at 100 V, a current path exists from one signal source to the other through the series combination of the two back-to-back diodes. D_2 breaks down when V(1) reaches 50 V and V(2) reaches −50 V, and the two diodes could be destroyed in reality. The solution to this problem is to add a parasitic series resistance, **RS**, to the diode model, which limits the current when D_1 or D_2 operate in the breakdown region.

Resimulation of the circuit with RS=100 added to the **.MODEL** line in Figure 10.25 results in the waveforms V(3) and V(4) shown in Figure 10.27. Note that because of the diode reverse conduction the voltage at node 3 is clamped at 50 V. In reality a careful sizing of the series resistors is necessary in order to limit the reverse current to the maximum value prescribed in the data book and avoid destruction of the diode.

This example has illustrated a failure condition that can develop during the transient analysis because of an ideal model and improper circuit design. The addition of series resistance for limiting the current solved the problem.

Analysis failure can also occur because of insufficient charge storage in a circuit. In the absence of charge storage elements a circuit tends to switch in zero time, which does not happen in reality and cannot be handled by the existing solution algorithms. This issue is presented in the following example.

EXAMPLE 10.10

Simulate the behavior of the NMOS relaxation oscillator shown in Figure 10.28 (Kelessaglou and Pederson 1989) using SPICE. The model parameters for the enhancement and depletion transistors are

```
LEVEL=1 VTO=0.7  KP=30U (enhancement transistor)
LEVEL=1 VTO=-0.7 KP=30U LAMBDA=0.01 (depletion transistor)
```

Solution

The input description for SPICE is listed in Figure 10.29. Note that the current source, I_1, is used only to kick-start the oscillations. Also note that the substrates of all transistors are connected at $V_{EE} = -9$ V. The transient response is requested to be displayed only from 1400 μs to 1800 μs. As pointed out in Sec. 6.2, SPICE computes the waveform starting from 0 but saves only the results after $t = 1.4$ ms.

In SPICE2 the analysis fails at $t = 46.23$ μs, when the time step is reduced below a minimum value set by the program to be 10^{-9} of the smaller of $2 \times TSTEP$ or $TMAX$.

The first observation is that the MOS transistors do not have any capacitances given in the **.MODEL** statement. It is necessary to correct the ideality of the MOSFET models by adding a gate-source capacitance, **CGSO**, to the two models, MOD1 and MOD2. After CGSO=1P is included in the **.MODEL** statement, the transient analysis finishes successfully; the resulting waveforms are shown in Figure 10.30. The period and pulse width can be verified by hand using the GDS values printed in the operating point information for depletion devices M_5 and M_8.

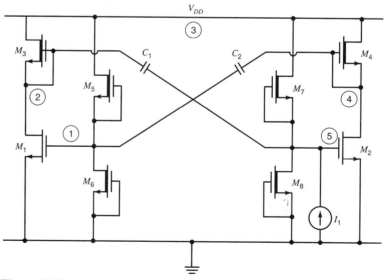

Figure 10.28   NMOS relaxation oscillator.

```
MOS RELAXATION OSCILLATOR
M1 2 1 0 6 MOD1 W=100U L=5U
M2 4 5 0 6 MOD1 W=100U L=5U
M3 3 2 2 6 MOD2 W=20U L=5U
M4 3 4 4 6 MOD2 W=20U L=5U
M5 3 1 1 6 MOD2 W=20U L=5U
M6 1 0 0 6 MOD2 W=20U L=5U
M7 3 5 5 6 MOD2 W=20U L=5U
M8 5 0 0 6 MOD2 W=20U L=5U
C1 2 5 100P
C2 4 1 200P
VEE 6 0 -9
VDD 3 0 5
I1 5 0 PULSE 10U 0 0 0 0 1
.MODEL MOD1 NMOS VTO=+0.7 KP=30U
*+ CGSO=1P
.MODEL MOD2 NMOS VTO=-0.7 KP=30U LAMBDA=0.01
*+ CGSO=1P
.OP
.TRAN 2U 1800U 1400U
.PLOT TRAN V(5) V(1) V(4)
.WIDTH OUT=80
.OPTION NOPAGE NOMOD LIMPTS=1001 ITL5=0
.OPTION ACCT ABSTOL=1N
.END
```

Figure 10.29 SPICE input for NMOS relaxation oscillator.

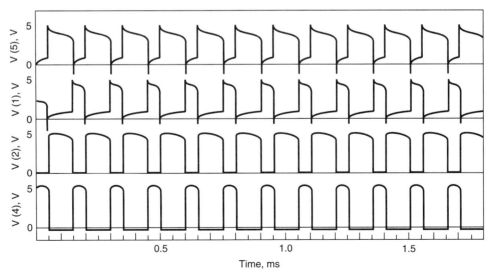

Figure 10.30 Waveforms V(1), V(2), V(4), and V(5) for the NMOS relaxation oscillator.

The above examples have focused on the main causes for the failure of transient analysis and on how to overcome them. Another important issue is the accuracy of the solution computed by the transient analysis.

Several observations were made in the previous chapter on the stability of numerical integration algorithms. The trapezoidal method was seen to oscillate around the solution for values of the time step that are larger than a limit related to the time constant of the circuit. This behavior can be noticed in the computed response of a circuit, especially when the circuit switches. This numerical oscillation can be avoided by selecting the Gear algorithm or controlling the tolerances for a tighter control of the truncation error, which results in a smaller time step.

The following example illustrates ringing around the solution.

EXAMPLE 10.11

Find the waveform of the current flowing through the diode D_1 in Figure 10.6 when a voltage step from 5 V to -5 V is applied to the circuit. For the first analysis use the default model for the diode with the following two additional parameters:

```
RS=100 TT=40N
```

The parameter **TT** is a finite transit time of the carriers in the neutral region of the diode; the effect of this parameter is that when a diode is switched off it takes a finite time to eliminate the carriers from the neutral region, and during this time the diode conducts in reverse direction.

Solution
```
DIODE SWITCHING
*
D1 1 0 DMOD
VIN 1 0 PULSE 5 -5 50N .1N .1N 50N 100N
.MODEL DMOD D RS=100
+ TT=40N
*+ CJO=10P
.TRAN .5N 100N
.PLOT TRAN I(VIN)
*.OPTION METHOD=GEAR MAXORD=2
.WIDTH OUT=80
.END
```

The SPICE input description is listed above. The waveform of the diode current resulting from the SPICE simulation is plotted in Figure 10.31. Current ringing around the value of zero can be noticed following the storage time, during which the diode conducts in reverse. This result can be explained only by the oscillatory nature of the trapezoidal integration method.

Whenever the simulated response of circuits displays oscillations that are not anticipated by design, the integration method should be changed to the second-order Gear

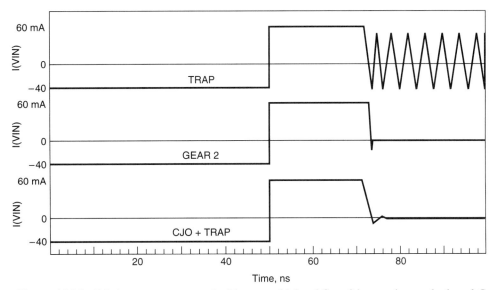

Figure 10.31   Diode current computed with trapezoidal and Gear 2 integration methods and C_J.

method by adding the following line to the SPICE input:

```
.OPTION METHOD=GEAR MAXORD=2
```

The second-order Gear method is characterized by numerical damping. After the addition of the above line to the input, repeating the simulation yields the second waveform shown in Figure 10.31, which exhibits the expected behavior.

The model of the diode used in this example is ideal except for the finite transit time, **TT**, and series resistance, **RS**. The charge storage of the diode is incomplete without the depletion region charge (see Sec. 3.2). If this charge is modeled by specifying a value for the parameter **CJO** on the **.MODEL** line, a gradual decay of the current is expected with a time constant equal to $C_J R_S$. A new simulation with **CJO** added and the default trapezoidal integration yields the third current waveform shown in Figure 10.31. It can be noticed that the current returns to zero with minimal numerical ringing.

This last observation leads to the conclusion that the more complete the model, the more accurate and stable is the solution. A common cause for numerical oscillation is sudden changes in a circuit variable or in the model equations. The latter can also be caused by imperfections in the built-in analytical models in SPICE in addition to the absence of certain model parameters.

Next, a few comments are necessary about the impact of transient analysis parameters *TSTEP* and *TMAX* on the accuracy of the result. In Sec. 6.4 the accuracy of the Fourier coefficients calculation was shown to depend on the value of *TMAX*, which controls the number of points evaluated for a signal each period. Numerical ringing can also be avoided by selecting smaller values for *TMAX* or reducing the relative tolerance **RELTOL**. In general, the more time points are computed, the more accurate the results. Sometimes the maximum time step must be set by the user in order to obtain an accurate solution.

EXAMPLE 10.12

Simulate the current flowing through the CMOS inverter shown in Figure 10.32. The SPICE input description and model parameters are listed in Figure 10.33. The input voltage is ramped from 0 V to 10 V over 10 μs.

Solution
Current flows through the inverter only for values of v_{IN} from 1 V to 9 V. The graphical solution of the inverter current in Figure 10.34, computed with the default trapezoidal method, incorrectly displays ringing of 10 μA after reaching the peak value.

The same current computed with the Gear 2 method is smooth and overlaps the trapezoidal solution in Figure 10.34. An alternate way to obtain the correct solution is to limit the maximum time step the program takes during the trapezoidal integration. A value of 50 ns added to the **.TRAN** line in Figure 10.33 also leads to a smooth waveform.

Modeling the subthreshold current with parameter **NFS** added to the device specification has an effect on the result similar to that of the Gear integration, that is, a smooth current waveform.

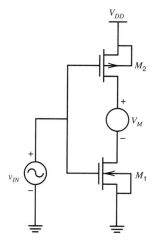

Figure 10.32   CMOS inverter circuit diagram.

```
CMOS INVERTER
*
M1 2 1 0 0 NMOS
+    L=10U W=20U AD=160P AS=160P PD=36U PS=36U
VM 21 2
M2 21 1 3 3 PMOS
+    L=10U W=40U AD=1600P AS=1600P PD=216U PS=216U
VIN 1 0 PWL 0 0 10U 10
VDD 3 0 10
*
.MODEL NMOS NMOS
+    LEVEL=2 VTO=.9 GAMMA=.6 PB=.8 CGSO=.28E-9 CGDO=.28E-9 CGBO=.25E-9
+    CJ=.3E-3 MJ=.49 CJSW=.66E-9 MJSW=.26 TOX=.5E-7 NSUB=.5E16 NSS=1E11
+    TPG=1 XJ=.5E-6 LD=.38E-6 UO=610 UCRIT=5.4E+4 UEXP=.1
+    VMAX=1.2E+5 NEFF=3 KF=1.0E-32 AF=1.0
+    NFS=1E11
*
.MODEL PMOS PMOS
+    LEVEL=2 VTO=-.8 GAMMA=.69 PB=.8 CGSO=.28E-9 CGDO=.28E-9 CGBO=.25E-9
+    CJ=.4E-3 MJ=.46 CJSW=.6E-9 MJSW=.24 TOX=.5E-7 NSUB=.65E+16 NSS=1E+11
+    TPG=-1 XJ=.7U LD=.52U UO=230 UCRIT=5.2E+4 UEXP=.18
+    VMAX=.4E+5 NEFF=3 KF=.2E-32 AF=1.0
+    NFS=1E11
*
*.TRAN 200N 10U 0 50N
.TRAN 200N 10U
.PLOT TRAN I(VM)
.WIDTH OUT=80
.END
```

Figure 10.33   SPICE description of CMOS inverter.

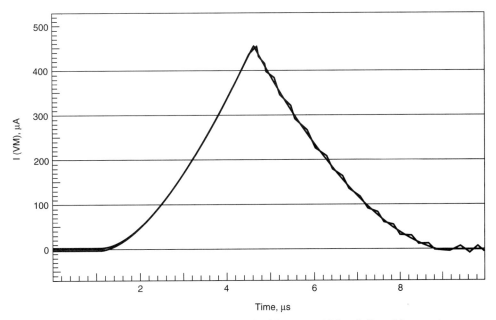

Figure 10.34   CMOS inverter current computed with trapezoidal and Gear 2 integration methods.

PSpice does not provide a choice in integration methods. The results of PSpice simulations for the above circuits show a certain level of damping, which points to the Gear method.

10.5 CIRCUIT-SPECIFIC CONVERGENCE

10.5.1 Oscillators

The analysis of oscillators can represent a challenge for the user of circuit simulation. As described in Example 6.3, the number of periods for oscillations to build up is proportional to the quality factor of the resonant circuit, Q. Another possible difficulty in simulating oscillators is related to the numerical integration methods. In the previous chapter and in the previous section it was shown that the Gear integration method has a damping effect on oscillations. Therefore, trapezoidal integration, the default in SPICE2 and SPICE3, but seemingly not in PSpice, should always be used when simulating oscillators. It may be necessary to set a value of $TMAX$, the maximum allowed integration step, in order to control the accuracy of the trapezoidal method.

An important aid for initiating oscillations in a simulator is either a single pulse at the input of the amplifier block or the initialization of the charge storage elements close to the steady-state values reached during oscillations.

The most representative example for the difficulty encountered in simulating oscillators is a crystal oscillator. Because of the high Q, of the order of tens of thousands, it would theoretically take a number of periods on the same order of magnitude to reach steady state. It is impractical to simulate a circuit for so many cycles. The following example provides insight into the analysis of a Pierce crystal oscillator.

EXAMPLE 10.13

Verify the behavior of the CMOS Pierce crystal oscillator shown in Figure 10.35 using SPICE2 and PSpice. The circuit drawing includes the equivalent schematic of the crystal, C_0, C_x, L_x, and R_s. The crystal has a resonant frequency of 3.5795 MHz, common in color television, and has the following equivalent circuit parameters:

$$C_x = 2.47 \text{ fF} \quad L_x = 0.8 \text{ H} \quad R_s = 600 \ \Omega \quad C_0 = 7 \text{ pF}$$

Use the following model parameters for the n- and p-channel MOSFETs:

```
NMOS: VTO=1   KP=20U LAMBDA=0.02
PMOS: VTO=-1  KP=10U LAMBDA=0.02
```

An attempt to run the circuit as is or using a start-up pulse does not produce the expected oscillations.

Solution

A first approach for simulating this circuit is to replace the crystal with an equivalent circuit with a reduced Q that allows for a rapid buildup of oscillations. The crystal has

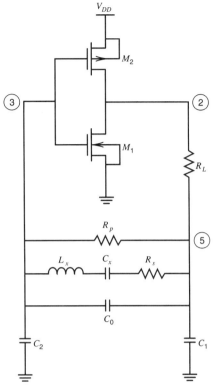

Figure 10.35 Pierce CMOS crystal oscillator.

a series and a parallel resonant frequency, f_s and f_p, respectively. At frequencies below f_s the reactance of the crystal is capacitive, whereas above f_s the reactance is inductive and L_x, C_x, and C_0 can be substituted by an effective inductance, L_{eff}. The oscillation frequency, ω_0, of the equivalent circuit is given by

$$\omega_0 = \frac{1}{\sqrt{L_{eff} \dfrac{C_1 C_2}{C_1 + C_2}}} \tag{10.6}$$

This frequency, ω_0, is very close to the crystal resonant frequency, ω_s, leading to the following value for L_{eff}:

$$L_{eff} = \frac{1}{\omega_s^2 C_1 C_2 / (C_1 + C_2)} = 0.18 \text{ mH} \tag{10.7}$$

The new equivalent resonant circuit has $Q \approx 7$, which is considerably smaller than the original value. The crystal equivalent circuit, C_x, L_x, C_0, must be replaced by L_{eff}. The SPICE input description is listed below; a **.SUBCKT** block represents the crystal.

```
PIERCE XTAL OSCILLATOR W/ CMOS
.OPTION LIMPTS=10000 ITL5=0
*
M1 2 3 0 0 NMOS W=40U L=10U
M2 2 3 1 1 PMOS W=80U L=10U
RL 2 5 10K
C1 5 0 22P
C2 3 7 22P
VDD 1 0 5
*
* XTAL
*
X1 3 5 XTAL
.SUBCKT XTAL 1 2
LEFF 1 3 .18M
RS 3 2 600
RP 1 2 22MEG
.ENDS XTAL
*
* KICKING SOURCE
*
VKICK 7 0 PULSE 0 10M .1N .1N .1N 1 1
*
.MODEL NMOS NMOS VTO=1 KP=20U LAMBDA=0.02
.MODEL PMOS PMOS VTO=-1 KP=10U LAMBDA=0.02
*
.OP
.TRAN 10N 50U 45U 10N
.PLOT TRAN V(2) V(3)
.END
```

Resimulation of the circuit results in the waveforms $V(2)$ and $V(3)$ shown in Figure 10.36. The circuit sustains oscillations centered around 2.5 V with an amplitude of 1.75 V at the output of the gain block, node 2, and 0.75 V across the crystal with a period of 280 ns, corresponding to the resonant frequency, ω_0.

Note that the SPICE circuit description includes a voltage source, VKICK, which provides a 0.1 ns pulse for triggering the oscillations. The **.TRAN** line requests the results to be saved only after 45 μs, that is, after approximately 150 cycles, when the steady state should have been reached. The same results are obtained with both SPICE2 and PSpice.

Once the correct operation of the circuit has been verified, the circuit can be simulated with the real crystal. As shown in Sec. 6.3, it is necessary to initialize the charge storage elements as close as possible to the steady-state values; the previous results can help in this task. Assume that at $t = 0+$ both $V(3)$ and $V(5)$ are at 2.5 V,

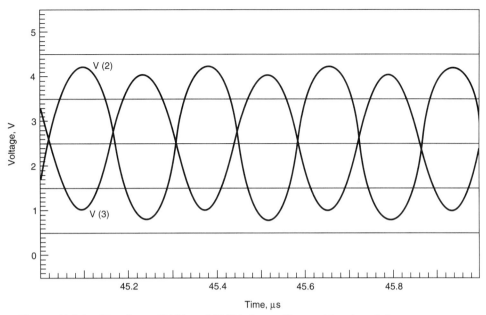

Figure 10.36 Waveforms V(2) and V(3) for oscillator with reduced Q.

which corresponds to the maximum current in the inductor L_x. The results of the simulation using the equivalent inductor, L_{eff}, instead of the crystal can be used for guidance. Note that it is important to initialize the state of one of the crystal components, such as L_x, in order to achieve steady-state oscillations in the solution. Therefore, the initial current through L_x and C_0 should approximate the value of the current amplitude through the equivalent crystal, L_{eff}. For a rigorous derivation of the oscillation amplitude consult the text by Pederson and Mayaram (1990).

```
PIERCE XTAL OSCILLATOR W/ CMOS
*
* THIS CKT OSCILLATES ONLY PSPICE 3.02 (OLD) BECAUSE IT USES TRAP
*
.OPTION LIMPTS=10000 ITL5=0
*
M1 2 3 0 0 NMOS W=40U L=10U
M2 2 3 1 1 PMOS W=80U L=10U
RL 2 5 10K
C1 5 0 22P IC=2.5
C2 3 0 22P IC=2.5
VDD 1 0 5
*
* XTAL
*
```

```
X1 3 5 XTAL
.SUBCKT XTAL 1 2
* LEFF 3 6 .18M
LX 1 3 0.8 IC=0.6M
CX 3 4 2.47FF
C0 1 2 7PF
RS 4 2 600
RP 1 2 220K
.ENDS XTAL
*
.MODEL NMOS NMOS VTO=1 KP=20U LAMBDA=0.02
.MODEL PMOS PMOS VTO=-1 KP=10U LAMBDA=0.02
*
.OP
.TRAN 10N 50U 45U 10N UIC
.PLOT TRAN V(2) V(3)
.END
```

The new SPICE input is shown above; note that the initial pulse is omitted and the keyword **UIC** is used in the **.TRAN** statement in order to start the analysis at steady state. The results are shown in Figure 10.37 and are identical to the previous solution.

The estimated steady state is verified by the analysis performed in SPICE2 or SPICE3, and the oscillations are sustained for the 200 cycles simulated. The result

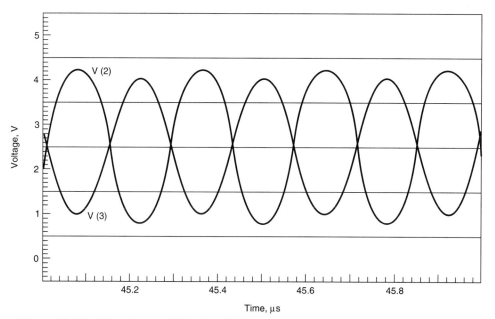

Figure 10.37   Waveforms $V(2)$ and $V(3)$ for oscillator with crystal.

of a PSpice analysis for this circuit shows the oscillations decaying, which could be explained by the use of Gear integration.

10.5.2 BJT versus MOSFET Specifics

MOSFET circuits have more convergence problems than bipolar circuits because of a number of differences between the two devices types. First, the physical structures of the two devices are different. The gate terminal of a MOSFET is insulated; that is, it is an open circuit in DC. The self-conductance of the gate is therefore zero in DC, often leading to ill-conditioned circuit matrices, as demonstrated by Example 10.1, and subsequently to the failure of SPICE to find a solution. By contrast, continuous current flows in or out of the base terminal of a bipolar transistor, independently of region of operation and analysis mode.

Second, the generality of analytical models used in SPICE to describe the two devices is not the same. Whereas the Ebers-Moll or Gummel-Poon formulation for the BJT transistor applies to all regions of operation, the MOSFET models combine different equations to describe distinct regions of operations and various second-order effects. The different formulations have different levels of continuity for the equivalent conductance at the transition points. The continuity of the conductance, which is the first derivative of the function, is important for the convergence of the iterative process.

Third, the implementation details in SPICE also affect convergence. There are differences in the operating points in which the program initializes the two types of devices and in the way new operating points are selected at each iteration. An important difference between BJTs and MOSFETs in the first iteration is that by default the former are initialized conducting whereas, generally, MOSFETs are initialized turned off. MOSFETs are initialized with $V_{GS} = VTO$, and the difficulty with MOSFETs is that the actual threshold voltage, V_{TH}, is usually increased by back-gate bias to a higher value than the zero-bias threshold voltage, VTO. Only MOSFET devices with subthreshold current, model parameter **NFS**, are initially in the conduction state. This second-order effect is supported only by the higher-level models, summarized in Appendix A and described in more detail in the references (Antognetti and Massobrio 1988; Sheu, Scharfetter, and Ko 1985; Vladimirescu and Liu 1981). This is one example where convergence can be improved by changing model parameters.

The default initialization of transistors has a different impact on convergence depending on the operation of the circuit. A smaller number of iterations has been noticed for analog (linear) bipolar circuits as compared to digital (logic) bipolar circuits because BJTs are initialized as conducting in SPICE. The explanation can be found in the mode of operation of analog circuits, which have the majority of the transistors turned on, in contrast with digital circuits, which have an important percentage of the devices turned off. Example 10.4 demonstrated that the initialization of two protective devices as **OFF** improves the convergence and leads to a solution without source ramping. Experiments in the SPICE code that initialized all MOSFETs in the conduction state have proven to speed up the convergence of analog circuits. The user can access initialization through the device initial conditions, $IC = v_{DS0}, v_{GS0}, v_{BS0}$, which are activated only in transient analysis in conjunction with the **UIC** option.

The importance of initialization and device specifics for the convergence of MOS analog circuits, especially those with high-impedance nodes, often present in cascode loads, can be exemplified by a CMOS differential amplifier (Senderowicz 1991).

EXAMPLE 10.14

Find the DC operating point of the circuit shown in Figure 10.38 using the **LEVEL**=2 parameters given in the SPICE description.

Solution
This circuit is a differential amplifier in a unity-feedback loop; the conversion from double- to single-ended output is achieved by PMOS transistors M_6 through M_9. The state of this circuit is set by connecting the appropriate bias transistors M_{11} through M_{14} to the gates of the transistor M_5, the differential pair transistor M_2, and the load transistor pairs M_3-M_4, and M_6-M_7.

Based on the knowledge acquired in Chap. 9, one can see the difficulty associated with solving the modified nodal equations when a number of nodes are of very low conductance; nodes 3 and 4 have very high impedances because of the cascode connection, and additionally, all transistors are biased very close to the threshold voltage, on one hand, and at the limit between saturation and linear region, on the other hand. This approach to biasing is common in analog CMOS circuits.

The attempt to find the DC operating point of the circuit as it is represented by the input description in Figure 10.39 fails in SPICE2 but succeeds in PSpice and SPICE3 after source ramping. Addition of the option `ITL6=40` to the SPICE2 input does not help for this circuit. It is important to note that SPICE2, which provides more feedback related to a solution failure, encounters problems in the matrix solution: the messages `PIVOT CHANGE ON THE FLY` and `*ERROR*: MAXIMUM ENTRY...IS LESS THAN PIVTOL` can be found in the output file.

When ramping methods fail for this type of difficult circuit, running a transient analysis while ramping the supplies from 0 to the DC value or leaving them unchanged may lead to a solution. If the supplies are ramped for part of the time interval, the DC value should be preserved for the rest of the time-domain analysis in order to allow the circuit to scttle. A time-domain analysis of an MOS circuit has the additional advantage of a well-conditioned matrix because charge storage elements provide finite conductance at the gates of MOSFETs. Transient ramping of the CMOS differential amplifier in Fig. 10.39 is performed when the `.TRAN` and `.PRINT` lines are activated. A DC solution is avoided by using the **UIC** keyword on the `.TRAN` line. A transient analysis also fails in SPICE2 for this circuit.

Several options can be modified for this circuit. Based on the observation of the condition of the circuit matrix, a first approach is to tighten the pivot selection criterion

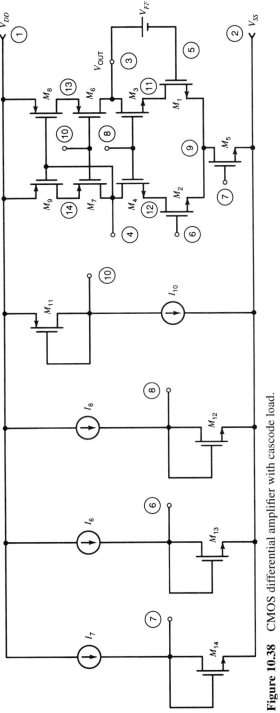

Figure 10.38 CMOS differential amplifier with cascode load.

369

```
CMOS DIFFERENTIAL AMPLIFIER
*
M1      11   5    9    2    N    W=120U   L=1.2U   AD=2N   AS=2N   PD=12U PS=12U
M2      12   6    9    2    N    W=120U   L=1.2U   AD=2N   AS=2N   PD=12U PS=12U
M3      3    8    11   2    N    W=120U   L=1.2U   AD=2N   AS=2N   PD=12U PS=12U
M4      4    8    12   2    N    W=120U   L=1.2U   AD=2N   AS=2N   PD=12U PS=12U
M5      9    7    2    2    N    W=240U   L=1.2U   AD=2N   AS=2N   PD=12U PS=12U
M6      3    10   13   1    P    W=120U   L=1.2U   AD=2N   AS=2N   PD=12U PS=12U
M7      4    10   14   1    P    W=120U   L=1.2U   AD=2N   AS=2N   PD=12U PS=12U
M8      13   4    1    1    P    W=120U   L=1.2U   AD=2N   AS=2N   PD=12U PS=12U
M9      14   4    1    1    P    W=120U   L=1.2U   AD=2N   AS=2N   PD=12U PS=12U
*
* BIAS CIRCUIT
* VOLTAGES AT NODES 6, 7, 8 AND 10 ARE SUPPLIED BY
* M13, M14, M12 AND M11
*
M11     10   10   1    1    P    W=120U   L=4.8U   AD=2N   AS=2N   PD=12U PS=12U
M12     8    8    2    2    N    W=120U   L=9.6U   AD=2N   AS=2N   PD=12U PS=12U
M13     6    6    2    2    N  W=120U     L=4.8U   AD=2N   AS=2N   PD=12U PS=12U
M14     7    7    2    2    N    W=120U   L=1.2U   AD=2N   AS=2N   PD=12U PS=12U
I10     10   2    200U
I8      1    8    200U
I6      1    6    200U
I7      1    7    400U
*
VDD 1   0    5.0
*+ PULSE(0.0 5.0 0 100U 10N 100U 200U)
VSS 2   0    0.0
*
VFF 3   5    2.0
*+ PULSE(0.0 2.0 0 100U 10N 100U 200U)
*
.WIDTH OUT=80
.OP
.OPT ACCT
*.OPT PIVREL=1E-2
*.OPT ABSTOL=1U
*.OPT ITL6=40
*.OPT ITL4=100
*.TRAN 1US 200US UIC
*.PRINT TRAN V(2) V(3) V(4) V(5) V(6)
*.PRINT TRAN V(7) V(8) V(9) V(10) V(11)
*.PRINT TRAN V(12) V(13) V(14)
*
* N_CHANNEL TRANSISTOR, SANITIZED MODELS
*
.MODEL N NMOS LEVEL=2
+ UO=600 VTO=700E-3 TPG=1.000 TOX=25.0E-9
+ NSUB=5.0E+16 UCRIT=2E+4 UEXP=0.1
+ XJ=1E-10 LD=100E-9 PB=0.8 JS=100.0E-6 RSH=100
+ NFS=200E+9 VMAX=60E+3 NEFF=2.5 DELTA=3
+ CJ=400E-6 MJ=300E-3 CJSW=2.00E-9 MJSW=1.0
+ CGSO=2.0E-10 CGDO=2.0E-10 CGBO=3.0E-10 FC=0.5
*
* P_CHANNEL TRANSISTOR
*
.MODEL P PMOS LEVEL=2
+ UO=200 VTO=-700E-3 TPG=-1.000 TOX=25.0E-9
+ NSUB=2.0E+16 UCRIT=2E+4 UEXP=0.1
+ XJ=1E-10 PB=0.8 JS=100.0E-6 RSH=200
+ NFS=100E+9 VMAX=20.00E+3 NEFF=.9 DELTA=1
+ CJ=400E-6 MJ=300E-3 CJSW=2.00E-9 MJSW=1.0
+ CGSO=2.0E-10 CGDO=2.0E-10 CGBO=3.0E-10 FC=0.5
*
.END
```

Figure 10.39 SPICE input for CMOS differential amplifier with cascode load.

by increasing the value of *PIVREL* by including the following statement:

```
.OPTIONS PIVREL=1E-2
```

Second, because of the possible operation near threshold of some transistors, the absolute current tolerance, *ABSTOL,* can be raised:

```
.OPTIONS ABSTOL=1U
```

These two options contribute to a successful time-domain solution in SPICE2. The voltages at the circuit nodes are listed for the last 10 time points in Figure 10.40; an average value is chosen to initialize the node voltages with a **.NODESET** statement.

```
.NODESET
+ V(3) = 3.29
+ V(4) = 3.71
+ V(5) = 1.29
+ V(6) = 1.28
+ V(7) = 1.1
+ V(8) = 1.54
+ V(9) = 0.19
+ V(10) = 3.32
+ V(11) = 0.39
+ V(12) = 0.39
+ V(13) = 4.63
+ V(14) = 4.64
```

When the node voltages are initialized, the DC operating point is obtained for this circuit in only 8 iterations.

The solution and the operating point information are listed in Figure 10.41. One can verify the bias point of the transistors in this circuit and understand that the convergence difficulty is caused by the proximity of the operating points to the limits of the subthreshold conduction, linear and saturation regions. MOS analog circuits are biased with V_{GS} close to *VTO* for maximum gain and at the edge of saturation for maximum output signal swing (Gray & Meyer 1985). VTH represents the bias- and geometry-adjusted values of the zero-bias threshold voltages, **VTO**, specified in the **.MODEL** statement.

An alternate way to find a solution for this circuit and for amplifiers in general is to cut the feedback loop and find a DC solution of the open-loop amplifier. Then, the node voltages obtained from the open-loop circuit can be used in a **.NODESET** statement to initialize the closed-loop amplifier. The open-loop solution for this circuit is nontrivial.

```
*******03/07/93 ******** SPICE  2G.6      3/15/83********20:18:56*****

CMOS DIFFERENTIAL AMPLIFIER

****     TRANSIENT ANALYSIS                   TEMPERATURE =  27.000 DEG C

*********************************************************************

       TIME     V(3)        V(4)        V(5)        V(6)
      1.910E-04  3.287E+00   3.709E+00   1.287E+00   1.283E+00
      1.920E-04  3.288E+00   3.711E+00   1.288E+00   1.284E+00
      1.930E-04  3.289E+00   3.712E+00   1.289E+00   1.285E+00
      1.940E-04  3.288E+00   3.711E+00   1.288E+00   1.285E+00
      1.950E-04  3.287E+00   3.710E+00   1.287E+00   1.283E+00
      1.960E-04  3.286E+00   3.708E+00   1.286E+00   1.282E+00
      1.970E-04  3.285E+00   3.707E+00   1.285E+00   1.281E+00
      1.980E-04  3.286E+00   3.708E+00   1.286E+00   1.282E+00
      1.990E-04  3.288E+00   3.710E+00   1.288E+00   1.284E+00
      2.000E-04  3.290E+00   3.712E+00   1.290E+00   1.286E+00


       TIME     V(7)        V(8)        V(9)        V(10)

      1.910E-04  1.102E+00   1.537E+00   1.883E-01   3.320E+00
      1.920E-04  1.102E+00   1.541E+00   1.893E-01   3.316E+00
      1.930E-04  1.102E+00   1.546E+00   1.903E-01   3.313E+00
      1.940E-04  1.102E+00   1.542E+00   1.896E-01   3.315E+00
      1.950E-04  1.102E+00   1.538E+00   1.886E-01   3.319E+00
      1.960E-04  1.102E+00   1.533E+00   1.876E-01   3.323E+00
      1.970E-04  1.102E+00   1.529E+00   1.866E-01   3.327E+00
      1.980E-04  1.102E+00   1.534E+00   1.877E-01   3.323E+00
      1.990E-04  1.102E+00   1.540E+00   1.890E-01   3.317E+00
      2.000E-04  1.102E+00   1.546E+00   1.904E-01   3.312E+00

       TIME     V(11)       V(12)       V(13)       V(14)

      1.910E-04  3.844E-01   3.874E-01   4.633E+00   4.642E+00
      1.920E-04  3.876E-01   3.906E-01   4.630E+00   4.639E+00
      1.930E-04  3.907E-01   3.939E-01   4.626E+00   4.635E+00
      1.940E-04  3.883E-01   3.914E-01   4.629E+00   4.638E+00
      1.950E-04  3.852E-01   3.882E-01   4.632E+00   4.642E+00
      1.960E-04  3.821E-01   3.850E-01   4.636E+00   4.645E+00
      1.970E-04  3.791E-01   3.817E-01   4.639E+00   4.649E+00
      1.980E-04  3.824E-01   3.853E-01   4.635E+00   4.645E+00
      1.990E-04  3.867E-01   3.897E-01   4.631E+00   4.640E+00
      2.000E-04  3.909E-01   3.942E-01   4.626E+00   4.635E+00
```

Figure 10.40   Transient solution of CMOS differential amplifier.

10.5.3 CONVERGENCE OF LARGE CIRCUITS

The larger a circuit, the more complex is its behavior. A large circuit usually consists of a number of individual functional blocks, or cells, which perform different functions, some analog, linear or nonlinear, and some digital. The large number of transistors operate in very different conditions depending on the functions of the blocks. The difficulty of finding a solution is directly related to the number of nonlinear elements. With no prior knowledge of the expected function, SPICE requires considerable more iterations than for a few transistors. For most common purposes a circuit with more than 100 semiconductor devices can be considered a large circuit.

It is a good practice to describe a large circuit hierarchically. The **.SUBCKT** definition capability of SPICE introduced in Chap. 7 should be used for this purpose. Before the entire circuit is stimulated, the component functional blocks should be characterized individually. Key nodes can be expressed as interface nodes and initialized. This is necessary since **.NODESET** and **.IC** cannot initialize nodes internal to a subcircuit. Specifying the state of transistors that are **OFF** also helps.

Because of the size of the circuit, the 100 iterations of the default *ITL1* are often insufficient for finding a DC solution. *ITL1* should be increased to 300 to 500 but not

```
*******03/08/93  ******** SPICE  2G.6  3/15/83********21:30:55*****

CMOS DIFFERENTIAL AMPLIFIER

****     SMALL SIGNAL BIAS SOLUTION     TEMPERATURE =  27.000 DEG C

*************************************************************************

 NODE    VOLTAGE     NODE    VOLTAGE     NODE    VOLTAGE     NODE    VOLTAGE

(   1)    5.0000    (   2)    0.0000    (   3)    3.2872    (   4)    3.7093
(   5)    1.2872    (   6)    1.2831    (   7)    1.1019    (   8)    1.5372
(   9)    0.1884    (  10)    3.3196    (  11)    0.3848    (  12)    0.3878
(  13)    4.6327    (  14)    4.6421

     VOLTAGE SOURCE CURRENTS

     NAME        CURRENT

     VDD       -1.467D-03
     VSS        1.467D-03
     VFF        0.000D+00

     TOTAL POWER DISSIPATION   3.67D-03   WATTS        (continued on next page)
```

Figure 10.41 DC operating point computed after initialization.

CMOS DIFFERENTIAL AMPLIFIER

**** OPERATING POINT INFORMATION   TEMPERATURE = 27.000 DEG C

****

**** MOSFETS

	M1	M2	M3	M4	M5	M6	M7
MODEL	N	N	N	N	N	P	P
ID	2.35E-04	2.32E-04	2.35E-04	2.32E-04	4.67E-04	-2.35E-04	-2.32E-04
VGS	1.099	1.095	1.152	1.149	1.102	-1.313	-1.322
VDS	0.196	0.199	2.902	3.322	0.188	-1.346	-0.933
VBS	-0.188	-0.188	-0.385	-0.388	0.000	0.367	0.358
VTH	0.845	0.845	0.931	0.932	0.765	-0.859	-0.857
VDSAT	0.172	0.170	0.156	0.154	0.202	-0.306	-0.314
GM	1.19E-03	1.23E-03	1.82E-03	1.82E-03	1.37E-03	9.70E-04	9.25E-04
GDS	7.69E-04	6.84E-04	9.14E-06	8.52E-06	4.02E-03	3.30E-05	4.06E-05
GMB	5.37E-04	5.53E-04	7.46E-04	7.48E-04	6.88E-04	2.47E-04	2.36E-04
CBD	7.32E-13	7.31E-13	4.96E-13	4.81E-13	7.82E-13	5.79E-13	6.13E-13
CBS	7.65E-13	7.65E-13	7.23E-13	7.22E-13	8.09E-13	7.22E-13	7.24E-13
CGSOVL	2.40E-14	2.40E-14	2.40E-14	2.40E-14	4.80E-14	2.40E-14	2.40E-14
CGDOVL	2.40E-14	2.40E-14	2.40E-14	2.40E-14	4.80E-14	2.40E-14	2.40E-14
CGBOVL	3.00E-16	3.00E-16	3.00E-16	3.00E-16	3.00E-16	3.60E-16	3.60E-16
CGS	1.10E-13	1.10E-13	1.11E-13	1.11E-13	2.01E-13	1.33E-13	1.33E-13
CGD	1.20E-14	9.21E-15	0.00E+00	0.00E+00	1.13E-13	0.00E+00	0.00E+00
CGB	0.00E+00	0.00E+00	0.00E+00	0.00E+00	0.00E+00	0.00E+00	0.00E+00

	M8	M9	M11	M12	M13	M14
MODEL	P	P	P	N	N	N
ID	-2.35E-04	-2.32E-04	-2.00E-04	2.00E-04	2.00E-04	4.00E-04
VGS	-1.291	-1.291	-1.680	1.537	1.283	1.102
VDS	-0.367	-0.358	-1.680	1.537	1.283	1.102
VBS	0.000	0.000	0.000	0.000	0.000	0.000
VTH	-0.751	-0.751	-0.751	0.757	0.756	0.762
VDSAT	-0.348	-0.348	-0.677	0.540	0.361	0.207
GM	5.26E-04	5.08E-04	4.08E-04	4.76E-04	6.89E-04	2.16E-03
GDS	5.98E-04	6.23E-04	6.03E-06	1.13E-06	2.46E-06	2.55E-05
GMB	1.64E-04	1.59E-04	1.13E-04	2.14E-04	3.23E-04	1.04E-03
CBD	7.40E-13	7.42E-13	5.80E-13	5.90E 13	6.11E-13	6.31E-13
CBS	8.09E-13	8.09E-13	8.11E-13	8.18E-13	8.18E-13	8.11E-13
CGSOVL	2.40E-14	2.40E-14	2.40E-14	2.40E-14	2.40E-14	2.40E-14
CGDOVL	2.40E-14	2.40E-14	2.40E-14	2.40E-14	2.40E-14	2.40E-14
CGBOVL	3.60E-16	3.60E-16	1.44E-15	2.82E-15	1.38E-15	3.00E-16
CGS	1.29E-13	1.29E-13	5.30E-13	1.04E-12	5.08E-13	1.11E-13
CGD	3.87E-14	4.20E-14	0.00E+00	0.00E+00	0.00E+00	0.00E+00
CGB	0.00E+00	0.00E+00	0.00E+00	0.00E+00	0.00E+00	0.00E+00

Figure 10.41   *(continued)*

beyond; the probability for a circuit to converge beyond this number is low. Source ramping offers a better chance of success than increasing *ITL1* above the limit of 500 iterations.

Another approach that may help large circuits converge is to relax the tolerances **RELTOL** and **ABSTOL**. *RELTOL* can be increased to 5×10^{-3}, and *ABSTOL* can be made as high as six orders of magnitude below the highest current of a single device in the circuit. Once a solution is available, it can be used for initialization and the tolerances can be tightened back for more rigorous results.

If all the above attempts fail, a transient analysis with **UIC** and all supplies ramped up from zero should help. The time interval should be chosen so that elements with the slowest time constants can scttlc to steady-state values. The results can then be used either to find an operating point or to initialize another transient analysis.

10.6 SUMMARY

Based on the information on the solution algorithms of the previous chapter, a number of approaches and simulator option parameters for overcoming convergence failure have been presented in this chapter.

The following is the sequence of actions to undertake when SPICE fails to find the DC solution.

1. Carefully check the circuit connectivity and element values for possible specification errors and typographical errors; the **NODE**, **LIST** and **MODEL** options provide useful information on connectivity, element values, and model parameters, respectively. Note that only the **MODEL** option is turned on by default in most SPICE versions.

2. If the circuit specification is correct and SPICE does not converge, try first increasing the maximum number of iterations *ITL1* to 300 to 500. PSpice and SPICE3 automatically exercise the ramping methods if the main iterative process fails; SPICE2, however, needs to be directed to apply source ramping by setting *ITL6* to a value of 20 to 100, the maximum number of iterations allowed for any one set of source values. If automatic convergence algorithms fail, the number of iterations taken at each step should be increased; for example, increase *ITL2* in PSpice from the default of 20 to 40 or more.

3. Initialization of node voltages and cutoff semiconductor devices can be of help in finding a solution. For large circuits it can be useful to initialize all nodes once a solution is available and reduce the time for subsequent DC solutions.

4. The next step toward finding a solution is to add or delete certain physical effects of the model. Examples in this chapter have demonstrated the importance of finite output conductance in saturation and subthreshold conduction for MOS circuits. In other situations a simpler model can lead to better convergence; see Example 10.7.

5. Relaxation of the convergence tolerances, **RELTOL** and **ABSTOL**, can also lead to a solution. *ABSTOL* can easily be raised to 1 nA for MOSFETs. In general, *ABSTOL* should not be more than 9 orders of magnitude smaller than the largest current of a nonlinear device.

6. The protective parallel junction conductance, *GMIN*, can be increased 2 or 3 orders of magnitude.

7. A very reliable method for finding a DC solution is to run a transient analysis with the **UIC** option. All supplies should be ramped from zero to the final value for part of the interval and then held constant for the nodes to settle.

8. The solution of large-gain blocks connected in a feedback loop can be found by opening the loop and then using the results to initialize the closed-loop circuit.

Convergence failure can occur not only in a DC analysis but also in the time-domain solution. The following steps should be taken.

1. Check the circuit for charge storage, which insures a finite transition time from state to state. Also check the circuit for an abnormally high range of values for a single component type; the ratio of the largest to the smallest value of a component type should preferably be within nine orders of magnitude. See Sec. 9.2.2 for the underlying explanation.

2. Check model parameters that can produce unrealistic conductance values, such as **FC**.

3. Increase the number of iterations, *ITL4*, allowed at each time point to 40 or higher from the default of 10.

4. Relax the tolerances **ABSTOL** and **RELTOL**.

5. Use a different integration method if available. **METHOD=GEAR** with **MAXORD**=2 or **MAXORD**=3 is recommended.

6. Reduce the maximum allowed time step, *TMAX*.

7. Disable control of truncation error by setting *TRTOL* to a large value, such as 10^3. The time step in this situation is controlled by the iteration count; SPICE2 grants the option of setting **LVLTIM**=1, which is equivalent to setting *TRTOL* to a very large value.

8. Reduce, rather than relax, *RELTOL*; a value of 10^{-4} or smaller can force a smaller time step and avoid the bypass of a seemingly inactive device (see Sec. 9.3.2). Bypass is sometimes the culprit in a failed time-domain analysis.

If more than one SPICE simulator is available, it is useful to try to solve a difficult circuit with another version. Once a solution is available, it can be transferred as a **.NODESET** initialization to the simulator of choice, which is usually linked to a graphical postprocessing package for documentation purposes. Not many users have this luxury, however.

REFERENCES

Antognetti, P., and G. Massobrio. 1988. *Semiconductor Device Modeling with SPICE.* New York: McGraw-Hill.

Gray, P. R., and R. G. Meyer. 1985. *Analysis and Design of Analog Integrated Circuits.* New York: John Wiley & Sons.

Grove, A. S. 1967. *Physics and Technology of Semiconductor Devices.* New York: John Wiley & Sons.

Jorgensen, J. M. 1976. (October 5). CMOS Schmitt Trigger. *U.S. Patent 3,984,703.*

Kelessoglou, T., and D. O. Pederson. 1989. NECTAR—A knowledge-based environment to enhance SPICE. *IEEE Journal of Solid-State Circuits.* SC-24 (April) 452–457.

Muller, R. S., and T. I. Kamins. 1977. *Device Electronics for Integrated Circuits.* New York: John Wiley & Sons.

Pederson, D. O., and K. Mayaram. 1990. *Integrated Circuits for Communications.* Boston: Kluwer Academic.

Senderowicz, D. 1991. Personal communication.

Sheu, B. J., D. L. Scharfetter, and P. K. Ko 1985. SPICE2 implementation of BSIM. Univ. of California, Berkeley, ERL Memo UCB/ERL M85/42 (May).

Sze, S. M. 1981. *Physics of Semiconductor Devices.* New York: John Wiley & Sons.

Vladimirescu, A, and S. Liu. 1981. The simulation of MOS integrated circuits using SPICE2. Univ. of California, Berkeley, ERL Memo UCB/ERL M80/7 (March).

APPENDIX A

SEMICONDUCTOR-DEVICE MODELS

A.1 DIODE

Table A.1 Diode Model Parameters

Name	Parameter	Units	Default	Example	Scale Factor
IS	Saturation current	A	1×10^{-14}	1E-16	*area*
N	Emission coefficient	—	1	1.5	—
RS	Ohmic resistance	Ω	0	100	*1/area*
TT	Transit time	s	0	0.1N	—
CJO	Zero-bias junction capacitance	F	0	2P	*area*
VJ	Junction potential	V	1	0.6	—
M	Grading coefficient	—	0.5	0.33	—
EG	Activation energy	eV	1.11	1.11 Si 0.69 Sbd 0.67 Ge	—
XTI	I_S temperature exponent	—	3	3 pn 2 Sbd	—
BV	Breakdown voltage	V	∞	40	—
IBV	Current at breakdown voltage	A	1×10^{-3}	10U	*area*
FC	Coefficient for forward-biased depletion capacitance formula	—	0.5		—
KF	Flicker noise coefficient	—	0		—
AF	Flicker noise exponent	—	1		—

A.1.1 DC, Transient, and AC Models

All SPICE2 and SPICE3 model parameters are listed in Table A.1.

$$
I_D = \begin{cases}
IS(e^{qV_D/NkT} - 1) + V_D GMIN & \text{for } V_D \geq -5\dfrac{NkT}{q} \\[3mm]
-IS + V_D GMIN & \text{for } -BV < V_D - 5\dfrac{NkT}{q} \\[3mm]
-IS(e^{-q(BV+V_D)/kT} - 1) - IBV & \text{for } V_D < -BV
\end{cases}
\tag{A.1}
$$

$$
C_D = \begin{cases}
TT\dfrac{dI_D}{dV_D} + \dfrac{CJO}{(1 - V_D/VJO)^M} & \text{for } V_D < FC \cdot VJ \\[4mm]
TT\dfrac{dI_D}{dV_D} + \dfrac{CJO}{(1 - FC)^{1+M}}\left[1 - FC(1+M) + \dfrac{MV_D}{VJ}\right] & \text{for } V_D \geq FC \cdot VJ
\end{cases}
\tag{A.2}
$$

A.1.2 Temperature Effects

Model parameters *IS, VJ, CJO,* and *EG* vary with temperature according to the following functions of temperature, $I_s(T)$, $\phi_J(T)$, $C_J(T)$ and $E_g(T)$:

$$
I_S(T_2) = I_S(T_1)\left(\frac{T_2}{T_1}\right)^{XTI/N} \exp\left[-\frac{qEG(300)}{NkT_2}\left(1 - \frac{T_2}{T_1}\right)\right]
\tag{A.3}
$$

$$
\phi_J(T_2) = \frac{T_2}{T_1}\phi_J(T_1) - 2\frac{kT_2}{q}\ln\left(\frac{T_2}{T_1}\right)^{1.5} - \left[\frac{T_2}{T_1}E_g(T_1) - E_g(T_2)\right]
\tag{A.4}
$$

$$
E_g(T) = E_g(0) - \frac{(7.02 \times 10^{-4}\ \text{eV/K})T^2}{(1108\ \text{K}) + T}
\tag{A.5}
$$

$$
C_J(T_2) = C_J(T_1)\left\{1 + M\left[400 \times 10^{-6}(T_2 - T_1) - \frac{\phi_J(T_2) - \phi_J(T_1)}{\phi_J(T_1)}\right]\right\}
\tag{A.6}
$$

Model parameters are assumed to be specified at the reference temperature, T_1, equal to *TNOM*, which defaults to 300 K. *TNOM* can be modified with a **.OPTIONS** statement.

A.1.3 Noise Model

The three noise contributions are due to the parasitic series resistance, *RS*, and the shot and flicker noise of the pn junction:

$$\overline{i_{RS}^2} = \frac{4kT\Delta f}{RS}$$

$$\overline{i_d^2} = 2qI_D\Delta f + \frac{KFI_D^{AF}}{f}\Delta f \tag{A.7}$$

A.2 BIPOLAR JUNCTION TRANSISTOR

Table A.2 BJT Model Parameters

Name	Parameter	Units	Default	Example	Scale Factor
IS	Saturation current	A	1×10^{-16}	1E-16	*area*
BF	Forward current gain	—	100	80	—
NF	Forward emission coefficient	—	1	2	—
VAF	Forward Early voltage	V	∞	100	—
IKF	β_F high-current roll-off corner	A	∞	0.1	*area*
ISE	BE junction leakage current	A	0	1E-13	*area*
NE	BE junction leakage emission coefficient	—	1.5	2	—
BR	Reverse current gain	—	1	3	—
NR	Reverse emission coefficient	—	1	1.5	—
VAR	Reverse Early voltage	V	∞	250	—
IKR	β_R high-current roll-off corner	A	∞	0.1	*area*
ISC	BC junction leakage current	A	0	1E-13	*area*
NC	BC junction leakage emission coefficient	—	1.5	2	—
RC	Collector resistance	Ω	0	200	1/*area*
RE	Emitter resistance	Ω	0	2	1/*area*
RB	Zero-bias base resistance	Ω	0	100	1/*area*
RBM	Minimum base resistance at high current	Ω	*RB*	10	1/*area*
IRB	Current where base resistance falls halfway to its minimum value	A	∞	0.1	*area*
TF	Forward transit time	s	0	1N	—
XTF	Coefficient for bias dependence of τ_F	—	0		—
VTF	Voltage for τ_F dependence on V_{BC}	V	∞		—
ITF	Current where $\tau_F = f(I_C, V_{BC})$ starts	A	0		*area*
PTF	Excess phase at $f = 1/(2\pi TF)$	degrees	0		—
TR	Reverse transit time	s	0	100N	—
CJE	BE zero-bias junction capacitance	F	0	2P	*area*
VJE	BE built-in potential	V	0.75	0.6	—
MJE	BE grading coefficient	—	0.33	0.33	—
CJC	BC zero-bias junction capacitance	F	0	2P	*area*
VJC	BC built-in potential	V	0.75	0.6	—
MJC	BC grading coefficient	—	0.33	0.5	—

Table A.2 *(continued)*

Name	Parameter	Units	Default	Example	Scale Factor
XCJC	Fraction of C_{JC} connected at internal base node B'	—	1	0.5	—
CJS	CS zero-bias junction capacitance	F	0	2P	*area*
VJS	CS built-in potential	V	0.75	0.6	—
MJS	CS grading coefficient	—	0	0.5	—
EG	Activation energy	eV	1.11	1.11 Si	—
XTI	I_S temperature exponent	—	3		—
XTB	β_F and β_R temperature exponent	—	0		—
FC	Coefficient for forward-biased depletion capacitance formula	—	0.5		—
KF	Flicker noise coefficient	—	0		—
AF	Flicker noise exponent	—	1		—

A.2.1 DC Model

$$I_C = \frac{IS}{q_b}\left(e^{qV_{BE}/NFkT} - e^{qV_{BC}/NRkT}\right) - \frac{IS}{BR}\left(e^{qV_{BC}/NRkT} - 1\right)$$
$$- ISC\left(e^{qV_{BC}/NCkT} - 1\right) \qquad (A.8)$$

$$I_B = \frac{IS}{BF}\left(e^{qV_{BE}/NFkT} - 1\right) + \frac{IS}{BR}\left(e^{qV_{BC}/NRkT} - 1\right)$$
$$+ ISE\left(e^{qV_{BE}/NEkT} - 1\right) + ISC\left(e^{qV_{BC}/NCkT} - 1\right) \quad (A.9)$$

where q_b is defined by

$$q_b = \frac{q_1}{2}\left(1 + \sqrt{1 + 4q_2}\right)$$
$$q_1 = \left(1 - \frac{V_{BC}}{VAF} - \frac{V_{BE}}{VAR}\right)^{-1} \qquad (A.10)$$
$$q_2 = \frac{IS}{IKF}\left(e^{qV_{BE}/NFkT} - 1\right) + \frac{IS}{IKR}\left(e^{qV_{BC}/NRkT} - 1\right)$$

Older versions of SPICE use parameters **C2** and **C4**, which correspond to the junction leakage currents **ISE** and **ISC**:

$$ISE = C2 \cdot IS$$
$$ISC = C4 \cdot IS$$

The effective series base resistance, $R_{BB'}$ is

$$
R_{BB'} = \begin{cases} RBM + \dfrac{RB - RBM}{q_b} & \text{if } IRB \text{ is not specified} \\[2ex] RBM + 3(RB - RBM)\dfrac{\tan z - z}{z \tan^2 z} & \text{if } IRB \text{ is specified} \end{cases}
$$

$(A.11)$

where

$$
z = \frac{-1 + \sqrt{1 + 1.44 I_B / \pi^2 IRB}}{24 / \pi^2 \sqrt{I_B / IRB}}
$$

A.2.2   Transient and AC Models

$$
\begin{aligned}
C_{BE} &= C_{DE} + C_{JE} \\
C_{BC} &= C_{DC} + C_{JB'C} \\
C_{CS} &= C_{JS}
\end{aligned}
$$

$(A.12)$

Diffusion capacitances are implemented as

$$
\begin{aligned}
C_{DE} &= \frac{\partial}{\partial V_{BE}}\left[\tau_F \frac{IS}{q_b}\left(e^{q V_{BE}/NFkT} - 1 \right) \right] \\[2ex]
C_{DC} &= TR \frac{qIS}{NRkT} e^{q V_{BC}/NRkT}
\end{aligned}
$$

$(A.13)$

where

$$
\tau_F = TF\left[1 + XTF e^{V_{BC}/1.44 VTF}\left(\frac{I_{CC}}{I_{CC} + ITF} \right)^2 \right]
$$

and

$$
I_{CC} = IS\left(e^{q V_{BE}/NFkT} - 1 \right)
$$

$(A.14)$

Junction capacitances are defined by

$$
C_J = \frac{\partial Q_J}{\partial V_J}\bigg|_{E,C,S} = \begin{cases} \dfrac{C_J}{(1 - V_J/\phi_J)^m} & \text{for } V_J < FC\phi_J \\[2ex] \dfrac{C_J}{(1 - FC)^m}\left[1 + \dfrac{m}{\phi_J(1 - FC)}(V_J - FC\phi_J) \right] & \\[1ex] & \text{for } V_J \geq FC\phi_J \end{cases}
$$

$(A.15)$

The BC junction capacitance has two components, one connected to the external and one to the internal base node:

$$C_{JB'C} = XJC \cdot C_{JC}$$
$$C_{JBC} = (1 - XJC)C_{JC} \tag{A.16}$$

At high frequencies a phase shift equal to

$$\theta = \omega PTF \cdot TF \tag{A.17}$$

is applied to the phasor $\mathbf{I}_c$:

$$\mathbf{I}_c = I_c e^{j\theta}$$

This effect is also present in the time-domain expression of $i_c(t)$.

A.2.3 Temperature Effects

The following quantities are adjusted for temperature variation:

$$I_S(T_2) = I_S(T_1)\left(\frac{T_2}{T_1}\right)^{XTI} \exp\left[-\frac{qEG(300)}{kT_2}\left(1 - \frac{T_2}{T_1}\right)\right] \tag{A.18}$$

$$\beta(T_2) = \beta(T_1)\left(\frac{T_2}{T_1}\right)^{XTB} \tag{A.19}$$

$$I_{SE}(T_2) = I_{SE}(T_1)\left(\frac{T_2}{T_1}\right)^{-XTB}\left[\frac{I_S(T_2)}{I_S(T_1)}\right]^{1/NE}$$
$$I_{SC}(T_2) = I_{SC}(T_1)\left(\frac{T_2}{T_1}\right)^{-XTB}\left[\frac{I_S(T_2)}{I_S(T_1)}\right]^{1/NC} \tag{A.20}$$

The temperature dependence of E_g, ϕ_J, and C_J for the diode is implemented as determined by Eqs. $A.4$ to $A.6$.

A.2.4 Noise Model

Noise is modeled as thermal noise for the parasitic series resistances and as shot and flicker noise for i_C and i_B:

$$\overline{i_R^2} = \frac{4kT\Delta f}{R} \tag{A.21}$$

$$\overline{i_b^2} = 2qI_B\Delta f + \frac{KF \cdot I_B^{AF}}{f}\Delta f \tag{A.22}$$

$$\overline{i_c^2} = 2qI_C\Delta f \tag{A.23}$$

A.3 MOSFET

Table A.3 MOSFET Model Parameters

Name	Parameter	Units	Default	Example	Scale Factor
LEVEL	Model index	—	1		
VTO	Threshold voltage	V	0	1.0	—
KP	Transconductance parameter	A/V^2	2×10^{-5}	1.0E-3	—
GAMMA	Bulk threshold parameter	$V^{1/2}$	0	0.5	—
PHI	Surface potential	V	0.6	0.7	—
LAMBDA	Channel length modulation parameter	V^{-1}	0	1.0E-4	—
RD	Drain ohmic resistance	Ω	0	10	—
RS	Source ohmic resistance	Ω	0	10	—
RSH	D and S diffusion sheet resistance	Ω	0	10	*NRD* *NRS*
CBD	Zero-bias BD junction capacitance	F	0	5P	—
CBS	Zero-bias BS junction capacitance	F	0	1P	—
CJ	Zero-bias bulk junction bottom capacitance	F/m^2	0	2.0E-4	*AD* *AS*
MJ	Bulk junction grading coefficient	—	0.5	0.5	—
CJSW	Zero-bias bulk junction sidewall capacitance	F/m	0	1.0E-9	*PD* *PS*
MJSW	Bulk junction grading coefficient	—	0.33	0.25	—
PB	Bulk junction potential	V	1	0.6	—
IS	Bulk junction saturation current	A	1×10^{-14}	1.0E-16	—
JS	Bulk junction saturation current per junction area	A/m^2	0	1.0E-8	*AD* *AS*
CGDO	GD overlap capacitance per unit channel width	F/m	0	4.0E-11	*W*
CGSO	GS overlap capacitance per unit channel width	F/m	0	4.0E-11	*W*
CGBO	GB overlap capacitance per unit channel length	F/m	0	2.0E-10	*L*
NSUB	Substrate doping	cm^{-3}	0	4.0E15	—
NSS	Surface state density	cm^{-2}	0	1.0E10	—
NFS	Fast surface state density	cm^{-2}	0	1.0E10	
TOX	Thin-oxide thickness	m	∞	0.1U	—
TPG	Type of gate material +1 opposite to substrate −1 same as substrate 0 Al gate	—	1		
XJ	Metallurgical junction depth	m	0	0.5U	—
LD	Lateral diffusion	m	0	0.2U	—
UO	Surface mobility	$cm^2/V \cdot s$	600	700	—

Table A.3 *(continued)*

Name	Parameter	Units	Default	Example	Scale Factor
UCRIT	Critical field for mobility degradation (`LEVEL=2`)	V/cm	1×10^4	1.0E4	—
UEXP	Critical field exponent in mobility degradation (`LEVEL=2`)	—	0	0.1	—
VMAX	Maximum drift velocity of carriers	m/s	0	5.0E4	—
NEFF	Total channel charge (fixed and mobile) coefficient (`LEVEL=2`)	—	1.0	5.0	—
XQC	Thin-oxide capacitance model flag and channel charge share for drain coefficient (0–0.5)	—	1	0.4	—
DELTA	Width effect on threshold voltage	—	0	1.0	—
THETA	Mobility modulation (`LEVEL=3`)	V^{-1}	0	0.1	—
ETA	Static feedback coefficient (`LEVEL=3`)	—	0	1.0	—
KAPPA	Saturation field factor (`LEVEL=3`)	—	0.2	0.5	—

A.3.1 DC Model

The **LEVEL** parameter differentiates and sets the analytical models describing the behavior of a MOSFET. There are 3 levels in SPICE2 and 6 in SPICE3. Only the equations for `LEVEL=2` and `LEVEL=3` are listed; these two models together with the `LEVEL=1` model, described in Chap. 3, are available in all SPICE versions.

`LEVEL=2`:

$$
I_{DS} = \beta \left\{ \left(V_{GS} - V_{BIN} - \frac{\eta V_{DS}}{2} \right) V_{DS} \right.
$$
$$
\left. - \frac{2}{3} \gamma_S \left[(PHI + V_{DS} - V_{BS})^{3/2} - (PHI - V_{BS})^{3/2} \right] \right\} \tag{A.24}
$$

where the transconductance and mobility factors are defined by

$$
\beta = \mu_s C_{ox} \frac{W}{L_{eff} - \Delta L} \tag{A.25}
$$

$$
\mu_s = UO \left[\frac{UCRIT \epsilon_{Si}}{C_{ox}(V_{GS} - V_{TH})} \right]^{UEXP} \tag{A.26}
$$

The built-in voltage including small-size effects is

$$
V_{BIN} = V_{FB} + PHI + DELTA \frac{\pi \epsilon_{Si}}{4 C_{ox} W} (PHI - V_{BS}) \tag{A.27}
$$

where the narrow-channel effect is represented by:

$$
\eta = 1 + DELTA \frac{\pi \epsilon_{Si}}{4 C_{ox} W}
$$

The adjusted and zero-bias threshold voltages are

$$V_{TH} = V_{BIN} + \gamma_S \sqrt{PHI - V_{BS}} \tag{A.28}$$

$$VTO = V_{FB} + PHI - GAMMA \sqrt{PHI} \tag{A.29}$$

where

$$\gamma_S = GAMMA(1 - \alpha_S - \alpha_D)$$

$$\alpha_S = \frac{1}{2} \frac{XJ}{L} \left(\sqrt{1 + 2\frac{W_S}{XJ}} - 1 \right)$$

$$\alpha_D = \frac{1}{2} \frac{XJ}{L} \left(\sqrt{1 + 2\frac{W_D}{XJ}} - 1 \right) \tag{A.30}$$

$$W_S = X_d \sqrt{PHI - V_{BS}}$$

$$W_D = X_d \sqrt{PHI - V_{BS} + V_{DS}}$$

$$X_d = \sqrt{\frac{2\epsilon_{Si}}{q \cdot NSUB}}$$

The geometric channel length, **L**, is adjusted due to lateral diffusion, *LD*, and channel shortening in saturation:

$$L_{eff} = L - 2LD \tag{A.31}$$

$$\Delta L = \begin{cases} L \cdot LAMBDA \cdot V_{DS} & \text{if } LAMBDA \text{ is specified} \\ f(V_{DSAT}) & \text{if } LAMBDA \text{ is not specified} \end{cases} \tag{A.32}$$

The pinch-off saturation voltage is defined by

$$V_{DSAT} = \frac{(V_{GS} - V_{BIN})}{\eta}$$

$$+ \frac{1}{2} \left(\frac{\gamma_S}{\eta} \right)^2 \left\{ 1 - \left[1 + 4 \left(\frac{\eta}{\gamma_S} \right)^2 \left(\frac{V_{GS} - V_{BIN}}{\eta} + PHI - V_{BS} \right) \right]^{1/2} \right\} \tag{A.33}$$

$$\Delta L = X_d \left[\frac{V_{DS} - V_{DSAT}}{4} + \sqrt{1 + \left(\frac{V_{DS} - V_{DSAT}}{4} \right)^2} \right]^{1/2} \tag{A.34}$$

The velocity-saturation-based model derives V_{DSAT} and ΔL from the following equations:

$$I_{DSAT} - VMAX \cdot W \cdot Q_{chan}(L) = 0 \qquad (A.35)$$

$$Q_{chan} = C_{ox}\left[V_{GS} - V_{BIN} - \eta V_{DSAT} - \gamma_S(PHI - V_{BS} + V_{DSAT})^{1/2}\right] \qquad (A.36)$$

$$\Delta L = X_d\left[\sqrt{\left(\frac{X_d VMAX}{2\mu_s}\right)^2 + (V_{DS} - V_{DSAT})} + \frac{X_d VMAX}{2\mu_s}\right] \qquad (A.37)$$

The channel shortening ΔL and therefore the output conductance in saturation can be adjusted using the total channel charge coefficient **NEFF** which multiplies **NSUB** in the expression of of X_d, Eq. A.30. This saturation model is used only when parameter **VMAX** is specified. Subthreshold conduction is modeled when **NFS** is present:

$$
\begin{aligned}
I_{DS} = \beta \Bigg\{ &\left(V_{ON} - V_{BIN} - \frac{\eta V_{DS}}{2}\right)V_{DS} \\
&-\frac{2}{3}\gamma_S\left[(PHI + V_{DS} - V_{BS})^{3/2} - (PHI - V_{BS})^{3/2}\right]\Bigg\} e^{q(V_{GS}-V_{ON})/nkT}
\end{aligned} \qquad (A.38)
$$

where

$$
\begin{aligned}
V_{ON} &= V_{TH} + \frac{nkT}{q} \\
n &= 1 + \frac{C_{fs}}{C_{ox}} + \frac{C_d}{C_{ox}} \\
C_{fs} &= q \cdot NFS \\
C_d &= \frac{\partial Q_B}{\partial V_{BS}} \\
&= \left(-\gamma_S \frac{d}{dV_{BS}}\sqrt{PHI - V_{BS}} - \frac{\partial \gamma_S}{\partial V_{BS}}\sqrt{PHI - V_{BS}} + DELTA\frac{\pi\epsilon_{Si}}{4C_{ox}W}\right)C_{ox}
\end{aligned} \qquad (A.39)
$$

LEVEL=3:

$$I_{DS} = \beta\left(V_{GS} - V_{TH} - \frac{1 + F_B}{2}V_{DS}\right)V_{DS} \qquad (A.40)$$

where

$$\beta = \mu_{eff}C_{ox}\frac{W}{L_{eff}} \qquad (A.41)$$

$$\mu_s = \frac{UO}{1 + THETA(V_{GS} - V_{TH})} \tag{A.42}$$

$$\mu_{eff} = \frac{\mu_s}{1 + \dfrac{\mu_s}{VMAX \cdot L_{eff}} V_{DS}} \tag{A.43}$$

The threshold voltage is defined by

$$V_{TH} = V_{FB} + PHI - \sigma V_{DS} + \gamma F_S \sqrt{PHI - V_{BS}} + F_N(PHI - V_{BS}) \tag{A.44}$$

where

$$\sigma = ETA \frac{8.15 \times 10^{-22}}{C_{ox} L_{eff}^3} \tag{A.45}$$

$$F_S = 1 - \frac{XJ}{L_{eff}} \left[\frac{LD + W_C}{XJ} \sqrt{1 - \left(\frac{W_P/XJ}{1 + W_P/XJ} \right)^2} - \frac{LD}{XJ} \right] \tag{A.46}$$

$$F_N = DELTA \frac{\pi \epsilon_{Si}}{2 C_{ox} W} \tag{A.47}$$

$$F_B = \frac{\gamma F_S}{4 \sqrt{PHI - V_{BS}}} + F_N \tag{A.48}$$

$$W_P = X_d \sqrt{PHI - V_{BS}}$$

$$\frac{W_C}{XJ} = d_0 + d_1 \frac{W_P}{XJ} + d_2 \left(\frac{W_P}{XJ} \right)^2 \tag{A.49}$$

$$d_0 = 0.0631353, \qquad d_1 = 0.8013292, \qquad d_2 = -0.01110777$$

The saturation model is based on velocity-limited carrier flow:

$$V_{DSAT} = \frac{V_{GS} - V_{TH}}{1 + F_B} + \frac{VMAX \cdot L_{eff}}{\mu_s} - \sqrt{\left(\frac{V_{GS} - V_{TH}}{1 + F_B} \right)^2 + \left(\frac{VMAX \cdot L_{eff}}{\mu_s} \right)^2} \tag{A.50}$$

$$\Delta L = X_d \left[\sqrt{\left(\frac{E_P X_d}{2} \right)^2 + KAPPA(V_{DS} - V_{DSAT})} - \frac{E_P X_d}{2} \right] \tag{A.51}$$

where

$$E_P = \frac{I_{DSAT}}{G_{DSAT} L_{eff}}$$

For the BSIM and BSIM2 models, **LEVEL**=4 and **LEVEL**=5 in SPICE3, see the University of California at Berkeley research reports (Jeng 1990; Sheu, Scharfetter, and Ko 1985).

A.3.2 Transient and AC Models

The gate capacitances defined by the Meyer model, C_{GS}, C_{GD}, and C_{GB}, shown graphically as a function of V_{GS} in Figure 3.17, are listed below for the three main regions of operation of a MOSFET.

In the cutoff region, $V_{GS} \le V_{TH}$, all three capacitances are constant:

$$
\begin{aligned}
C_{GB} &= C_{OX} + CGBO \cdot L_{eff} \\
C_{GS} &= CGSO \cdot W \\
C_{GD} &= CGDO \cdot W
\end{aligned}
\qquad (A.52)
$$

In saturation, $V_{TH} < V_{GS} \le V_{TH} + V_{DS}$, the expressions are

$$
\begin{aligned}
C_{GB} &= CGBO \cdot L_{eff} \\
C_{GS} &= \frac{2}{3}C_{OX} + CGSO \cdot W \\
C_{GD} &= CGDO \cdot W
\end{aligned}
\qquad (A.53)
$$

In linear operation, $V_{GS} > V_{TH} + V_{DS}$,

$$
\begin{aligned}
C_{GB} &= CGBO \cdot L_{eff} \\
C_{GS} &= C_{OX}\left\{1 - \left[\frac{V_{GS} - V_{DS} - V_{ON}}{2(V_{GS} - V_{ON}) - V_{DS}}\right]^2\right\} + CGSO \cdot W \\
C_{GD} &= C_{OX}\left\{1 - \left[\frac{V_{GS} - V_{ON}}{2(V_{GS} - V_{ON}) - V_{DS}}\right]^2\right\} + CGDO \cdot W
\end{aligned}
\qquad (A.54)
$$

where

$$
\begin{aligned}
C_{OX} &= C_{ox}W \cdot L_{eff} \\
C_{ox} &= \frac{\epsilon_{ox}\epsilon_0}{TOX}
\end{aligned}
$$

The charge conservation model derives asymmetrical capacitances according to the following definitions:

$$
Q_{chan} = Q_D + Q_S = -(Q_G + Q_B) \qquad (A.55)
$$

$$Q_D = XQC \cdot Q_{chan} \tag{A.56}$$

$$C_{xy} = \frac{\partial Q_x}{\partial V_y} \neq \frac{\partial Q_y}{\partial V_x} = C_{yx} \tag{A.57}$$

A.3.3   Temperature Model

In addition to I_S, ϕ_J, and C_J which have the temperature dependence described for the diode, the intrinsic concentration n_i and the mobility are adjusted for temperature:

$$n_i(T) = n_i(300)\left(\frac{T}{300}\right)^{1.5} \exp\left[\frac{q}{2k}\left(\frac{1.16\,\text{eV}}{300\,\text{K}} - \frac{E_g}{T}\right)\right] \tag{A.58}$$

$$n_i(300) = 1.45 \times 10^{10}\ \text{m}^{-3}$$

$$\mu(T) = \mu(300)\left(\frac{300}{T}\right)^{1.5} \tag{A.59}$$

A.3.4   Noise Model

The noise contributed by the drain-source current is

$$\overline{i_{ds}^2} = \frac{8kTg_m}{3}\Delta f + \frac{KF \cdot I_{DS}^{AF}}{f\,C_{ox}L_{eff}^2}\Delta f \tag{A.60}$$

REFERENCES

Jeng, M.-C. 1990. Design and modeling of deep-submicrometer MOSFETs. University of California, Berkeley, ERL Memo UCB/ERL M90/90 (October).

Sheu, B. J., D. L. Scharfetter, and P. K. Ko. 1985. SPICE2 implementation of BSIM model. University of California, Berkeley, ERL Memo UCB/ERL M85/42 (May).

APPENDIX B

ERROR MESSAGES

A large percentage of aborted simulation runs are due to erroneous input specifications. This section enumerates all the SPICE2 error and warning messages related to input specifications. Other SPICE programs flag the same problems as the ones listed below, but the wording may differ.

B.1 GENERAL SYNTAX ERRORS

ERROR: UNRECOGNIZABLE DATA CARD
This message follows an input statement that starts with a number in the first field. This error may arise when the continuation character, +, is omitted in the first column of a continuation statement. This error is fatal.

ERROR: UNKNOWN DATA CARD: *Name*
Name, that is, the first field of the statement, does not start with any of the accepted key characters in the first column. An example is the occurrence of two title statements in the same circuit specification. This error is fatal.

ERROR: ELEMENT TYPE NOT YET IMPLEMENTED
This error message should hardly ever occur, since it duplicates the previous message. It is an added protection that ensures that additional element types are implemented correctly at future times. This error is fatal.

ERROR: NEGATIVE NODE NUMBER FOUND
This statement is printed immediately following an input statement that contains a negative node number. This error is fatal.

`*ERROR*: NODE NUMBERS ARE MISSING`

This error message is printed immediately following an input statement that does not contain the correct number of nodes for a particular element type. This error message would occur if, for example, only three nodes are specified for a MOSFET; even if all bulk terminals are connected to ground, the fourth node must be defined for a MOSFET. This error is fatal.

`*ERROR*: .END CARD MISSING`

`*ERROR*: ILLEGAL NUMBER--SCAN STOPPED AT COLUMN` *number*

A number with an absolute value outside the interval from 10^{-35} to 10^{35} has been specified.

B.2   MULTITERMINAL ELEMENT ERRORS

`*ERROR*: MUTUAL INDUCTANCE REFERENCES ARE MISSING`

A mutual inductance name must be followed by two inductor names starting with the letter *L*. This error is fatal.

`WARNING: COEFFICIENT OF COUPLING RESET TO 1.0`

A coupling coefficient in excess of 1 must have been specified in the above statement. The analysis continues with a coupling coefficient of 1.

`*ERROR*: Z0 MUST BE SPECIFIED`

A value must be specified for the characteristic impedance of a transmission line. This error is fatal.

`*ERROR*: EITHER TD OR F MUST BE SPECIFIED`

A value must be specified for either the delay time, *TD*, or the frequency, *F*, of a transmission line. This error is fatal.

B.3   SOURCE SPECIFICATION ERRORS

`*ERROR*: VOLTAGE SOURCE NOT FOUND ON ABOVE LINE`

A current-controlled source requires the name of the voltage source through which the controlling current flows to be specified. The voltage source name must follow the node numbers of the source. This error is fatal.

`*ERROR*: UNKNOWN SOURCE FUNCTION:` *source_function*

A transient source function is specified. Check the types and abbreviations for transient source functions. This error is fatal.

ERROR: ELEMENT *Name* **PIECEWISE LINEAR SOURCE TABLE NOT INCREASING IN TIME**
The time values of the (t_i, V_i) coordinates must be in increasing order.

B.4   ELEMENT, SEMICONDUCTOR-DEVICE, AND MODEL ERRORS

ERROR: VALUE IS MISSING OR IS NONPOSITIVE
This error message can follow an element definition statement that is expected to contain a value. The following elements belong in this category: resistors, capacitors, inductors, mutual inductors, and controlled sources. Each of these elements must be accompanied by a positive value. This message may be encountered also following a semiconductor device definition statement that contains negative geometry parameters, such as *area, W,* or *L*. This error is fatal.

ERROR: VALUE IS ZERO
This message follows a zero-valued resistor. The solution of nodal circuit equations in SPICE precludes the use of zero-valued resistors, which lead to infinite conductances. For convergence reasons it is not advisable to use resistor values less than 1 mΩ. This error is fatal.

ERROR: UNKNOWN PARAMETER: *Name*
A parameter name used in an element statement is not valid. Check element statement for valid parameter names. This error is fatal.

ERROR: EXTRA NUMERICAL DATA ON MOSFET CARD
A MOSFET can have up to eight device parameters. Up to eight values can follow the model name, which are assigned in order to *L, W, AD, AS, PD, PS, NRD,* and *NRS*. The number of values in the above line exceeds the maximum number of parameter values. This error is fatal.

ERROR: MODEL NAME IS MISSING
A model name is expected to follow the node specification on diode, BJT, JFET, and MOSFET statements. This error is fatal.

ERROR: MODEL TYPE IS MISSING
Every **.MODEL** statement must contain a model type. This error is fatal.

ERROR: UNKNOWN MODEL TYPE: *model_type*
A model type was specified in the above statement that is not one of the eight types recognized by SPICE2. This error is fatal.

ERROR: UNKNOWN MODEL PARAMETER: *Name*
The above parameter name is not supported by the model. This error is fatal.

WARNING: MINIMUM BASE RESISTANCE (RBM) IS LESS THAN TOTAL (RB)FOR MODEL *MODname* RBM SET EQUAL TO RB

WARNING: THE VALUE OF LAMBDA FOR MOSFET MODEL, *MODname*, IS UNUSUALLY LARGE AND MIGHT CAUSE NONCONVERGENCE

WARNING: IN DIODE MODEL *MODname* IBV INCREASED TO *value* TO RESOLVE INCOMPATIBILITY WITH SPECIFIED IS

WARNING: UNABLE TO MATCH FORWARD AND REVERSE DIODE REGIONS, BV = *value* AND IBV = *value*

B.5 CIRCUIT TOPOLOGY ERRORS

ERROR: CIRCUIT HAS NO NODES
The circuit needs to contain at least one other node than ground.

ERROR: LESS THAN 2 CONNECTIONS AT NODE *number*
At least two elements must be connected at any node.

ERROR: NO DC PATH TO GROUND FROM NODE *number*
From every node there must be a path to ground in order to find a DC solution.

ERROR: INDUCTOR/VOLTAGE SOURCE LOOP FOUND, CONTAINING *Vname*
Such a loop would contradict Kirchhoff's voltage law.

WARNING: ATTEMPT TO REFERENCE UNDEFINED NODE *number*—NODE RESET TO 0

B.6 SUBCIRCUIT DEFINITION ERRORS

ERROR: SUBCIRCUIT DEFINITION DUPLICATES NODE *number*
Two terminals (nodes) on the subcircuit definition line have the same number.

ERROR: NONPOSITIVE NODE NUMBER FOUND IN SUBCIRCUIT DEFINITION
All node numbers must be positive numbers.

ERROR: SUBCIRCUIT NAME MISSING
A name starting with a character must follow the word .SUBCKT on a subcircuit definition line or the node numbers on an **X** element, subcircuit instantiation, line.

ERROR: SUBCIRCUIT NODES MISSING
Node numbers are expected to follow the subcircuit name on a SPICE2 subcircuit definition line.

ERROR: UNKNOWN SUBCIRCUIT NAME: *SUBname*
A **.SUBCKT** definition for *SUBname* referenced by an **X** element cannot be found in the circuit deck.

ERROR: .ENDS CARD MISSING
The end of the circuit deck, **.END** line, has been encountered before all **.SUBCKT** lines have been matched by **.ENDS** lines; every subcircuit definition must be completed with an **.ENDS** line.

ERROR: *Xname* **HAS DIFFERENT NUMBER OF NODES THAN** *SUBname*
The number of nodes on an **X** line must match the number of nodes on the subcircuit definition line it references.

ERROR: SUBCIRCUIT *SUBname* **IS DEFINED RECURSIVELY**
A subcircuit definition contains an **X** element that references the subcircuit being defined.

WARNING: ABOVE LINE NOT ALLOWED WITHIN SUBCIRCUIT--IGNORED

WARNING: NO SUBCIRCUIT DEFINITION KNOWN--LINE IGNORED

B.7 ANALYSIS ERRORS

ERROR: MAXIMUM ENTRY IN THIS COLUMN AT STEP *number* *value* **IS LESS THAN PIVTOL**
Circuit matrix is singular; see Sec. 10.2.1 for detailed examples.

ERROR: NO CONVERGENCE IN DC ANALYSIS. LAST NODE VOLTAGES: *list*
Failure to find a DC solution; see Chap. 10 for advice on overcoming convergence problems.

ERROR: NO CONVERGENCE IN DC TRANSFER CURVES AT *Name = value* **LAST NODE VOLTAGES:** *list*
Solution failure during a **.DC** analysis at a specific *value* of the variable *Name;* in SPICE2 increase *ITL2* and consult Example 10.8.

ERROR: INTERNAL TIMESTEP TOO SMALL IN TRANSIENT ANALYSIS
The smallest acceptable time step has been reached after repetitively cutting the time step without converging to a solution; see Chaps. 9 and 10 for more insight.

ERROR: TRANSIENT ANALYSIS ITERATIONS EXCEED LIMIT OF *number*
THIS LIMIT MAY BE OVERRIDDEN USING THE ITL5 PARAMETER ON THE
.OPTION CARD
The transient analysis is stopped after a preset number of iterations, equal to 5000 in SPICE2, to give the user the opportunity to judge the correctness of results and whether analysis should be continued; set $ITL5 = 0$ to remove this limit. Many SPICE programs do not have this built-in limit, and when interactivity is available, such as in SPICE3, the user can monitor the correctness of the solution.

ERROR: MEMORY REQUIREMENT EXCEEDS MACHINE CAPACITY MEMORY
NEEDS EXCEED *value1, value2*
The analysis of the circuit requires more internal memory than available.

ERROR: TEMPERATURE SWEEP SHOULD BE THE SECOND SWEEP
SOURCE, CHANGE THE ORDER AND RE-EXECUTE
In a **.DC** statement the temperature variable must always be the second variable if another sweep variable is specified.

WARNING: UNDERFLOW OCCURRED *number* **TIME(S)**

WARNING: UNDERFLOW *number* **TIME(S) IN AC ANALYSIS AT FREQ**
freq **HZ**

WARNING: UNDERFLOW *number* **TIME(S) IN DISTORTION ANALYSIS AT**
FREQ *freq* **HZ**
A smaller number than can be represented on the computer was generated during a **.DC** **.TRAN**, **.AC**, or **.DISTO** analysis. SPICE has trapped and minimized the effect of the problem; note that in IEEE format, floating-point arithmetic computation may continue after an underflow condition leading to the creation and proliferation of out-of-range or NaNs (not-a-number), quantities.

WARNING: MORE THAN *number* **POINTS FOR** *Name* **ANALYSIS, ANALYSIS**
OMITTED. THIS LIMIT MAY BE OVERRIDDEN USING THE LIMPTS
PARAMETER ON THE .OPTION CARD
Some SPICE programs limit the number of print/plot points that can be output by **.DC**, **.TRAN**, or **.AC** analysis. Use the **LIMPTS** options parameter to override this limit.

WARNING: NO *Name* **OUTPUTS SPECIFIED...ANALYSIS OMITTED**
In SPICE2 a **.DC**, **.TRAN**, or **.AC** analysis must be accompanied by a **.PLOT** or **.PRINT** request.

WARNING: TOO FEW POINTS FOR PLOTTING
Too few points have been computed or requested; the line-printer plot has been omitted.

`WARNING: MISSING PARAMETER(S)...ANALYSIS OMITTED`
Incorrect `.DC` specification.

`WARNING: UNKNOWN FREQUENCY FUNCTION:` *Name*`...ANALYSIS OMITTED`
Frequency variation is limited to `LIN`, `OCT`, or `DEC`; see Chap. 5.

`WARNING: FREQUENCY PARAMETERS INCORRECT...ANALYSIS OMITTED`

`WARNING: START FREQ > STOP FREQ...ANALYSIS OMITTED`

`WARNING: TIME PARAMETERS INCORRECT...ANALYSIS OMITTED`

`WARNING: START TIME > STOP TIME...ANALYSIS OMITTED`

`WARNING: ILLEGAL OUTPUT VARIABLE...ANALYSIS OMITTED`
Incorrect output variable specification in `.TF` analysis.

`WARNING: VOLTAGE OUTPUT UNRECOGNIZABLE...ANALYSIS OMITTED`
Incorrect noise analysis output specification.

`WARNING: INVALID INPUT SOURCE...ANALYSIS OMITTED`
Incorrect input noise source specification.

`WARNING: DISTORTION LOAD RESISTOR MISSING...ANALYSIS OMITTED`

`WARNING: DISTORTION PARAMETERS INCORRECT...ANALYSIS
OMITTED`

`WARNING: FOURIER PARAMETERS INCORRECT...ANALYSIS OMITTED`

`WARNING: FOURIER ANALYSIS FUNDAMENTAL FREQUENCY IS INCOMPAT-
IBLE WITH TRANSIENT ANALYSIS PRINT INTERVAL...FOURIER
ANALYSIS OMITTED`
Transient analysis interval ($TSTOP - TSTART$) is less than the period of the fundamental specified on the `.FOUR` line.

`WARNING: OUTPUT VARIABLE UNRECOGNIZABLE...ANALYSIS OMITTED`
Incorrect sensitivity analysis output specification.

`WARNING: UNKNOWN ANALYSIS MODE:` *Name*`...LINE IGNORED`
The analysis type on a plot or print statement must be one of `DC`, `AC`, `TRAN`, `NOISE` or `DISTO`.

`WARNING: UNRECOGNIZABLE OUTPUT VARIABLE ON ABOVE LINE`
Incorrect output variable specification on a plot or print statement.

B.8   MISCELLANEOUS ERRORS

`*ERROR*: CPU TIME LIMIT EXCEEDED...ANALYSIS STOPPED`
Some SPICE programs allow the user to set a limit on how long an analysis can run.

`WARNING: NUMDGT MAY NOT EXCEED` *number*`;MAXIMUM VALUE ASSUMED`

`WARNING: UNKNOWN OPTION:` *Name* `...IGNORED`

`WARNING: ILLEGAL VALUE SPECIFIED FOR OPTION:` *Name* `...IGNORED`

The following warnings are issued for an incorrect `.NODESET` or `.IC` line.

`WARNING: OUT-OF-PLACE NON-NUMERIC FIELD` *Name* `SKIPPED`

`WARNING: INITIAL VALUE MISSING FOR NODE` *number*

`WARNING: ATTEMPT TO SPECIFY INITIAL CONDITION FOR GROUND IGNORED`

APPENDIX C

SPICE STATEMENTS

This appendix lists all the SPICE statements introduced in this text. Statements followed by [SPICE3] appear only in SPICE3.

C.1 ELEMENT STATEMENTS

R*name node1 node2 rvalue* $<$**TC**=*tc1*$<$*tc2*$>>$

R*name node1 node2* $<$*rvalue*$><$*Mname*$><$**L**=*L*$><$**W**=*W*$>$ [SPICE3]

C*name node1 node2 1value* $<$**IC**=v_{C0} $>$

C*name node1 node2* **POLY** c_0 c_1 $<$ c_2 ... $><$**IC**=v_{C0} $>$

C*name node1 node2* $<$*cvalue*$><$*Mname*$><$**L**=*L*$><$**W**=*W*$><$**IC**=v_{C0} $>$
  [SPICE3]

L*name node1 node2 1value* $<$**IC**=i_{L0} $>$

L*name node1 node2* **POLY** l_0 l_1 $<$ l_2 ... $><$**IC**= i_{L0} $>$

V*name node1 node2* $<<$**DC**$>$*dc_value*$><$**AC** $<$*ac_mag* $<$*ac_phase*$>>>$
+ $<$*TRAN_function* $<$*value1* $<$*value2* ... $>>>$

I*name node1 node2* $<<$**DC**$>$ *dc_value*$><$**AC** $<$*ac_mag* $<$*ac_phase*$>>>$
+ $<$*TRAN_function* $<$*value1* $<$*value2* ... $>>>$

TRAN_function can be one of the following:

  PULSE(*V1 V2* $<$*TD* $<$*TR* $<$*TF* $<$*PW* $<$*PER*$>>>>>$)
  SIN(*VO VA* $<$*FREQ* $<$*TD* $<$*THETA*$>>>$)
  SFFM(*VO VA* $<$*FC* $<$*MDI* $<$*FS*$>>>$)

EXP ($V1\ V2\ <TD1\ TAU1\ TD2\ <TAU2>>$)

PWL ($t_1 V_1 < t_2 V_2 < t_3 V_3 \ldots >>$)

K*name* L*name1* L*name2* *k*

G*name* *n+ n− nc+ nc− gvalue*

G*name* *n+ n−* <**POLY** (*ndim*) > *nc1+ nc1−* < *nc2+ nc2− . . .* > $p_0 < p_1$

+   < $p_2 \ldots >>$ <**IC**=$v_{nc1+,nc1-,v_nc2+,nc2-,\ldots}$>

E*name* *n+ n− nc+ nc− evalue*

E*name* *n+ n−* <**POLY** (*ndim*) > *nc1+ nc1−* < *nc2+ nc2− . . .* > $p_0 < p_1$

+   < $p_2 \ldots >>$ <**IC**=$v_{nc1+,nc1-}, v_{nc2+,nc2-},\ldots$>

F*name* *n+ n− Vname fvalue*

F*name* *n+ n−* <**POLY** (*ndim*) > *Vname1* <*Vname2 . . .* > $p_0 < p_1 < p_2 \ldots >>$

+   <**IC**=*i(Vname1), i(Vname2) . . .* >

H*name* *n+ n− Vname hvalue*

H*name* *n+ n−* <**POLY** (*ndim*) > *Vname1* <*Vname2 . . .* > $p_0 < p_1 < p_2 \ldots >>$

+   <**IC**=*i(Vname1), i(Vname2) . . .* >

B*name* *node1 node2* <**V/I**> = *expr*   [SPICE3]

S*name* *n+ n− nc+ nc− Model* <**ON/OFF**>   [SPICE3]

W*name* *n+ n− Vname Model* <**ON/OFF**>   [SPICE3]

T*name* n_1 n_2 n_3 n_4 **Z0**=*z0* <**TD**=*td*><**F**=*freq* <**NL**=*n1*>><**IC**=$v_{n1,n2}, i_1, v_{n3,n4}, i_2$ >

D*name* *n+ n− MODname* <*area*><**OFF**><**IC**=v_{D0} >

Q*name* *nc nb ne* <*ns*> *MODname* <*area*><**OFF**><**IC**=v_{BE0}, v_{CE0} >

J*name* *nd ng ns MODname* <*area*><**OFF**><**IC**=v_{DS0}, v_{GS0} >

M*name* *nd ng ns nb MODname* <<**L**=>*L*><<**W**=>*W*><**AD**=*AD*><**AS**=*AS*>

+   <**PD**=*PD*><**PS**=*PS*><**NRD**=*NRD*><**NRS**=*NRS*><**OFF**>

+   <**IC**=$v_{DS0}, v_{GS0}, v_{BS0}$ >

Z*name* *nd ng ns MODname* <*area*><**OFF**><**IC**=v_{DS0}, v_{GS0} >   [SPICE3]

X*name* *xnode1* <*xnode2 . . .* > *SUBname*

C.2   GLOBAL STATEMENTS

.MODEL *MODname MODtype* <*PARAM1=value1* <*PARAM2=value2 . . .* >>

MODtype can be one of the following:

D   Diode model

NPN   npn BJT model

PNP	pnp BJT model
NJF	n-channel JFET model
PJF	p-channel JFET model
NMOS	n-channel MOSFET model
PMOS	p-channel MOSFET model
R	Diffused resistor model [SPICE3]
C	Diffused capacitor model [SPICE3]
URC	Uniform-distributed RC model [SPICE3]
SW	Voltage-controlled switch model [SPICE3]
CSW	Current-controlled switch model [SPICE3]
NMF	n-channel MESFET model [SPICE3]
PMF	p-channel MESFET model [SPICE3]

.SUBCKT *SUBname node1* <*node2 . . .*>

.TEMP *temp1* <*temp2 . . .*>

C.3 CONTROL STATEMENTS

.OP

.DC **V/I***name1 start1 stop1 step1* <**V/I***name2 start2 stop2 step2*>

.TF *OUT_var* **V/I***name*

.SENS *OUT_var1* <*OUT_var2 . . .*>

.AC **DEC/OCT/LIN** *numpts fstart fstop*

.NOISE **V**(*n1*<*,n2*>) **V/I***name nums*

.DISTO *RLname* <*nums* < $f2/f1$ < P_{ref} < S_2 >>>>

.PZ *node_in1 node_in2 node_out1 node_out2* <**CUR/VOL**><**POL/ZER/PZ**>
  [SPICE3]

.TRAN *TSTEP TSTOP* <*TSTART* <*TMAX*>><**UIC**>

.FOUR *freq OUT_var1* <*OUT_var2 . . .*>

.PRINT/PLOT *Analysis_TYPE OUT_var1* <*OUT_var2 . . .*><*plot_limits*>

  Analysis_TYPE can be one of the following:

DC, AC, TRAN, NOISE, DISTO

.NODESET **V**(*node1*)=*value1* <**V**(*node2*)=*value2 . . .*>

.IC **V**(*node1*)=*value1* <**V**(*node2*)=*value2 . . .*>

.ENDS <*SUBname*>

.END

APPENDIX D

GEAR INTEGRATION FORMULAS

Gear 1: $\quad x_{n+1} = x_n + h\dot{x}_{n+1} - \dfrac{h^2}{2}\dfrac{d^2x}{dt^2}(\xi)$

Gear 2: $\quad x_{n+1} = \dfrac{4}{3}x_n - \dfrac{1}{3}x_{n-1} + \dfrac{2h}{3}\dot{x}_{n+1} - \dfrac{2h^3}{9}\dfrac{d^3x}{dt^3}(\xi)$

Gear 3: $\quad x_{n+1} = \dfrac{18}{11}x_n - \dfrac{9}{11}x_{n-1} + \dfrac{2}{11}x_{n-2} - \dfrac{6h}{11}\dot{x}_{n+1} - \dfrac{3h^4}{22}\dfrac{d^4x}{dt^4}(\xi)$

Gear 4: $\quad x_{n+1} = \dfrac{48}{25}x_n - \dfrac{36}{25}x_{n-1} + \dfrac{16}{25}x_{n-2} - \dfrac{3}{25}x_{n-3} + \dfrac{12h}{25}\dot{x}_{n+1} - \dfrac{3h^5}{125}\dfrac{d^5x}{dt^5}(\xi)$

Gear 5: $\quad x_{n+1} = \dfrac{300}{137}x_n - \dfrac{300}{137}x_{n-1} + \dfrac{200}{137}x_{n-2} - \dfrac{75}{137}x_{n-3} + \dfrac{12}{137}x_{n-4}$

$\qquad\qquad + \dfrac{60h}{137}\dot{x}_{n+1} - \dfrac{10h^6}{137}\dfrac{d^6x}{dt^6}(\xi)$

Gear 6: $\quad x_{n+1} = \dfrac{360}{147}x_n - \dfrac{450}{147}x_{n-1} + \dfrac{400}{147}x_{n-2} - \dfrac{225}{147}x_{n-3} + \dfrac{72}{147}x_{n-4} - \dfrac{10}{147}x_{n-5}$

$\qquad\qquad + \dfrac{60h}{147}\dot{x}_{n+1} - \dfrac{180h^6}{3087}\dfrac{d^7x}{dt^7}(\xi)$

SPICE INPUT DECK

This is an example SPICE input for a differential amplifier from the University of California, Berkeley, SPICE benchmarks. Most types of statements and analyses are used.

```
DIFFPAIR CIRCUIT - SIMPLE DIFFERENTIAL AMPLIFIER
*
* CIRCUIT DESCRIPTION
* ELEMENT STATEMENTS
*
Q1 4 2 6 QNL
Q2 5 3 6 QNL
Q3 6 7 9 QNL
Q4 7 7 9 QNL
RS1 1 2 1K
RS2 3 0 1K
RC1 4 8 10K
RC2 5 8 10K
RBIAS 7 8 20K
VCC 8 0 12
VEE 9 0 -12
VIN 1 0 SIN( 0 0.1 5MEG) AC 1
*
* GLOBAL STATEMENTS
*
.MODEL QNL NPN BF=80 RB=100 CCS=2P TF=0.3N TR=6N
+ CJE=3P CJC=2P VAF=50
*
* ANALYSIS REQUESTS
```

```
*
.OP
.TF V(5) VIN
.DC VIN -0.25 0.25 0.005
.AC DEC 10 1 10GHZ
.TRAN 5N 500N
*
* OUTPUT REQUESTS
*
.PLOT DC V(5)
.PLOT AC VM(5) VP(5)
.PLOT TRAN V(5,4)
*
.END
```

INDEX